Springer Geography

The Springer Geography series seeks to publish a broad portfolio of scientific books, aiming at researchers, students, and everyone interested in geographical research.

The series includes peer-reviewed monographs, edited volumes, textbooks, and conference proceedings. It covers the major topics in geography and geographical sciences including, but not limited to; Economic Geography, Landscape and Urban Planning, Urban Geography, Physical Geography and Environmental Geography.

Springer Geography—now indexed in Scopus

More information about this series at http://www.springer.com/series/10180

Hua Peng

China Danxia

Higher Education Press

Springer

Hua Peng
School of Geography and Planning
(1956–2018) Sun Yat-sen University
Guangzhou, China

ISSN 2194-315X ISSN 2194-3168 (electronic)
Springer Geography
ISBN 978-981-13-5958-3 ISBN 978-981-13-5959-0 (eBook)
https://doi.org/10.1007/978-981-13-5959-0

Jointly published with Higher Education Press
The print edition is not for sale in China (Mainland). Customers from China (Mainland) please order the print book from: Higher Education Press.
ISBN of the China (Mainland) edition: 978-7-04-053369-9

This Springer imprint is published by the registered company Springer Nature Singapore Pte Ltd.
The registered company address is: 152 Beach Road, #21-01/04 Gateway East, Singapore 189721, Singapore

Foreword By Zhijiu Cui

The author of this book was the academic leader of China Danxia landform research, the team leader in China Danxia Application for World Natural Heritage and the chairman of Red Beds and Danxia Geomorphology Working Group of International Association of Geomorphologists (IAG). Since the writing purpose of this book is international, this book is the most professional and authoritative work with higher scientific value in this field.

Danxia landform is a type of distinctive landform which was first named and developed in China. It is considered as a major contribution from Chinese scholars to the world earth science by international experts. A Danxia landform working group was established by IAG in order to call for international recognition and global research. Nowadays, China Danxia has been included in the World Natural Heritage. The six heritage sites introduced in this book are the most in-depth research cases. This book systematically introduces the results of Danxia landform and the typical cases of six heritage sites. If the English version could be published, it will exert significant guiding influence on future international Danxia landform research, which is conducive to the international promotion of China Danxia landform. In fact, the results of Danxia have resonated with international scholars in many ways.

This book is a high-level scientific work, which is of high readability for its rich illustration and fluent words. Also, this is a good book for its high-level science popularization. There are few articles on China sandstone landform research that are published internationally, and Danxia landform is a kind of sandstone landform. As a result, we can say that this book has filled this gap since the research on China Danxia landform has received great attentions by international scholars.

The author of this book dedicated his life to Danxia landform research. He has explored Danxia landform in most of China, and he has much communication and resonance with me. His expiration makes me sad and heart-broken. Writing this prologue could be a spiritual communication between us. May we memorize and learn from him.

November 2018 Zhijiu Cui

Foreword by Piotr Migoń In memory of Prof. Hua Peng

It is my great privilege to write the foreword to the "China Danxia" book—an important contribution to geomorphology and the testament of lifelong dedication of Prof. Hua Peng to the Danxia landscapes of China. I feel both honoured and very sad, wishing that we could discuss plans of further collaboration, rather than I reflect upon the loss of a distinguished scientist who left us so early.

I recall meeting Prof. Hua Peng for the first time in the year 2000, during my first visit to China, when I was privileged to attend the Thematic Conference of the International Association of Geomorphologists in Nanjing. Knowing very little about China at that time, I looked with awe at photographs contained in his presentation, showing unbelievably beautiful landforms from the Guangdong Province. Then I learnt that they are predominantly from sandstone and denoted as Danxia in China. I was lucky to receive a publication from him, with his signature, which I consult until today as researching sandstone tablelands and cuestas is among my own main research topics.

Then we met again in autumn 2007 when I was involved in an international team of experts helping the Chinese colleagues to develop the World Heritage application for China Danxia. Our contacts intensified and two years later a kind invitation came to me to attend the 1st International Conference on Danxia Landform, in Hua Peng's own playground—the beautiful Danxiashan. This conference and associated field trip exposed me to more of Danxia and both of us shared the view that such awesome sceneries do not only belong to China, but the entire world, and from scientific point of view are significant globally.

Thus, the next step was to submit a joint proposal to the International Association of Geomorphologists to create a Working Group dedicated to Danxia geomorphology. It came into effect in summer 2009 and has continued over the following years. The impressive list of activities of the Group is the best evidence of personal dedication, energy and vision that typified Prof. Hua Peng. I was impressed how successful he was in coordinating research and organizational efforts of Chinese colleagues, in gathering keen students around him, in establishing international co-operations across the world, and in constantly enlarging the scope of interest, from humid subtopics of south-east China to the arid north-west, and from monolithic towers of Langshan and Taining to fragile red beds.

In 2012, I was happy to host Prof. Hua Peng in Poland. We discussed further plans for comparative research between Danxia and sandstone formations across the world. Then we met again in Paris, attending the Eighth International Conference on Geomorphology. Unfortunately, I was unable to take part in his major Northwest China research expedition in 2016 for which he kindly invited me. We thought that the plans are simply postponed, but now we know they will never come true with Hua Peng involved. However, I believe the research on Danxia geomorphology, carried out by his students and successors, will continue in his memory. He surely would wish so!

However, although Prof. Hua Peng himself is no longer with us, his spirit is still here and his contribution to Danxia research will not be forgotten. I feel very privileged to have had the personal relationship with Hua Peng and his collaborators and I owe him very much, as it was him who showed me so many hidden gems of Chinese nature. I believe that this book, strong through both the scientific content and impressive images, will contribute to fulfill his dream—that more and more people will know about Danxia, will appreciate Danxia, and will undertake an effort to understand Danxia better.

December 2018

Prof. Piotr Migoń
Institute of Geography and Regional Development
University of Wrocław, Poland

Co-opted Executive member of IAG

Foreword by Fusheng Guo

In the Late Mesozoic in Southeastern China, many continental fault basins were formed with the extension of crustal tectonics. Under the generally dry and hot paleoclimate condition, a set of red beds characterized by red sandstone and conglomerate was deposited, with its thickness reaching several thousand meters.

Since Cenozoic, under the force of tectonic uplift, fault cutting, gravity collapse, weathering and erosion, these red beds have evolved into a series of special landscapes such as red cliffs and distinctive peaks and rocks. In the 1920s and 1930s, when Prof. Jinglan Feng and Prof. Guoda Chen, two famous geologists in China, conducted geological surveys in Jiangxi and Guangdong Provinces, they were attracted by these red cliff landscapes, and that was the opening of Danxia landscapes research.

Over the past 90 years, more than 1000 Danxia landforms have been found nationwide. In recent years, comparable landforms have been found in North America, Europe and Australia. In 2004, 2005 and 2007 respectively, UNESCO approved the establishment of three World Geological Parks, namely Danxiashan in Guangdong, Taining in Fujian and Longhushan in Jiangxi. In August 2010, the "China Danxia", composed of Chishui in Guizhou, Taining in Fujian, Langshan in Hunan, Danxiashan in Guangdong, Longhushan in Jiangxi and Jianglangshan in Zhejiang, was officially admitted in the World Heritage List. With the joint efforts of Chinese Danxia geomorphologists and local governments, the Danxia geomorphology named in China has finally been introduced to the world as a special geomorphological type, which has been recognized by the international geomorphological circles.

Danxia landform generally has gentle peaks, iconic cliff slopes and gently inclined foothills. As a result, it is briefly defined as "red layer landform with steep cliff slopes". The word "Danxia" is both visualized and elegant, and the description of "color as Lilium concolor Salisb, bright as sunset" gives people infinite beautiful imaginations. Therefore, since ancient times, Danxia geomorphic areas have attracted Taoist and Buddhist believers to build temples in these places. At the same time, it has been favored by the vast number of poets and writers, leaving a large number of poems behind. The scientific research of red beds and Danxia landforms has important theoretical significance and practical value. The red bed sedimentary rock sequence records the important geological information of the earth's evolution in the Late Mesozoic, and the study of the red beds can provide an important basis for explaining the surface sedimentary environment and Paleoclimate in this period.

At present, many Geoparks in China have become teaching and research bases of universities and scientific research institutes. A large number of undergraduates and graduates have become the promotors of Danxia landform, and some of them will definitely become researchers of Danxia landform in the future. The research on Danxia geomorphology can not only improve the level of scientific understanding of geomorphological evolution, but also enrich the scientific content of geoparks, promoting adolescents' love for nature, enhancing their awareness of protecting the environment.

Professor Hua Peng from Sun Yat-sen University devoted his whole life to the study of Danxia landform. His contribution can be summarized as follows, (1) he has improved the scientific theoretical system of Danxia landform and established a more systematic research method; (2) he has summarized the distribution pattern of red beds and Danxia landform in China; (3) he has developed in-depth research on Danxia landform in humid region in Southeastern China, organizing the World Heritage Application among six provinces in South China; (4) he has initiated the study of red-bed desertification and the prelude of large-scale investigation of Danxia landform in arid areas in Western China; (5) as the academic leader of Danxia landform research, he has launched 12 national academic conferences on Danxia landform; (6) he has established the Red Beds and Danxia Geomorphology Working Group in the International Association of Geomorphologists, promoting the international comparative study of Danxia Landform.

Professor Hua Peng was a close friend of mine. I was very sad to know this sudden expiration, which is undoubtedly a great loss to Danxia geomorphology. As a member of Danxia Landform Working Group, we should turn grief into strength, continue to devote ourselves to Danxia landform research, cultivate more young scholars, and promote Danxia landform discipline, which are the things that Prof. Hua Peng would love to see, I suppose. Professor Hua Peng has made outstanding contributions to the application for World Heritage of Danxia landform in China. He is a practitioner of "Clear waters and green mountains are as good as mountains of gold and silver", whose name is not only well known by his colleagues, but also widely spread among the people. I am very glad to learn that his book *China Danxia* will be published soon. This book is a comprehensive summary of Danxia landform in China, much of the content is the innovative achievements first introduced by the author. This book will lead the direction of Danxia landform research in China in the future. At the same time, it will be helpful for international colleagues to understand the special Danxia landform, promoting international comparative study of Danxia landform. I have read it first and would love to recommend it to my colleagues engaged in geological and geomorphological research.

November 2018

Fusheng Guo
Former Vice President of Red Beds and
Danxia Geomorphology Working Group of IAG

Vice President of East China University of Technology

Preface

Danxia landforms were initially named in China. It has been studied for nearly 90 years. After the establishing stage, formative stage, and developmental stage, Danxia landforms have become a relatively mature discipline in China. In May, 2009, the First International Symposium on Danxia Landform was held at Mt. Danxiashan of Shaoguan City in Guangdong Province, the place after which Danxia Landform has been named. Danxia landform then began to be known by international academic circles. During the 7th International Conference on Geomorphology in Melbourne, July, 2009, the Council of IAG (International Association Geomorphologists) approved the establishment of Danxia Geomorphology Working Group, marking the research of Danxia landforms stepped into international stage.

During the 10th National Conference of Danxia Landforms and Tourism Development in July, 2006, suggested by the Department of Langshan Mountain Scenic Area Administration of Hunan Province, a proposal about organizing typical Danxia Landform areas in China for a joint inscription on the list of World Natural Heritage was formed, which got active response from 13 Danxia landform areas in 10 provinces. In December, 2006, with the support from the former Ministry of Housing and Urban-Rural Development and the lead of the construction department of Hunan Province, the work of applying for the award of World Natural Heritage was started. With strict selection, 9 national parks (National Scenic Area) in 6 provinces of South China were nominated, followed by research and demonstration work. After four-year hard work, during the 34th convention of the World Heritage on August 1st, 2010, 6 famous Danxia Landform areas in South China were inscribed on the list of World Heritage Sites.

During this period, the nominated Danxia landform sites organized over 130 experts from various disciplines to work for the inscription. According to the criterions of World Heritage, they unified plans and goals, and conducted fundamental research and integrated demonstration on geology, geomorphology, biology, natural environment, social economy, history and culture of Danxia area. More than 30 experts from IUCN, IAG, IUGS and UNESCO world heritage center were invited to visit and make field investigation at the nominated places. Over 50 symposiums and meetings were organized, more than 20 of which were international meetings. Thus, the application process for the award of World Heritage witnessed the most in-depth research on Danxia landforms. It was also a process with highest degree of participation of the international experts.

Based on the 80-years research on Danxia Landforms in China, this book integrates the latest achievements during the process for the application process of World Heritage. In view of the introduction of the latest scientific system of Danxia landforms systematically, through the case study of six heritage sites, this book explains Danxia landforms from the aspect of geological setting, morphologic characteristics and evolution, biology and ecology, aesthetics value and earth science value, etc. It also briefly introduces the six heritage sites. We expect to provide a systematic introduction of Danxia landforms and typical base sites for colleagues both at home and abroad, and provide reference for the international comparison research on the Danxia landforms. We also hope to provide some information from all works of life for people who have interest in China Danxia World Natural Heritage Sites.

Guangzhou, China Hua Peng

Contents

Part I

1 The Basic Questions on Danxia Landform

1.1 Introduction of Danxia

Danxia, a terminology, similar to "Karst," is a unique type of landform and a name of the natural landscape. Literally in Chinese, Danxia is "reddish rays" or "rosy cloud" during sunrise or sunset. In Chinese geoscience circles, the term Danxia was initially designated by Chinese geologists in 1920s to classify specific stratigraphic sequence of red beds that forms hoodoo peaks at a gorgeous mountain named Mt. Danxiashan in Southern China (Feng and Zhu 1928) (Fig. 1.1). What is more, this name also appropriately expressed the natural beauty of the landscape.

After more than 80 years of study and popularization, and as a result of its special landscape esthetic value, Danxia has become almost a household name in China. It has evolved into an independent type of landform by Chinese geoscientists, and it is defined as a kind of landform developed on red beds and featured by steep slopes. This definition can be further explained as an erosional landscape that is formed from thick red beds that have been regionally uplifted, intensively faulted, and deeply dissected by fluvial erosion, weathering process, and mass movement, producing a variety of cliffs and bounded peaks surrounded by deep canyons.

1.2 Research History of Danxia Landform in China

1.2.1 Initial Stage (1928–1949)

Danxia served as a sort of unique landform that has been paid attention at late 1920s. In 1928, Feng Jinglan[1] and Zhu Huisheng for the first time named red beds and related stratum in North Guangdong as "Danxia Formation" and made a vivid description of its lithology and terrain formed. This sort of stratum and terrain classification drew attention of academics, which lay the cognitive foundation of a unique landform.

In 1939, Chen Guoda used the term of "Danxia Topography" for the first time in geology in Gongshui River Basin, Jiangxi. From then on, academics continued to use concepts of Danxia Formation, Danxia Topography, and Chen Stratum Classification.

In this period, some scholars including Zeng (1943) and Wu Shangshi et al. made study and discourse on stratum, lithology, structure, and landform development of South China red beds, which marks that Danxia landform has been advanced from simple concept to academic research.

The research scope mainly centered in Southeast China during this time, and the noun "Danxia Formation" was established. Scholars of this time demonstrated and defined the stratum

[1] All the Chinese names mentioned in the full text are surname first.

H. Peng, *China Danxia*, Springer Geography,
https://doi.org/10.1007/978-981-13-5959-0_1

Fig. 1.1 Danxia landforms at Mt. Danxiashan, China (photographer: Xie Jinshu)

level of Danxia and started this study as an independent landform.

1.2.2 Formative Development Stage (1950–1990)

With the regional geological and comprehensive survey after the founding of the People's Republic of China, the concept of Danxia landform has been widely applied. Zeng Zhaoxuan, especially, continued to process his study on red bed and landform features of South China and published the book of *Rock Topography* which for the first time regarded red bed landform as independent landform type to draw a conclusion. Huang Jin took Danxia terrain as a type of Guangdong Geomorphologic Map and defined "Danxia Topography" for the first time.

During the "Cultural Revolution," the study on Danxia landform came to a halt. After this period, Zeng Zhaoxuan and Huang Shaomin continued to discuss the cultivation and landform features of Southeast China red bed landform. They especially discussed the distribution, petrology features, landform development process, and landform features in the book of *Physical Geography of China: Geomorphology* in 1980. Huang (1982) summarized the mode and features of nearly horizontal red beds slope development under humid climate conditions, which further the study on development mechanism of Danxia landform.

During this period, other scholars had also conducted some study and investigation. These scholars studied Danxia landforms from Hebei, Gansu, Fujian, Anhui, and Guangdong. Meanwhile, tourism geographers started the study on Danxia tourism resource, which directs further study on landscape construction and utilization of Danxia landform.

The regional geological and comprehensive surveys in this period had collected large amount of firsthand references for Danxia landform studies, which helps make more scientific conclusions for defining the age and horizon of red beds. Zeng Zhaoxuan, et al., systematically concluded the Danxia research accomplishments, Huang Jin conducted further discussion on development mechanism of Danxia landform, and Chen Chuankang made studies on landscape resources and utilization of Danxia landform, all of which help make the terms of Danxia landform more widespread. As an independent landform type, academic studies on Danxia have formed a system and started a new branch field.

1.2.3 The Great Development Stage (1991–2008)

Before 1991, studies on Danxia landform were concentrated on individual and sporadic issues. In 1990, Huang Jin took charge the first National Natural Science Foundation Program on Danxia landform studies and started his comprehensive investigation among China Danxia areas. In

1991, the First Symposium on National Danxia Landform Tourism Development was held in Guangdong Mt. Danxiashan, and the establishment of "Danxia Geomorphology and Tourism Development Study Society of China" marked that study on Danxia landform advanced into great development stage.

In this period, Huang Jin worked on Danxia investigation for 3–6 months per year, and his field survey had covered 1000 sites of Danxia landforms area in 28 provinces. His actions enlarged the scope of research area and relevant studies with all his life.

From 1991 to 2008, the study society has held 11 sessions of symposium and published more than 380 papers, and more than 300 papers published in various academic journals. The study contents involve basic theories (definition, classification, feature, sediment, external force, development mechanism), approaches (dating, stress analysis, remote sensing, mapping), historic culture (the relationship between Danxia landform and religion, cliff carving, sculpting landscape, ancient human, cave tomb, cliff coffin, and ancient mountain village), conservation and utilization (resource assessment, planning, conservation and development), and popular science education. In 2006, Danxia landform was for the first time compiled into an independent chapter of *Modern Geomorphology*—a textbook serving for the tenth Five-Year Plan.

However, due to the lack of external exchange on the past Danxia landform research, the international does not understand the situation of China. Until 2000, at the International Association of Geomorphologists (IAG) Nanjing symposium, the bilingual version book *Danxia Geomorphology of China and its Progress in Research Work* was published and the exchange in the conference held by Peng Hua, made the international counterparts a preliminary understanding. In 2004, Mt. Danxiashan was granted the first batch World Geopark as "Danxia Landform Type." In 2006 and 2008, Taining in Fujian and Mt. Longhushan in Jiangxi were endowed with this denomination successively. In 2006, ten Danxia landform areas in six provinces of South China started a binding application for World Natural Heritage as "China Danxia." During these processes, many international experts on geology and geomorphology have taken part in the investigation, consulting and research and organized more than 20 symposiums or consulting in different scales, which promoted the further study on Danxia and the initial international communication.

In the process, promoted and organized by Chen Chuankang and Huang Jin, the studies on Danxia landform and its tourism development have been widespread among the country, presenting a prosperous state of development. Danxia landform, as a branch of geomorphology, has become a crucial growing point of modern geomorphology, and a fundamental framework has formed as an independent subject. Also, international study and communication on Danxia landform start and directly and indirectly serve for economy construction, which attract much attention from society and academic circles.

1.2.4 Internationalization Stage (Since 2009)

While on the subject of this stage, I would like to tell you a story first. At the International Association of Geomorphologists (IAG) Nanjing symposium, in 2000, I gave a parcel of books entitled *Danxia Geomorphology of China and Its Progress in Research Work* to all of the international colleagues. Before long, Robert Young from the University of Wollongong, Australia, mailed me his first edition of *Sandstone Landforms* and expressed his willingness to pay more attention to Danxia landform with the hope that we would have further cooperation. He had promised to attend the feasibility meeting for "China Danxia" which applies for World Natural Heritage with his wife Ann Young in Chishui City, Guizhou Province, in 2007. Due to visa issues, he and his wife did not attend but his Ph. D. student Robert Wray did. Then, due to his poor health, he also did not attend the First International Symposium on Danxia Landform, but again entrusted Robert Wray to express his regret and bring the latest edition *Sandstone Landforms*. To my great relief, that edition of the book, written jointly by Robert Young, Ann Young, and Robert Wray, increased introductory content of Danxia landform and referred to the main theory system. For the first time, international foremost experts and a book of authority introduced China's Danxia landform and its research. And in that very year, Danxia landform began to step into the international academic stage after 80 years of research in China.

In May 2009, the first international academic conference on Danxia landform was held in Mt. Danxiashan, Guangdong Province, issued a "declaration of Danxia." The conference affirmed the contribution of Chinese scholars, called the international recognition for Danxia landform and international comparative study, as an independent type, Danxia landform really has caught the attention of the international academic community.

On July 7, 2009, in the Seventh International Conference on Geomorphology Melbourne, Australia, the IAG Council approved the establishment of Danxia Geomorphology Working Group, symbolizing Danxia research has been placed onto the board. The scientific aims of the working group are to discuss and confirm the scientific definition that suited to worldwide Danxia landform; to define the essential development conditions of Danxia landform; to demonstrate its developing process; to build up the scientific classification system; to fully understand the global distribution rule of Danxia landform.

Then, in August 2010, the China Danxia world natural heritage series were nominated and included in the World Heritage List. From Oct 2010, two international conferences were held independently or Jointly. In October 2011, IAG Danxia Geomorphology Working Group first meeting and the second International Academic Symposium of Danxia Landform were held in Mt. Danxiashan. In Apr 2012, the 3rd conference of Sandstone Landscapes was jointly held in Europe; In Aug 2013, The Eighth international conference of Geomorphology, IAG Danxia Geomorphology Working Group focused discussions on Red beds, Danxia landform and Sandstone.

During 2013–2014, joint scientific research project was established between China and Slovakia "Landslide hazard and risk assessment for China Danxia World Heritage Sites"; four researchers were sent to Saint Louis University for exchange and started comparative study of red beds landform between Southeast China and Western USA; eight experts from Slovakia, Canada, and Australia were invited to China Danxia heritage sites to investigate Danxia landform.

In recent years, research on Danxia landforms in China has begin to focus on the quantitative study of weathering processes on red bed rock surfaces, as rock weathering may be considered to be the starting point of many dynamic systems and is also a dominant force in shaping the appearance of Danxia slopes. By using modern lab test techniques to analyze structures, mineralogical and chemical properties of rock samples, Zhu et al. (2009a, b, 2010), Chen et al. (2010) discussed many of the processes responsible for the weathering of red bed rocks. They found that different types of red bed rocks vary in weathering resistance. Sandstones and conglomerates, which contain few clay minerals and pores, are more resistant to weathering processes than those siltstones and argillaceous red bed rocks. These works were among the first to make quantitative analysis on the differential weathering of various lithotypes in Danxia landform area. In terms of biological weathering, Huang addressed that the action of moss, algae, and lichen can accelerate the exfoliation of red bed rocks on Danxia slopes, but the microscale weathering process of Danxia rock layers caused by the physical and chemical action of plants remains unclear.

Additionally, quantitative calculation on the rate of tectonic uplift and cliff retreat in Danxia landform areas also aroused concern from some researchers.

In the recent years, great changes have taken place in the field and direction of research. The Geographical Society of China (GSC) Red beds and Danxia Research Working Group was founded, with the approval of the Geographical Society of China in 2011. The Red Beds and Danxia Research Working Group of GSC finished the construction of organizational structures and adopted the work program at the 13th National Danxia Landform conference in 2013. Then, at the Eighth International Conference on Geomorphology in Paris, France, IAG made the necessary adjustment to "Danxia Geomorphology Working Group" according to China's conditions. The name of the working group was changed to "IAG Red Beds and Danxia Geomorphology Working Group". This was not just a change of name, but a qualitative promotion and extension of the research field, meaning the research field was extended to red beds and problems in red beds. The qualitative promotion means the change from landscape research and forms research to a comprehensive study of underground surface above ground which blends geology, geomorphology, and ecology. The extension also indicates the extension from landscape research to the comprehensive study of the geological environment, geomorphological hazards, ecological security, and land conservation. Also, it increases the contact with the national economy and the people's livelihood, and the development space of discipline and promotes international cooperation and exchange.

In addition, the special program of the Key Basic Research of the Ministry of Science and Technology of the People's Republic of China "National basic data survey of Danxia landform," operated by Professor Peng Hua, was approved by the Ministry of Science and Technology of the People's Republic of China in 2013. This project is led by Sun Yat-sen University and is participated jointly by six universities or institutions, including Shanghai Jiao Tong University and Nanjing University, etc. Although, Danxia landform, born in China, has been studied for more than 80 years in China, there is a lack of systematic basic research, and the research results are not intact. This five-year project is expected to be the foundation of the construction of the scientific system of Danxia landform. A display system of the basic data of Danxia landform, an exchange platform of academic research and applied research, a real-time update, and a powerful capability of concurrency for worldwide users will be formed. For the first time, at national scale, the collection of basic data, the construction of Danxia landform's basic database, and the scientific construction of a sharing platform mark that the "National basic data survey of Danxia landform" was integrated into state scientific planning. It has had a revolutionary impact on the study of China Danxia.

In general, the research history of Danxia landforms in China mentioned above presents a trajectory from early description and classification of features, to process explanation based on data analysis. During this transformation of research approaches, the investigation scales are becoming increasingly smaller and detailed; the traditional qualitative methods of field observations, measurement, and description of weathering features were shifted toward quantitative assessment.

1.3 Basic Concept

1.3.1 The Early Definition

Since Feng Jinglan puts forward "layer Danxia" in 1928, then in 1938–1939 Chen Guoda developed the concept of "Danxia terrain". Later on there was a long historical period nobody discussed its definition. The research mainly focused on physical, geological structure, color, shape, etc., but there was no lack of combination of description and concept explanation. Not until 1961 was Huang Jin in the preparation of Guangdong Province topographic maps who proposed the definition "Danxia landform is composed by the level or slight changed thick red sandstone, conglomerate, due to rock massive structure and abundant easy for water percolation vertical joints, and after down by running water erosion and the role of gravitational collapse, then forming the steep peaks or mountain terrain."

Geography Dictionary proposed that the Danxia landform refers to "landforms of Square

Mountain, odd peak, red cliff, grotto and megalith, and so on formed on the great thick red conglomerate" in 1983. The same year *Geological Dictionary* defines as "the thick, gentle attitude, joint development, iron calcium mixed uneven cement red conglomerate, on the combined effect of the differential weathering, gravitational collapse, erosion, corrosion, etc., resulted in the castle-like, pagoda-like, acicular, columnar, rod-like, square-shaped, or hill-like peaks terrain."

Zeng Zhaoxuan and Huang Jin in 1990 proposed Danxia was "the general name of each kind of red cliffs and strange peaks developed along the vertical joints in thick red sand, conglomerate" in the volume of *Encyclopedia of China Geography*.

Since 1990s, along with people becoming increasingly concerned about Danxia, the discussions about the concepts of Danxia also gradually increased. The concepts can be divided into two broad categories roughly:

First is the descriptive definition, which more comprehensively expressed Danxia landform's connotation and denotation, such as its material composition, geological structure, growth conditions, and geomorphological features. More representative is Huang Jin, etc., in 1992 amends the concept of descriptive explanation: "Mainly developed above the Mesozoic to Tertiary level or lightly inclined red thick, rigid, vertical joints developed conglomerate, sandstone and other clastic rocks, in the area of intermittently uplift during neotectonic movement, along the rock joints, bedding down by running water erosion, collapse back off, weathering and wave erosion, karstification and wind erosion and other external force, formed flat peaks (or inclined peaks, roof peaks), body steep, flat piedmont with cliff, square mountain, castle rock, rock towers, rock walls, rock peak, rock column, rock mushrooms, cave, rock pile, rock block and so on the strange, dangerous, elegant and beautiful red cliff landscapes, known as Danxia."

Second is the abstract definition, which more briefly expressed the essence of the attributes of Danxia landform but no explanation of its concept. For example, Huang Jin in 1992 defined "the red cliff and red wall and its related landform formed by the red clastic rock formation called Danxia landform"; Huang Keguang and Chen Zhijun in 1992 defined Danxia landform as "each kind of landform with steep slope face formed by the red clastic rock layer"; Peng Hua in 1993 thought it should add "continental facies" restriction to the red clastic rock, proposed "the Danxia landform is a kind of landform grown on the foundation of red continental clastic rock, and characterized as red wall and red cliff"; afterward, Huang (1995, 1996), etc., all added "continental facies" restriction in the revised definition. The above definitions stipulated Danxia landform the "red continental clastic rock" material composition and the "red cliffs slope" of the essential attribute. And the definition is currently accepted and used by the major part scholars.

Not only with the above differences in the interpretation of the definition, differences also exist in the recognition of Danxia landform. There are three different definitions: the first one is a broad definition, including red limestone and red volcanic rock and so on, no matter what kind of rock it is, every landform with red cliff can be classed as Danxia landform. The second one is a narrow definition, and only the landforms developed in Cretaceous red fluvial and lacustrine facies conglomerate can be labeled as Danxia landform. The third definition falls between the first two, and the landform developed in red beds mainly consists of red continental clastic rock (maybe including non-continental layers) is Danxia landform. The material of this landform is limited, but no limitation to the age of the red beds. Today, this definition is accepted widely by most Danxia researchers.

1.3.2 New Definitions and Explanations

(1) **Definition**

At present, the Danxia landform is concisely defined as "red beds landform characterized as scarp slope." Here, it expressed two essential attribute of the concept, first, the Danxia landform belongs to the red beds landform; second, the Danxia landform's characteristic is scarp slope.

The reason that has substituted the red continental clastic rock in original definition for red beds is because later studies have discovered that in the arid area and the semi-arid region, some siltstone even argillaceous rock may also grow the very good Danxia landform; also, there are some red rock layer with biological or chemical rocks, some with marine-terrigenous facies across, transition phase, and littoral–neritic facies interbedded or laminated. However, the main body stratum of the typical developed Danxia remains red continental coarse clastic rocks, especially sandstone and conglomerate.

The reason that replaces red scarp slope or red wall and red cliff with scarp slope is regional red beds have the big difference in color; some red beds color is not always a distinctive red, even with some other colors sandwich; the red beds scarp slope suffered long-term exposure by weathering, running water erosion, biological role, calcium precipitated in water, or organic matter taken from the top of the slope dyeing, and its surface color will be great changes. Therefore, the scarp slope does not necessarily require the scarp slope must be red, but mainly in the red.

(2) **Explanation**

Increase growth conditions for the explanation of Danxia landform concept and express as "Danxia landform was the uplifted and fracture cutting red beds in the movement of the earth's crust in recent geological period, mainly suffering water erosion, weathering, erosion, corrosion, gravity, and so on external force effect and shaped into such landform with the characteristics of scarp slope." This concept contains two basic elements—the material and shape, and two growth conditions—structure and external force.

- **The material basis for Danxia development is red beds**. Red beds for Danxia development mainly red continental clastic rocks, especially coarse clastic rock which are the conglomerate and sandstone as the main. In right conditions, the non-continental red beds (such as the littoral–neritic facies) or siltstone and mudstone (mudstone-arid areas even containing evaporites) may also develop Danxia. Therefore, the concept of sandstone landform concept cannot properly reflect the taxonomic status of Danxia.
- **Danxia is a prominent scarp slope**. Cliff with variety of scales and patterns is the basic element for forming Danxia Mountain block and canyon valley wall and also the most prominent individuality characteristic of Danxia monomer landform and the community landscape. Some experts think that Danxia scarp slope should be limited to be greater than 60° and more than 10 m. But the reality is that some of the early developed Danxia or old age might not reach the slope. Therefore, Danxia scarp slope can be seen as the bare bedrock's steep slopes.
- **Danxia development was controlled by the geological structure**. Firstly, Danxia developed in the region that the red beds were uplifted to above the local erosion base; secondly, basin geological structure controlled the spatial pattern of block; thirdly, rock attitude and faults, joints, and other structures controlled the shape of Danxia slope; the fourth is the new tectonic movement created the multi-phase and multi-sexual development of topography.
- **External force is the direct Danxia shaper**. Danxia is a landscape formed by water erosion and the gravity collapse as the main force. After red beds were uplifted, it started to accept the role of physical and chemical weathering, erosion, water erosion, slope gravity, etc., and then carved out a thousand pictures landscape. In addition, the wave action of coastal zone, sandstorm function of arid area, biological action and so on are the

Danxia's shape factors and show the different importance in the different area.

In some former definitions, some scholars' viewpoint contains a description of Danxia landscape, such as cone, castle shape, wall shape, and columnar. The monomer landform landscape expression, such as cluster-like, shaped peaks, peaks, the groups landform landscape expression also have red beds deposition time, the landform landscape esthetics expression, and so on. However, these restrictions do not apply to all Danxia or only belong to Danxia low-level features, and the author thinks that there is nonessential to be included in definition, but can be used as a basis for further classification. In addition, esthetics belong to a perceptual evaluation of the subjective consciousness it should not be the limiting factor to define the landform.

1.4 Essential Characteristics of Danxia Landforms

Red scarp slopes are the most significant feature of Danxia landforms. Huang (1982) summarized the basic slope shapes of nearly horizontal Danxia landforms as "flat top, steep cliff body, gentle foothill." These descriptors, respectively, indicate flat top surfaces controlled by nearly horizontal rock layers; scarp slopes controlled by vertical joints; and gentle colluvial foothills controlled by the internal friction angle of the colluvial rocks (Fig. 1.2). For those Danxia landforms developed at basin margin or near fault zone, they had cuesta characteristics, and the inclined top surfaces were essentially consistent with bedding planes (Fig. 1.2). In some places with intensive tectonic activity, the dip angle of rock layers can reach over 60°, and as a result, the bedding planes can form scarp slopes. Liu and Liu (2003) proposed a minimum relative height of 10 m and a minimum gradient of 55° for the identification of Danxia slopes in field investigation. In addition, Gao et al. (2004a, b) applied the King's hillslope development model to explain the shape and evolution pattern of Danxia slopes. He suggested that the Danxia slope surfaces can be divided into four parts: convex slope, free slope, debris slope, and concave slope. The major evolution patterns of Danxia slopes are parallel retreat and slope replacement.

Nevertheless, there are a variety of slope morphologies observed in the field. Different bedding inclinations may result in various shapes of Danxia top surface, such as "inclined top, round top, and sharp top"; due to water erosion and lithological differences of rock layers in vertical direction, a great mount of caves and grooves with various shapes are formed on the cliff surface (Peng 2004).

Red cliffs are the most significant feature of Danxia landform.

1.4.1 Basic Slope Morphology

- Sub-horizontal strata: flat top, steep cliffs, and gentle foothill slope.
- Gentle dipping strata: gentle inclined top, steep cliffs, and gentle foothill slope.
- Steep dipping strata: if dip angles of layers are more than 60°, the bedding plane can form scarp slope.

1.4.2 The Complexity of the Slope Morphology

The morphology of Danxia slopes can be very complicated.

- Water erosion type, collapsed slope type, multifactor (e.g., rainwater, wind, ripple) influenced type.
- Undulating slope surfaces due to lithology discontinuity along the bedding.
- Vertical grooves developed on Danxia slopes.
- Spheroid weathering on Danxia slope.
- Rock surface is not always red colored.

Fig. 1.2 Danxia landforms developed on nearly horizontal strata (Left); Cuesta Danxia landforms with monoclinic terrain (Right) (photographer: Hou Rongfeng)

1.5 The Taxonomic Position and Classification System

1.5.1 The Discipline Ownership

In the landscape classification system, the Chinese geologists and geomorphologists give Danxia an appropriate position, namely the Danxia belongs to red beds landform in rock geomorphology. In red beds landform, Danxia as a red beds landform with a characteristic of scarp slopes. In other words, red beds landform without scarp slopes may be the red beds hill or red beds mountain region or red beds mountain hills, so it cannot be called Danxia. Therefore, Danxia is a special kind of red beds landform (Fig. 1.3).

However, the international community has not taken it as an independent type to pay attention, Danxia is often included in the sandstone landform, red beds landform even in tectonic landform, sometimes, it is said to be "false-karst" landform. However, these classifications and views seem to be very difficult scientifically expressing its taxonomy status.

Danxia landforms are one of the most spectacular landscapes globally. However, due to the lack of international academic communication, the great amount of work that has already been done by Chinese scholars on Danxia landforms is poorly known outside China. Actually, according to the gained data so far, Danxia landforms are not unique in China. Literature shows that red beds, the sedimentary strata that are predominantly red in color due to the presence of ferric oxides and are responsible for the formation of Danxia landforms are widely distributed throughout the world. Many stunning landscapes developed on red beds outside China display a similar appearance with Danxia landforms, such as the sparse, towering buttes in Monument Valley, Arches National Park, Southwestern USA, sandstone peaks Purnululu National Park, Western Australia. Although there is lack of research on "Danxia," there is not rare research on those similar landforms. These Danxia-like landscapes have not been identified as an independent landform in the international geoscience community, and relevant studies were conducted in the scope of sandstone landforms, or named as "red bed landform", "pseudo Karst," or "quartzose Karst"; most of red beds are distributed globally in rainless region, and they are usually researched as arid area landform.

For the development of geomorphology subdiscipline, it is still very weak in the research on this kind of landform. The typical Danxia landform in China is mainly developed from terrestrial coarse clastic rocks (conglomerate or sandy

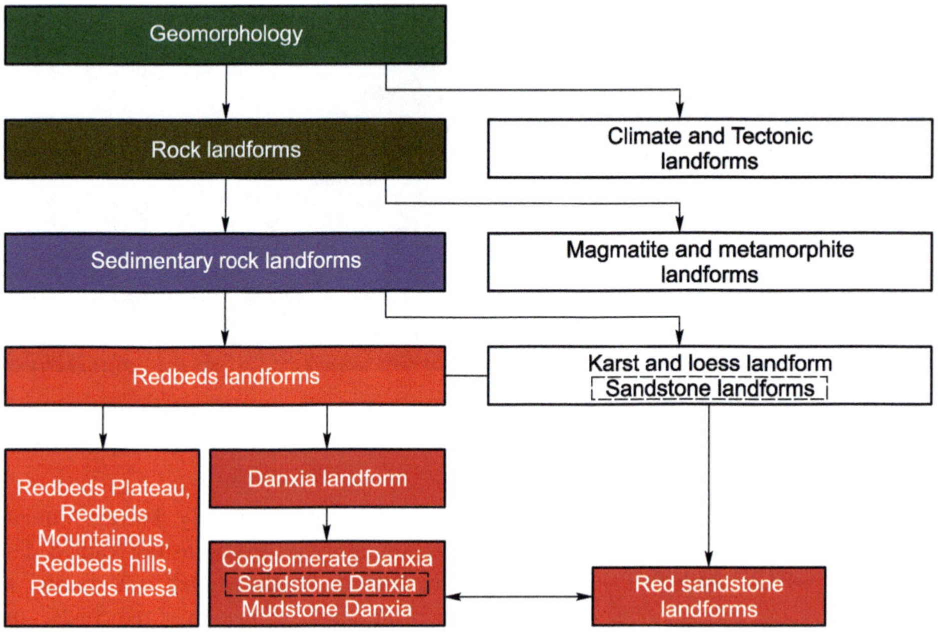

Fig. 1.3 Discipline ownership on Danxia Landform

conglomerate). Thus, it is not suitable to simply classify Danxia to sandstone landform. Landforms vary greatly from red beds; Danxia is just one of them, so that it is also inappropriate to roughly categorize Danxia into red bed landform. Some Danxia landforms involve karst phenomenon; however, it is not a scientific classification to attribute Danxia to "pseudo karst" or "quartzose karst." Therefore, it is inaccurate to simply classify Danxia into any of the above-mentioned category.

Admittedly, geoscientists have done a lot of research from different perspectives, although terminologies and ways of classification vary. However, we have to acknowledge that despite differences in terminologies, sandstone landforms, red bed landforms, or Danxia landforms are relatively under-researched if compared with other rock geomorphology such as Karst and granite landforms. The research on Danxia landforms is still in its fledgling stage internationally. Therefore, it is necessary for us to join our efforts in the research of it.

1.5.2 The Taxonomic Position and Classification System

As an independent subdiscipline of modern geomorphology, Danxia landform needs a scientific classification system to promote further research. Huang et al. (1992) proposed the first classification system of Danxia landform, on the basis of bedding inclination, climate, development stages, morphology, and existence of overlaid layers and solution weathering. From then on, some other classification systems of Danxia landforms were addressed by scholar (Luo 1994). However, most of them remained focused on the description of individual morphology features; few quantitative indexes have been included in those classification systems.

Based on the previous research, Hua Peng proposed a more integrated classification system of Danxia landforms (Table 1.1). He suggests the first-order classification criteria, including material foundation, geological structure, dominant exogenic forces, landform geometry, and

Table 1.1 Classification system of Danxia

Criteria			Type
Material	Detritus type		Conglomerate Danxia, sandstone Danxia, mudstone Danxia
	Special element		Karstic Danxia, earthy forest-type Danxia, mud-coating-type Danxia
Attitude of rock layer	–		Nearly horizontal Danxia, gently slope Danxia, steeply dipped Danxia
Dominate force	Climate		Danxia in humid region; Danxia in arid region; Danxia in Alpine region
	Exogenic force		Water erosion Danxia; weathered Danxia; gravity collapse Danxia; wind erosion Danxia; sea erosion Danxia; Karstic Danxia; artificial Danxia
Landform geometry	Single geometry	Positive landform	Escarpment face; mesa; cuesta; stone wall; stone column; stone peak; upland; hill; stone single peaks; stone balls; colluvial pile and colluvial rock block
		Negative landform	Danxia valley (linear ravines, narrow gorge, canyon, incised meander, broad valley); Danxia rock groove (bedding notch, bedding groove, vertical groove); Danxia cave (bedding cave, horizontal cave, periclinal cave, superposed cave, perforated cave, honeycomb cave, niche-like perforated cave, vertical cave); Danxia arch; Danxia pothole
	Combination structure		Plateau–canyon-type Danxia; mountainous Danxia; peak forest-type Danxia; single-peak-type Danxia; hilly Danxia
Developmental stages	–		Young stage; mature stage; old stage

development stages. Each criterion is subdivided into the secondary criteria, which form a secondary classification system.

1.6 Distribution of Danxia Landforms in China

Danxia landforms are widely distributed throughout China, and more than 1000 locations had been identified by the end of 2014. They are found in tropical and subtropical humid region, temperate humid and semi-humid region, semi-arid and arid region, as well as on Tibetan Plateau. Geographically, Danxia landscapes in China can be classified into three regions of densest occurrence (Fig. 1.4). They are Southeastern China, where it is featured by peak-hoodoo-cluster Danxia developed in the hilly region of South Yangtze River; Southwestern China, where it is featured by plateau–mountain–valley Danxia developed in the transitional zone between the margin of the Sichuan Basin and its surrounding plateau, as well as in the Yunnan–Guizhou Plateau; Northwestern China, where it is featured by arctic-alpine Plateau and arid mountain type of Danxia developed in Gansu Province and surrounding provinces.

1.7 Development Mechanism of Danxia Landforms

1.7.1 Material Basis

(1) **Distribution of red beds**

As mentioned above, red beds are the material basis for the formation of Danxia landforms in China. Many researches suggested that red beds in China belong to continental sedimentary rocks. The age of most red beds in China is no earlier than Mesozoic, and main were formed in Cretaceous, followed by Jurassic and Paleogene (Zeng and Huang 1980; Peng 2000a, b; Peng and Wu 2003; Guo et al. 2007a, b).

As shown in Fig. 1.5, red beds are widespread across China. In East China (I, II, III), red beds were mainly formed in Cretaceous and Paleogene. In Middle China (IV, V), they were mainly formed in

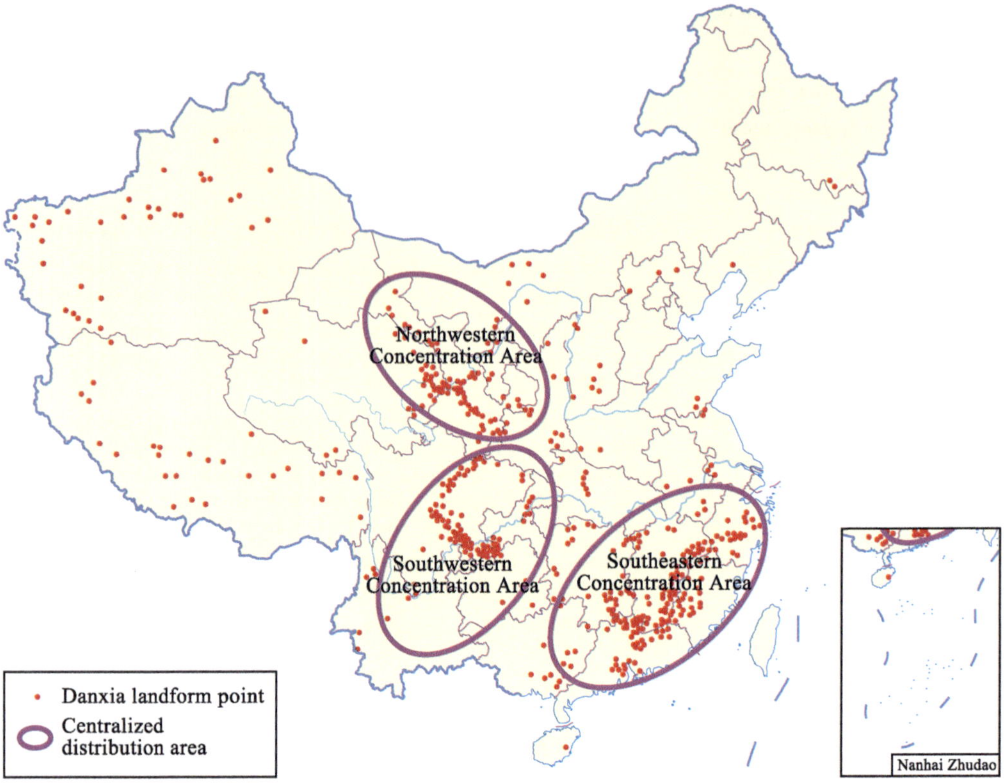

Fig. 1.4 Distribution of Danxia landforms in China

Cretaceous and Jurassic. In Northwest China (VI), they were mainly formed in Paleogene and Neogene; the Cretaceous red beds are mainly exposed at the margin of large basins, such as the Tarim Basin and Junggar Basin. Red beds in Qinghai–Tibet Plateau (VII) are mainly formed in Cretaceous; the Paleogene and Neogene continental deposits are mostly grayish yellow or brownish gray.

(2) **Material composition**

With regard to the composition of red beds, the dominant minerals are quartz, feldspar, and clay, which are mainly cemented by silica, calcite, ferric oxides, and clay minerals especially calcite, which is the most common cements in red bed rocks (Liang et al. 1992). The dissolution of calcite cement plays a significant role in the solution weathering of red bed rocks in humid South China (Liu 1994). The ferric oxides, which are responsible for the red color of red beds, typically occur as a coating on the grains of sediments comprising red beds.

However, because of various depositional environment and the later geological effect, the color of red beds is not always in bright red, but it could be presented as a range of brownish yellow, brown, purplish red, brownish red, and gray purple (Peng 2000a, b).

In addition, red beds also vary in lithology, including conglomerates, sandy conglomerates, sandstones, siltstones, and mudstones. Peng (2000a, b) addressed that Danxia escarpments were commonly developed on the relatively thick layers of hard conglomerate, sandy conglomerate, and sandstone, which often accumulated at basin margin. However, the siltstone and argillaceous red beds deposit in the center of a basin contain much soluble material and clay mineral, with poor water permeability, and rich

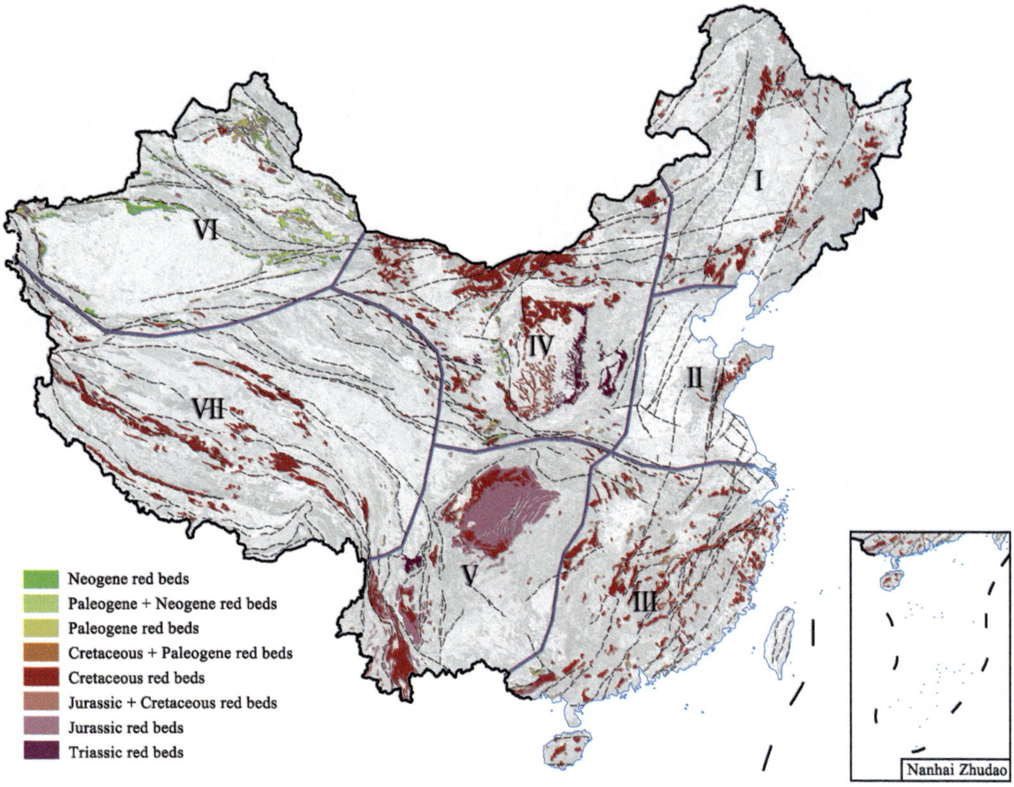

Fig. 1.5 Ages and distribution of red beds in China

in water, and they are easily damaged by erosion and weathering processes, forming red bed hills with low altitude.

1.7.2 Tectonic Control

According to the research of China geologic history, the unified China continental plate was formed in Triassic, by the combination of South China, North China, Indo-China, and other neighboring continental segments. Since the Jurassic, the eastern part of South China was in an underthrust zone of the Pacific plate, forming a NNE-oriented magma tectonically active zone. The eastern red bed basins were strongly controlled by this structural zone; the west part of China was extruded by the Indian plate, forming several nearly EW-oriented red bed basins from north to south. In the middle part, there was a NE-oriented transpressional zone, forming several big red bed basins, like Ordos Basin and Sichuan Basin, and numerous small intermontane basins (Peng and Wu 2003; Guo et al. 2007a, b).

In red bed basins, tectonic line pattern is the basic factor that controls the distribution patterns and shapes of Danxia hill blocks. Large tectonic lines control the general arrangement of hill blocks, while small structures control the trends, density, and the plane pattern of hill blocks. For instance, the arrangement of hill blocks at Mt. Danxiashan extends in a NNE direction along the major fault (Fig. 1.6). The trend of hill blocks and the arrangement of stone pillars were mainly along secondary faults and large joints of the nearly EW trend (Peng 2000a, b).

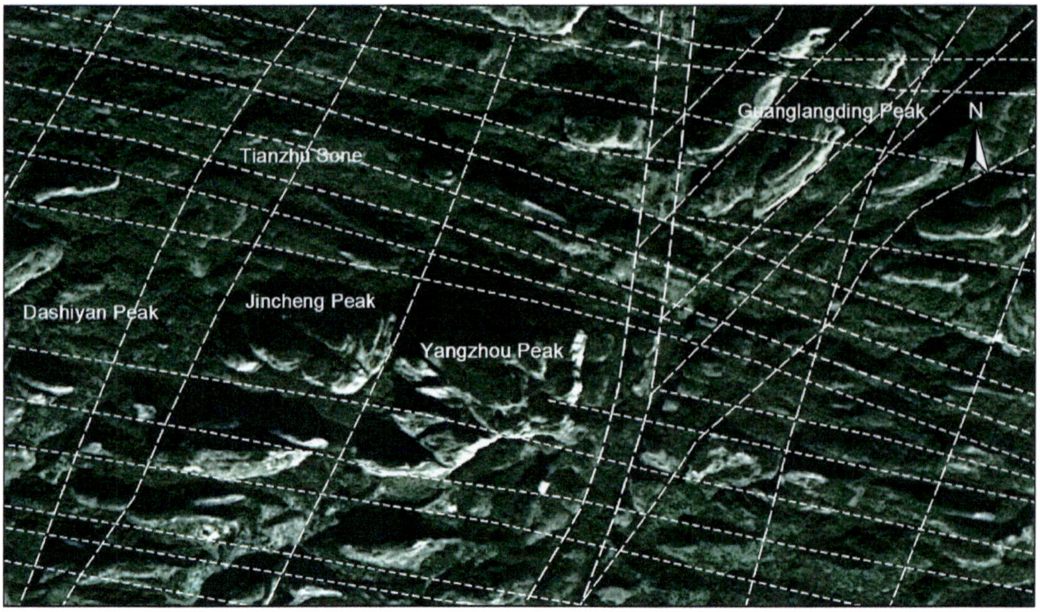

Fig. 1.6 Main structural line of eastern part of Danxiashan

In addition to the effect of tectonic structure and tectonic line pattern within red bed basins, tectonic uplift during the neotectonic movement also plays a significant role in the formation of Danxia landforms in China. For instance, at Mt. Langshan Danxia landform area, the overall terrain descends from west to east, and the regional tectonic uplift varied from strong in the west to weak in the east. This produced three levels of platform gradually descending toward the north, with each platform separated by about 200 m in altitude.

1.7.3 Exogenic Forces

After red bed basins were uplifted to a certain level, the curst would keep a relatively long-term stable state, which provides the necessary conditions for the erosion of red beds by exogenic forces. The main exogenic agents responsible for this erosion process include flowing water, weathering, and gravity effect. Among which, flowing water plays the most significant role in sculpting this landform, especially in the humid Southeastern China (Peng 2000a, b), flowing water incised the uplifted red beds along faults and vertical joints, forming deep and narrow gorges. Then, with lateral erosion, the gorges were widened to become broad canyons. During this process, the exposed red bed rock layers are also constantly being attacked by weathering (Fig. 1.7). In particular, on some steep slopes, there is little water erosion, while weathering action plays a more crucial role. The lithology difference in vertical direction may result in differential weathering on red bed rocks. The resistant hard sandy conglomerate weathers slowly, and it protrudes above the softer clay and silt layers, which weather rapidly and recede to form bedding grooves or caves on Danxia slopes.

Fig. 1.7 Grooves formed by the weathering of argillaceous siltstones interlayers at Hunyuan Cave of Mt. Danxiashan (**a**); Disintegration of siltstones at Chuanyan Cave of Taining Danxia landform area (**b**); Layered exfoliation of fine sandstones on cliff surfaces of Chishui Danxia landform area (**c**); Weathering of argillaceous cemented sandy conglomerates at the foot of Baota Peak of Mt. Danxiashan (**d**)

In terms of gravity effect, usually occurs on overhanging slopes that are formed by the lateral erosion of running water or the weathering action of soft interlayers. The overhanging rock blocks may collapse along the primary joints and unloading joints. They collapsed rocks pile up on the foot of scarp slopes, forming gentle colluvial foothills (Fig. 1.8).

1.7.4 Evolution Process

The evolution process of Danxia landform begins at the uplift of red bed basins. Then, with the repeated erosion and collapse processes, the area of the mountain top will be gradually reduced, and the original red bed plateau will be sculpted into peak clusters with castle shape, stone walls, or isolated stone pillars (Fig. 1.9). Based on the model of the fluvial cycle of erosion developed by Davis, Huang (1982) divided the evolution process of Danxia landform into four stages: young, mature, old, and rejuvenation. From then on, by taking Stranler's "area-altitude analysis of erosional topography" as reference, Peng Hua made some improvement to the evolution model of Danxia landform, and he classified the six serial nomination sites of China Danxia for

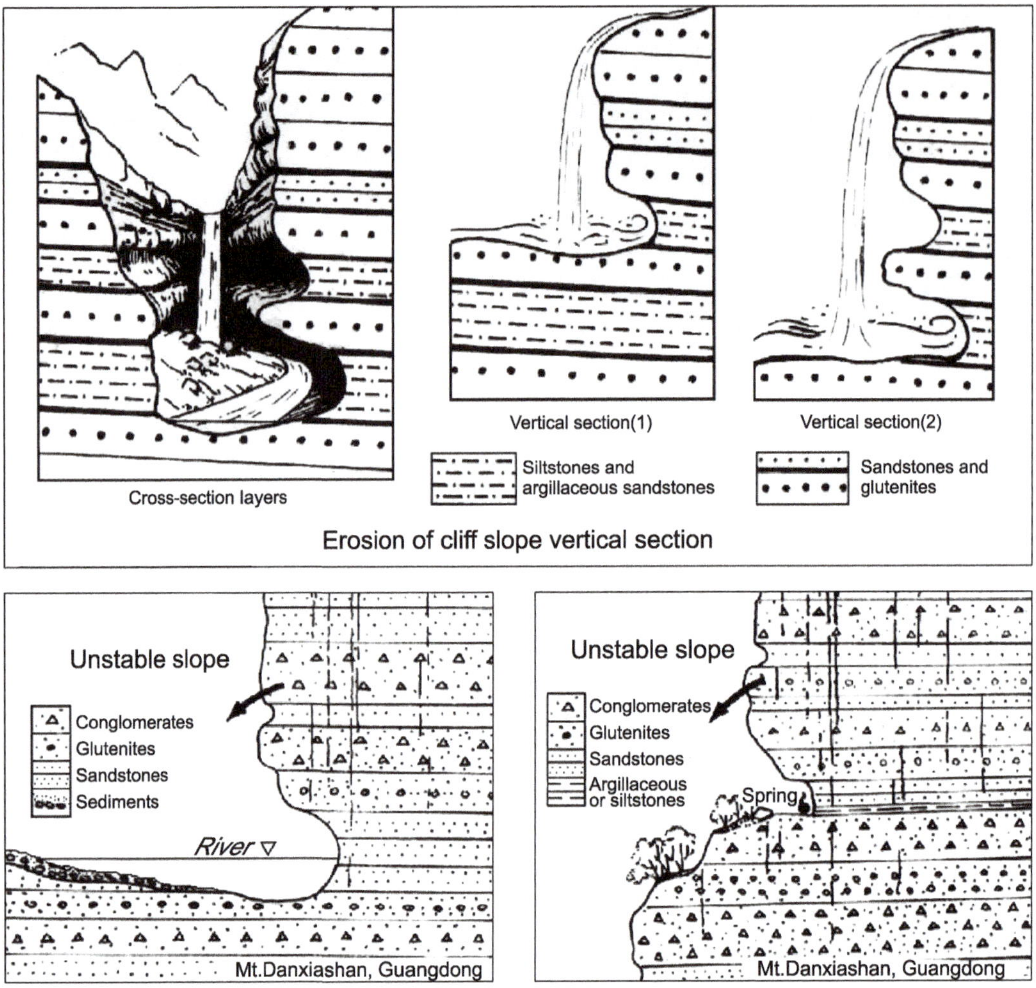

Fig. 1.8 Incising and lateral erosion of running water (above); Two main conditions of cliff collapse (below)

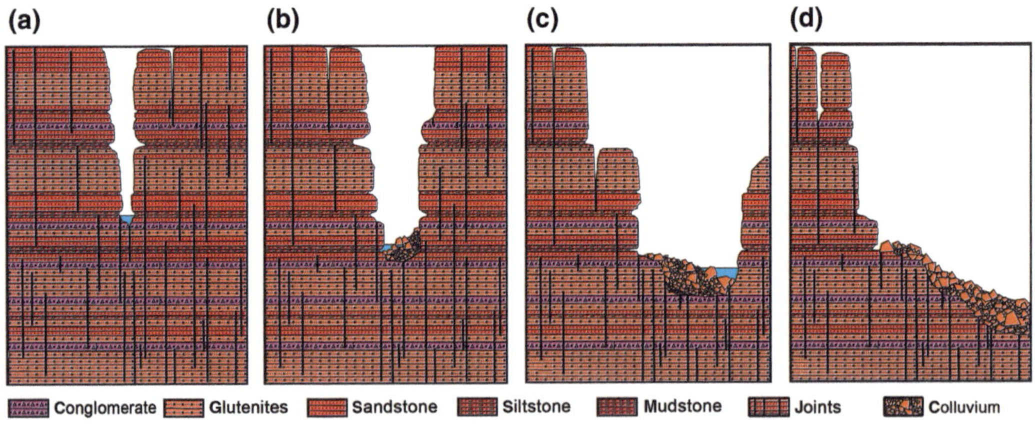

Fig. 1.9 Evolution process of Danxia landforms

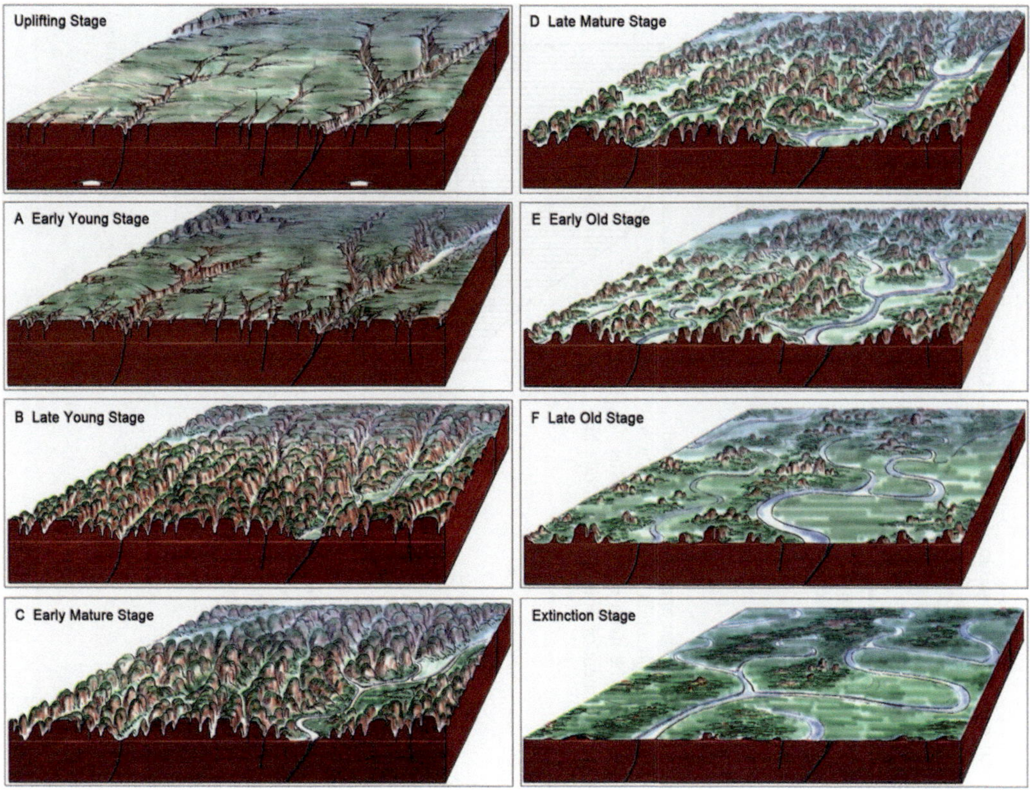

Fig. 1.10 Evolution processes of Danxia landforms

World Natural Heritage into six stages: early young, late young, early mature, late mature, early old, and late old (Fig. 1.10). He also proposed a quantitative index of denudation degree for this classification system (Table 1.2). In addition, researcher found that there are different stages for the evolution process of Danxia landform developed in the margin and center of a red bed basin, a finding supported by a research in Xinjiang Basin.

Table 1.2 Evolutionary Stages of Danxia landform

Type	Indicator and foundation	Feature (supposing the crust is long-term stable after uplifting)	Graphic
Early Young Stage	Original sedimentary surface or ancient denudated planation kept continuous (>50%); erosion amount was less than 20%	Red bed uplifting area and plateau basin geomorphologic combination that belong to crustal uplifting—water incising—lane valley and gorge developing stage. A large number of original sedimentary surfaces, ancient denudated plantations, or weak erosion platforms were kept in the upper part; the longitudinal profile of gorge was stepladder like with many waterfalls and water drops	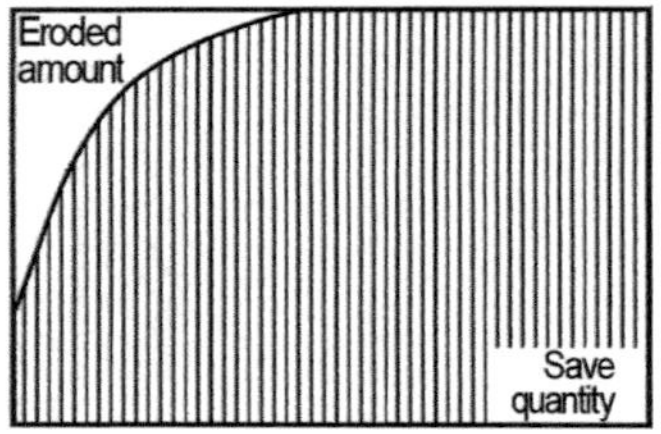
Late Young Stage	Summit surface reduced and original gentle surface was less than 30%; erosion amount was 20%–40%	It showed a dense peak cluster—gorge appearance. Original plateau was broken into discrete mountain blocks which were mainly like mesa or Loess Hill; the main river valley was mainly gorges with the development of point bar or flood plain in parts	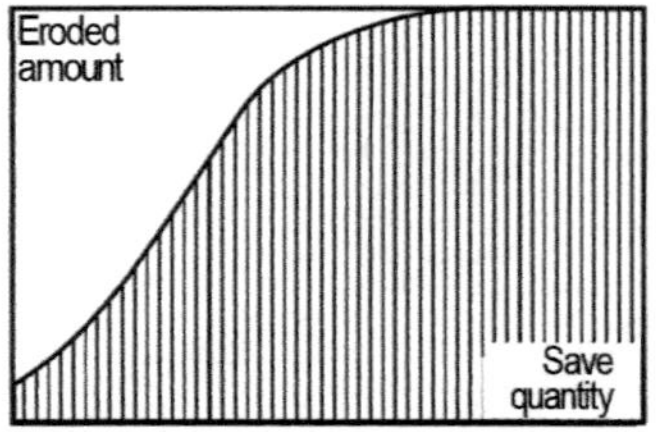
Early Mature Stage	The mountain blocks were discrete. The proportion of original surface was 10%–20%, and the erosion amount was 40%–55%	The mountain blocks were discrete and showed themselves like peak cluster. The ancient denudated planation was clear but the surface was narrow; the main river valley is distributed with gorges and straths alternating appearing. The main riverbed was close to regional erosion base level; there were peak forest near river valley with peak cluster far away from river valley. The surface was rough, and the height difference was largest	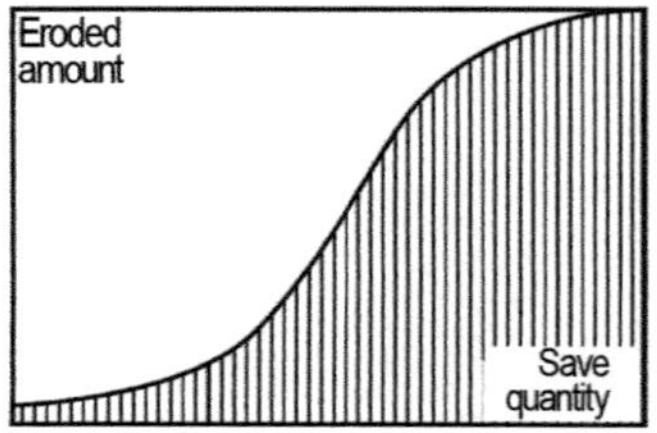

(continued)

Table 1.2 (continued)

Type	Indicator and foundation	Feature (supposing the crust is long-term stable after uplifting)	Graphic
Late Mature Stage	It showed a groups of peak cluster—peak forest appearance. The gentle summit surface was less than 10%, and erosion amount is 55%–70%	The main river valley turned into strath by erosion, and the main branch valley was also close to regional erosion base level. It showed a group of peak cluster—peak forest appearance. The summit surface reduced to less than 10% with the continuous reduction by weathering erosion	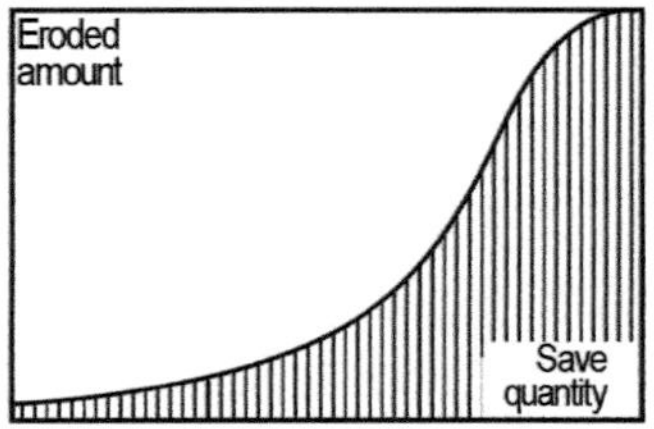
Early Old Stage	Combination of strath—peak forest—isolated peak. Denudated planation disappeared, and the erosion mount was 70%–85%	The main river valleys were all straths, and the main branch valleys reached the regional erosion datum plane. It showed a combination of strath—peak forest—isolated peak with small range peak cluster in parts. The mountain blocks reduced, and summit surface decreased with no original surface	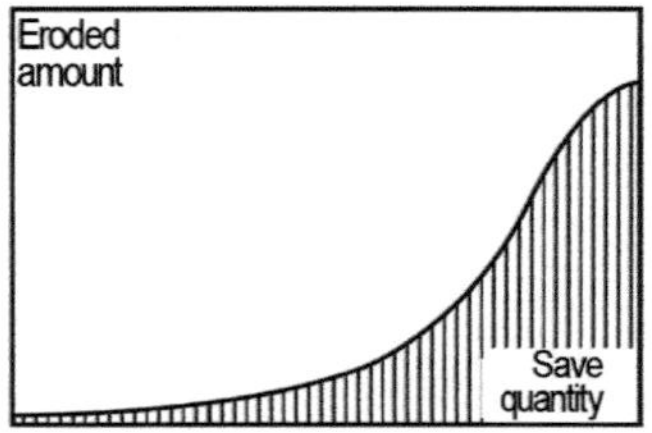
Late Old Stage	Peneplain erosion above the main riverbed was more than 85%	There was undulatory peneplain surfaces with small rang of peak forest, isolated peak, isolated hill, or isolated stone	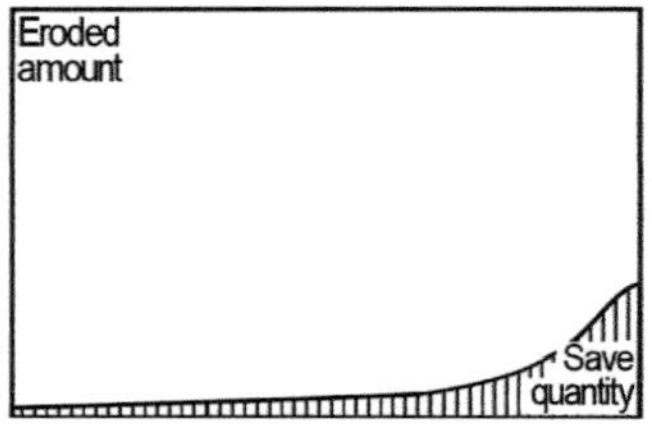

2 Identification of the Property

2.1 Introduction

2.1.1 What's China Danxia?

"China Danxia" is a general name for the Serial Heritage for World Natural Heritage. It contains six world natural heritage sites, which are developed by the Danxia landform.

It is not just a term of geomorphology, but also refers to a perfect scenery combination consisting of red cliffs, blue waters, green forests, and white clouds, even relying on the environment developed by the Danxia culture.

2.1.2 Why Is China Danxia an Appropriate Serial World Natural Heritage?

According to Paragraph 137 of the Operational Guidelines for the Implementation of the World Heritage Convention (shortened hereafter to Operational Guidelines), the six heritage sites of China Danxia meet the required criteria as follows:

(b) *the same type of property which is characteristic of the geographical zone*; The heritage sites are all located wetted area in southern and southeastern China and fully representative of China Danxia in those regions.

The Danxia landscapes in the humid regions of China, especially the peak cluster/peak forest type Danxia landscapes in southeast China, have a close association of mountains, forests, and water, which gives them the most spectacular, attractive, and colorful display of Danxia landscapes anywhere in China or the world. Although Chishui is located in southwest China, it lies within the North Guizhou–South Sichuan Danxia landscape region, which has the largest continuous extent of Danxia landscapes in China. It is also the site that best represents the plateau–canyon–waterfall–forest combination in a Danxia landscape.

(c) *the same geological, geomorphological formation, the same biogeographic province, or the same ecosystem type*; The heritage sites are all located within a Mesozoic activated geosyncline system in Southern China, with the same or similar evolutionary history and geotectonic background. They all developed from terrigenous red clastic rocks of Cretaceous age, within a subtropical humid monsoon climatic zone. Today they form distinctive red-colored landscapes, with associated typical terrestrial ecosystems dominated by evergreen broad-leaved forest vegetation.

The heritage sites represent different development stages in geomorphological evolution of Danxia landscapes from the youngest stages to

H. Peng, *China Danxia*, Springer Geography,
https://doi.org/10.1007/978-981-13-5959-0_2

oldest stages, and each has a distinctive array of landforms typical of that evolutionary stage. In combination, they provide a full, intact, and logical series of Danxia landscapes and landforms in southeast China.

In addition, the heritage sites are all protected areas. They are also among the most intensively researched of all Danxia landscape areas in China. This excellent basis of protection and knowledge means they are ideal places for promoting and developing science, education, tourism, and the sustainable conservation and development of natural resources.

Thus, the heritage serial properties as a whole represent the peak cluster/peak forest type Danxia landscapes in low-altitude humid regions of southeast China and the plateau–canyon type Danxia landscapes in humid regions of southwest China. They include the most beautiful Danxia landscapes in China and reflect the systemic value of China Danxia intactly.

The selection of sites for the serial heritage of China Danxia, first is based on the requirement that they have outstanding universal value, including the most representative Danxia landscape areas. They show the comprehensive characteristics and key diagnostic natural values of China Danxia. They are all in red sedimentary beds from Mesozoic and Cenozoic time, and best represent the earth science values of China Danxia. Second, they are the most representative sites demonstrate different stages of geomorphological evolution, they can constitute an intact evolutionary series of Danxia landform, and they must be areas where ecological environments are under best protection. At the same time, the heritage serial properties also include the most beautiful Danxia landscapes in China, which also have unparalleled outstanding scenic and aesthetic value compared to any other similar place in the world.

China Danxia, therefore, complies with the two conditions (b) and (c) for World Heritage serial properties and it has outstanding universal value at aspects of earth science, biology, ecology, and landscape aesthetics.

2.1.3 The Planned Sequence of China Danxia World Heritage

"China Danxia" world natural heritage includes a total of 6 sites in 6 regions. They are: Chishui, Guizhou region; Taining, Fujian region; Jianglangshan, Zhejiang region; Langshan, Hunan region; Danxiashan, Guangdong region; Longhushan, Jiangxi region.

The following are the heritage sites to compile the information. The properties are composed of sites that have been carefully selected and arranged according to how they demonstrate different stages of geomorphological evolution, from the youngest stages to oldest stages. The following ordering of sites, therefore, has a geological basis only, and it is not a ranking of the relative heritage value of the candidate sites.

- Chishui, Guizhou Province: Young Stage A/B—representative of plateau-canyon Danxia landscapes with intensive uplift and deep incision.
- Taining, Fujian Province: Young Stage C—representative of deeply incised river meanders in a mountain-plateau (Prototype of the peak cluster) and canyon landscape, with cliffs and caves of varying origin and formation.
- Langshan, Hunan Province: Mature Stage A—representative of Danxia peak clusters and peak forests, with a dense array of dome- and needle-shaped forms.
- Danxiashan, Guangdong Province: Mature Stage B/C—The named prototype site of Danxia landform, representative of classical Cluster type peak clusters and peak forests.
- Longhushan, Jiangxi Province: Old Stage A/B—representative of Danxia landscapes with scattered peak forests and single-peak groups of diverse origins, separated by lower altitude areas.
- Jianglangshan, Zhejiang Province: Old Stage C—representative of Danxia landscapes with prominent and isolated single-peak landforms surrounded by lower terrain.

2.2 Identification of the Property

Country	People's Republic of China
Name	General Name: China Danxia
State, Province or Region	Heritage sites of "China Danxia" (the first period), located in 7 cities of 6 provinces in Southern China: Guizhou Province Zunyi City (Chishui) Fujian Province Sanming City (Taining) Hu'nan Province Shaoyang City (Langshan) Guangdong Province Shaoguan City (Danxiashan) Jiangxi Province Yingtan City, Shangrao City (Longhushan) Zhejiang Province Quzhou City (Jianglangshan)

(1) Geographic coordinates

General coordinates: 24°51′48″N–28°33′03″N, 105°47′39″E–118°35′02″E

Coordinates of the serial heritage sites

Section	Heritage site	Center coordinates	
Guizhou Province	Chishui	West Section	28°22′11″N, 105°47′39″E
		East Section	28°25′19″N, 106°02′33″E
Fujian Province	Taining	North Section	27°00′37″N, 117°13′07″E
		South Section	26°51′56″N, 117°02′22″E
Hunan Province	Langshan	26°20′24″N, 110°46′45″E	
Guangdong Province	Danxiashan	24°57′55″N, 113°42′12″E	
Jiangxi Province	Longhushan	Longhushan Section	28°04′15″N, 116°59′05″E
		Guifeng Section	28°19′03″N, 117°25′10″E
Zhejiang Province	Jianglangshan	28°31′44″N, 118°33′43″E	

(2) Description of the boundaries of the property

The core zones of the heritage sites in the proposed serial property of China Danxia are all designated as National Park, National Natural Reserve or National Natural Heritage. The boundaries of the sites are carefully surveyed on the ground and identified in planning documents. The boundaries are based on the natural features such as rivers, valleys, and ridges. The boundaries are located to exclude major residential areas, mining areas, and any industrial or other large building or construction sites.

The boundaries of buffer zones are also distinct and well-surveyed and are clearly shown in planning documents. These, too, are primarily based on natural features but in places are roads or land-use boundaries. In setting the boundaries of the buffer zones, consideration was given to the need to protect the heritage core zones from external human influences as much as possible. Relevant regulations are enforced within village areas and special protecting and monitoring teams exist to implement effective protection.

The sites in the serial heritage are continuous natural areas focused on and including all main Danxia landscapes and their associated biological features and communities.

(3) Area

The total area of the nine sites in the serial property is 82,151 ha. The total area of the buffer zones is 136,206 ha. The total area of the entire the Heritage Sites is 218,357 ha.

(4) Distribution maps

Distribution maps of the Heritage Sites are shown in Figs. 2.1 and 2.2.

Area of the Heritage Sites (hectares)

Section name	Heritage site		Area of the Heritage site	Area of the buffer zone	Total area
Guizhou Province	Chishui	West Section	10,142	25,341	72,178
		East Section	17,222	19,473	
Fujian Province	Taining	North Section	5277	4247	23,488
		South Section	5810	8154	
Hunan Province	Langshan		6600	6200	12,800
Guangdong Province	Danxiashan		16,800	12,400	29,200
Jiangxi Province	Longhushan	Longhushan Section	16,950	41,030	79,510
		Guifeng Section	2740	18,790	
Zhejiang Province	Jianglangshan		610	571	1181
Total	9		82,151	136,206	218,357

Fig. 2.1 General map of distribution of the Heritage Sites of China Danxia in China

Fig. 2.2 Distribution of the Heritage Sites of China Danxia in Southeast China

2.3 Value of Heritage on China Danxia

According to paragraph 137 of the Operational Guidelines it is the series as a whole that must be of outstanding universal value. On this basis, the nomination text report the heritage sites of China Danxia meets the following four criteria for outstanding universal value, as specified in Section II.D paragraph 77 of the Operational Guidelines. Among them, (vii) and (viii) was identified by the World Heritage Committee.

(vii) **Contain superlative natural phenomena or areas of exceptional natural beauty and aesthetic importance**: China Danxia forms a majestic landscape of high mountain peaks, steep cliffs, and deep canyons with spectacular waterfalls, rivers, and lakes, and an extensive cover of natural forest vegetation. The six heritage sites are those that best demonstrate the most beautiful Danxia landscapes in southern China. They have scenic and aesthetic qualities that are of outstanding universal value.

(viii) **Be outstanding examples representing major stages of earth's history, including the record of life, significant ongoing geological processes in the development of landforms, or significant geomorphic or physiographic features**: The Danxia landscape is symbolic of a special phase in the regional development of the planet's continental crust. Basin evolution of the heritage sites reflects the evolutionary history of the earth's crust since the Cretaceous. China Danxia is also a model for demonstrating a long history of landform evolution in uplifted and intensively faulted sedimentary basins, including both past and ongoing geomorphological processes. Research over a period of more than 80 years has shown that the Danxia landscape is a special natural earth science phenomenon, regional in scale and global in its significance.

(ix) **Be outstanding examples, representing significant ongoing ecological and biological processes in the evolution and development of terrestrial, fresh water, coastal, and marine ecosystems and communities of plants and animals**: The property is located in two Biogeographical Provinces—"Chinese Subtropical Forest" and "South Chinese Rainforest", and in two of the world's eight Biogeographical Realms—the Palaearctic Realm and the Indo-Malayan Realm. The series of sites present an outstanding example of the composition of the biological community and the successional processes within evergreen broad-leaved forests developed under a southeast monsoonal climate. It also typifies the post-glacial ecological succession within its biogeographic setting.

(x) **Contain the most important and significant natural habitats for in situ conservation of biological diversity, including those containing threatened species of outstanding universal value from the point of view of science or conservation**: The property is an outstanding example of biodiversity of an ecotope within the "Chinese Subtropical Forest" and "South Chinese Rainforest". There are a total of 5722 higher plant species, 836 vertebrate species, and 3073 insect species. The property is located in the Southeast China-Hainan Moist Forests according to the "WWF Global 200 Biota regions". It contains an ancient biota with biological communities in their natural state. There are some 400 rare species of flora and fauna, and more than 40 endemic species within the heritage sites, which means it is of global significance for biological conservation.

It is important to note that each of the heritage sites of the China Danxia serial heritage meets at least two of the above criteria (see Part II for descriptions of each heritage site).

Value of the Heritage	
Description of boundaries of the heritage	The core zones of the heritage sites in the proposed serial property of China Danxia are all designated as National Park, National Natural Reserve or National Natural Heritage. The boundaries of the sites are carefully surveyed on the ground and identified in planning documents. The boundaries are based on the natural features such as rivers, valleys, and ridges. The boundaries are located to exclude major residential areas, mining areas, and any industrial or other large building or construction sites The boundaries of buffer zones are also distinct and well-surveyed, and are clearly shown in planning documents. These, too, are primarily based on natural features but in places are roads or land-use boundaries. In setting the boundaries of the buffer zones, consideration was given to the need to protect the heritage core zones from external human influences as much as possible. Relevant regulations are enforced within village areas and special protecting and monitoring teams exist to implement effective protection The sites in the serial heritage are continuous natural areas focused on and including all main Danxia landscapes and their associated biological features and communities
Statement of justification of outstanding universal value	(1) **Landscape Aesthetic Values** China Danxia forms an exquisite natural landscape, highlighted by massive mountain blocks and majestic cliffs. Red coloration in the rocks is a prominent landscape feature but is complemented by other natural elements such as mountains, forests, and water features—waterfalls, rivers, lakes, and wetlands. In total, the landscape elements produce sites of great scenic and aesthetic value and interest. The sites are prime examples of Danxia landscapes and landforms. These are renowned scenic spots in China and among the best of their type in the world The heritage sites of the China Danxia property form a series spanning the full range of landform development from youth to maturity and old age within the humid region of south China. Danxia landscapes in the youthful stage are characterized by incised meanders with steep gorges and deep glens. Maturely developed Danxia landscapes are dominated by forest-covered peak clusters. The old-age stage of Danxia landscape development has isolated peaks surrounded by more gently sloping low-lying land and rivers. The majestic and colorful Danxia landscape evokes a sense of wonder and awe among those who witness it. The great height and steepness of the red walls and cliffs, the supernatural shapes of the landforms, the beauty of the forested land surfaces, the quiet and serenity of the valleys and forests, and the mystery and fantasy of clouds and mists, all give Danxia landscapes an extraordinary beauty that is of world class (2) **Earth Science Value** China Danxia is an outstanding example representing elements of the evolution of the Earth's continental crust since the Mesozoic. It includes a range of geological phenomena and ongoing landform evolution. It contains a significant record of life on earth, and important and distinctive landforms. Overall, it can be considered to be of outstanding universal value for earth science **China Danxia demonstrates the character of the earth's continental crust at a specific phase of development**. The extensive Chinese red beds were formed in the late Mesozoic, and contain key geological information about that period, such as the character of continental fault basins, and ancient geography, climate and environment. The red beds were uplifted in the Cenozoic, which initiated the development of the Danxia landscapes. They reveal evolutionary elements of the continental crust, including the formation of large-scale crustal plates in a relatively late geological era. They also show the complete process of regional crust formation with alternating periods of activity and stability. In particular, Danxia landscape evolution is globally significant for revealing the history and processes of geographical and environmental changes and of climate changes on earth since the Cretaceous period

(continued)

Value of the Heritage	
	China Danxia displays the ongoing geomorphological evolution. The heritage sites of China Danxia display a great diversity of landscapes and landforms and ongoing land-forming processes are clearly shown. Danxia landscapes are important in global geology as an outstanding example of the ongoing geological changes in the earth's surface. The property is a natural museum for displaying geomorphological features and dynamic geomorphological process and is a natural laboratory for the study of geological science in the evolution of continental basins **Evolution of China Danxia reflects global changes and major events in the earth's land surface system since the Late Mesozoic–Cenozoic, which may be significant for research on current global changes**. China Danxia is valuable for global comparative research in geology and geomorphology. Research on Danxia development processes, its climatic environment and geographical features may also have value for understanding global changes in the earth's land surface system, and geological events since the Mesozoic **The property is an irreplaceable Danxia landform and landscape system**. The serial heritage of China Danxia includes landscapes at different stages in their geomorphic evolution, with different landform types and different combinations of landscape features. All stages in landscape evolution are present from youth through maturity to old age. Thus, the property serves as an in situ museum and a textbook for understanding the evolution of Danxia landscapes. The sites within the property are type localities for comparative study of Danxia geology. The overall scientific, aesthetic and ecological values of the property are not found in any other comparable heritage site in the world (3) **Biological and Ecological values** There is a complex and varied pattern of natural habitats in the property. Included are eight first-grade habitat types recognized by IUCN/SSC, accounting for 61.5% of these habitat classes in the world. The typical zonal vegetation in the property is evergreen broad-leaved forests, comprising 70 different formations and 102 associations. Driven by the southeast monsoon, these forest types are globally representative of intact subtropical evergreen broad-leaved forests. They display both primary forests and secondary successional forests. The particular geological and landform setting produces a great diversity of ecological systems and a complex mosaic in the spatial variation of communities at smaller scales. Discreteness of Danxia landscapes leads to intensive fragmentation of habitats. This produces special "island effects" on hilltops and in ravines in particular. The property is therefore a natural laboratory for the study of the dynamics of biological communities and for the understanding of conservation biology The property, located within the humid regions of eastern Asia, is representative of biological diversity in the Palaearctic Realm and Indo-Malayan Realm of recognized global biogeographical systems (Udvardy 1975). The property is also representative of the southeast China-Hainan moist forest ecotope within the system of 200 global biotas recognized by WWF, and it extends across three biological diversity centers in Southern China, Central China and Southwest China. Its biological composition has strong features of ancient flora and original communities. There are approximately 400 rare and endangered species at all levels, as well as more than 40 locally endemic species. So, the Danxia property is a key area and the world for protecting wildlife diversity and endangered species and is of primary significance for the preservation and in situ conservation of the world's natural habitats and biological diversity

2.4 Summary to the Heritage Sites

Province	Name	General description
Guizhou Province	Chishui	Chishui is located in an area combining the Sichuan Basin and Yunnan-Guizhou Plateau, which is the largest region of continuous Danxia landscapes in China. The combination of vigorous uplift of a plateau and strong dissection by flowing water has produced a landscape of very strong relief. It has a typical stepped river valley system with spectacular waterfalls. It is a representative of the plateau–canyon type Danxia landscape in the early youthful stage of landform development. Chishui also has the most intact central subtropical forest ecosystem and rich species diversity. The landscape is characterized by a combination of "red mountain," "blue water," "grand waterfall," and "forest sea"
Fujian Province	Taining	Taining Basin records the evolutionary history of a continental margin mobile belt in the east of the South China tectonic plate since the Cretaceous. The landscape has remnants of an archaic denudation-planation surface. This surface is dissected by dense rectilinear canyons and narrow gorges. There are well-developed cliff caves, deeply incised meanders, and original ravine ecosystems. Taining is representative of the youthful stage of Danxia geomorphic development. A combination of canyons, dense peak clusters, mountains, rivers, and lakes produces beautiful scenery. The landscape retains intact natural ecosystems with high biological diversity
Hunan Province	Langshan	Langshan is located in a zone between the South China Plate and the Yangtze Plate and has experienced intermittent crustal uplift. The landscape of Langshan features dome-shaped landforms, with dense peak clusters and peak forests. There are linear valleys, large-scale natural bridges, and unusual Karst-like features. An "isolated ecological island effect" is found in the vegetation and extremely narrow natural habitats are very prominent. Langshan has the most integral plant communities, with natural biological succession and co-evolution of animals and plants, found in any Danxia landscape. The landscape here also has a rare degree of natural beauty. In the serial heritage, it is representative of a landscape in the mature stage of development and is also an excellent example of a classic Danxia landscape
Guangdong Province	Danxiashan	Danxiashan developed in the geotectonic basin of the central fold belt of Nanling. It is the type area for Danxia geology, from which the name derives. It is a model site for Danxia landscapes and landforms, exhibiting a great diversity of all the primary types and features. It is representative of the cluster-group type of Danxia landscape, with peak clusters and peak forests and is in the mature or late mature stage of geomorphic development. Compared to other heritage sites, it has the most tropical plant and animal species and the most outstanding ravine rainforest. It is the type region of some special biological communities such as "isolated island effects"
Jiangxi Province	Longhushan	Longhushan is located in Xinjiang Basin. The area records key geological events of the early Cretaceous, such as volcanic eruption, gypsum–salt deposition, aeolian deposition, and dinosaur extinction. The landscape is notable for the diversity of landforms including peak clusters, peak forests, isolated peaks, and monadnocks. The landscape is in the late mature and early old-age stage of geomorphic development. Rare low-altitude, central subtropical evergreen broad-leaved forests remain in this region. These provide significant habitat for rare and endangered species of plants and animals. Notably, this is

(continued)

Province	Name	General description
		the birthplace of Taoism in China, and there are many important cultural associations with the landscape, as well as cultural relics such as archaic cliff coffin sites in high caves
Zhejiang Province	Jianglangshan	Xiakou Basin in which Jianglangshan is located is a tectonic basin located on a deep fault. The red beds here consist of firm pyroclastic material of the Fangyan Formation. Differential erosion of this resistant rock type has led to prominent isolated peaks standing above the landscape. Much of this lower terrain is an ancient denudation-planation surface. In addition to the isolated peaks, the landscape features are narrow alleys in deep ravines and huge near-vertical rock walls. The landscape here is in the old-age stage of geomorphic development

3 Physical Geography

3.1 Geology and Geomorphology

The heritage sites of China Danxia are located in six provinces in South China—Guizhou, Hunan, Jiangxi, Zhejiang, Fujian, and Guangdong. While Chishui belongs to the transition zone between the Sichuan Basin and Yunnan-Guizhou Plateau, the other nominated properties are all located in the Jiangnan hill region, distributed on both sides along the main ridge of Nanling Mountain and Wuyishan Mountain. Nanling Mountain and Wuyishan Mountain form the topographic skeleton with lower terrain on both sides. They are within an uplifted zone, dating from the Cenozoic.

Location of the Heritage Sites of China Danxia in China

H. Peng, *China Danxia*, Springer Geography,
https://doi.org/10.1007/978-981-13-5959-0_3

Except for some exceptionally high peaks, the altitude is usually below 2000 m. Low hills with altitudes lower than 1000 m are dominant. The Danxia Landscape areas are within basins, the average height being 300–500 m. Some mountains reach 600 m in altitude, the highest one being Swan Pool Huluping in Xishui in Guizhou, with the altitude of 1750 m. Rivers continued to dissect these regions while the Cretaceous red bed basins were uplifted, and the Danxia peaks were formed between the rivers. Various types of mountain and water bodies form a beautiful natural landscape with high aesthetic value.

General geomorphology in heritage sites

No.	Section	Candidate site	Basin	Red beds	Ages of red beds	Topography	Highest peak (m)	Lowest point (m)	Elevation-difference (m)	Main terrain level (m)	Geomorphology	Developmental Stages
1	Guizhou	Chishui	Sichuan Basin	Jiaguan Formation K_2f	Late Cretaceous	Transmitting zone between Sichuan Basin and Yunnan-Guizhou Plateau	1750	240	1510	1100–1200 1400–1500	Plateau, Low mountain canyon	Early Young Stage
2	Fujian	Taining	Zhukou-Meikou Basin	Chongan Formation K_2c	Mid Cretaceous	East Basin in the mid-section of Mt. Wuyi	674	200	474	400–500	Peak-group, Canyon shape	Young Stage
3	Hunan	Langshan	Zixin Basin	Lanlong Formation $K_{1-2}s$	Early Cretaceous	Basin in the west-section of Mt. Nanling	818	302	516	500–600	Hoodoo, Peak-group shape	Early Mature Stage
4	Guangdong	Danxiashan	Danxia Basin	Danxia Formation K_2d	Late Cretaceous	Basin in the mid-section of Mt. Nanling	625	58	567	300–400	Hoodoo, Peak-group shape	Late-Mature Stage
5	Jiangxi	Longhushan	Xinjiang Basin	Tangbian Formation K_2t, Hekou Formation K_2h	Late Cretaceous	West Basin of the North-section of Mt. Wuyi	401	48	353	120–280	Hoodoo peak group Single peak Remnant hill	Late-Mature and old stage Old Stage
6	Guizhou	Jianglangshan	Xiakou Basin	Fangyan Formation K_1f	Early Cretaceous	East Basin of the North-section of Mt. Wuyi	824	170	645	500, 800	Single peaks	Old Stage

3.2 Climate

The heritage sites are within the middle subtropical humid monsoon climatic zone. They have abundant sunshine and rainfall, with a short frost-free period. The winters are short and summers are relatively long. Winter and autumn are dry while spring and summer are rainy.

Climatic statistics for the heritage sites

No.	Location	Candidate site	Annual average temperature (°C)	Average temperature of January (°C)	Average temperature of July (°C)	Annual rainfall (mm)	Annual hours of sunshine (h)
1	Guizhou	Chishui	18.1	7.9	28.0	1287	1297
2	Fujian	Taining	17.1	5.9	33.7	1788	1607
3	Hunan	Langshan	15.5	4	26	1450	1495
4	Guangdong	Danxiashan	19.7	9.3	28.4	1715	1725
5	Jiangxi	Longhushan	18.0	5.5	29.7	1878	1820
6	Zhejiang	Jianglangshan	14.0	4.5	29.4	1650	2063

3.3 Hydrology

The heritage sites of Chishui, Langshan, and Longhushan are within the Yangtze River watershed system; Danxiashan is in the Zhuajiang watershed system; Taining is in the Minjiang watershed system; Jianglangshan is in the Qiantangjiang watershed system. The water resources in the rivers flowing through the heritage sites are rich and the water quality is good. There is a high seasonal difference in river flows, which change markedly according to the monsoon rainy season. Generally, spring and summer are high water seasons while autumn and winter are dry seasons.

3.4 Biotic Community and Ecological Environment

The heritage sites are located in the middle region of subtropical area in eastern China, where it is warm and humid with ample sunshine and rainfall. There are various soil types in the heritage sites, such as yellow forest soil, forest red loam, lawn red loam, fragmental red loam, mountain meadow soil, and calcareous soil.

According to the biogeographical system of Udvardy (Udvardy 1975), the nominated site is located in the “Chinese Subtropical Forest” and “South Chinese Rainforest” (which are two of the 193 Biogeographical provinces), within the Palaearctic Realm and the Indo-Malayan Realm (two of the eight Biogeographical Realms). The vegetation cover overall is 78%, and it is more than 90% in the core areas. The heritage sites have ancient biota, preserving many ancient components that are relics for the Cretaceous and Tertiary.

In terms of the WWF Global 200 Biological Areas system, the heritage sites are located in the “Southeast China-Hainan Moist Forests” of the Indo-Malaysian Kingdom, which is part of the “Tropical and Subtropical Moist Broadleaf Forest”. Species diversity is rich due to the favorable water and temperature conditions and the special ecological niches provided within the Danxia landscapes. There are 5772 higher plants, belonging to 293 families and 1271 genera; 836 vertebrates, belonging to 129 families and 37 orders; 3073 insects; and more than 400 rare and endangered species.

There is a great variety of biotic communities in the heritage sites, including 23 vegetations types, with 261 formations and 424 associations. Evergreen broad-leaf forest is the typical region-specific vegetation in Danxia landscapes, including 69 formations and 102 associations. Evergreen broad-leaf forest is an outstanding example of the biological community determined by the southeast monsoon. There are different communities at different spatial scales and vertical levels in the landscape. For example, among the plants there are xeric plant communities on the summits of hills and crags, mesic communities in the ravines, and other endemic communities. Among the animal communities are: forest animals, shrub animals, farmland animals, cultivated zone animals, aquatic animals, flood plain animals, and cave animals, etc.

4 Geological Structure

4.1 Geotectonic Background

All the heritage sites are located on the South China Plate, formed by the collision of the China ancient plate and the Yangtze Plate in the Neoproterozoic, which became combined during the Silurian. South China, North China, Indo-China, and other neighboring continental segments were combined in the Triassic, forming a unified China continental plate. Since the Jurassic, the eastern part of South China was in an underthrust zone of the Pacific Plate, forming a NNE-oriented magma tectonically active zone. The eastern basins of red beds were strongly controlled by this structural zone. In the west of South China was the Qinghai–Tibet orogenic belt, formed between the China Plate and the Gangdise, Indian Plate which moved rapidly northward after the Late Jurassic. The foreland basins of red beds in the western part of the heritage sites (including the Sichuan Basin) were formed within this background of tectonic activity.

During the Early Yanshan Movement in the Middle Jurassic–Early Cretaceous, an inner part of the continental margin tectonic–magma belt formed in the eastern sector of the South China Plate, and large-scale intermediate–acidic magma activities occurred in the Nanling-Wuyi regions. Compression changed into extension in the south of South China Plate, forming a series of NE-NNE extension-fault basins along the regional fault, and the extensive development of red bed deposition of interior basins facies.

There was a huge difference in the deep physical structure and state between the eastern and western sectors of the South China Plate. To the west of Helanshan-Longmenshan, the crust was underthrusted and thickened, and the resulting strong uplift formed the Tibetan Plateau in the Cenozoic. To the east of Taihangshan-Xuefengshan, the crust experienced divergence

H. Peng, *China Danxia*, Springer Geography,
https://doi.org/10.1007/978-981-13-5959-0_4

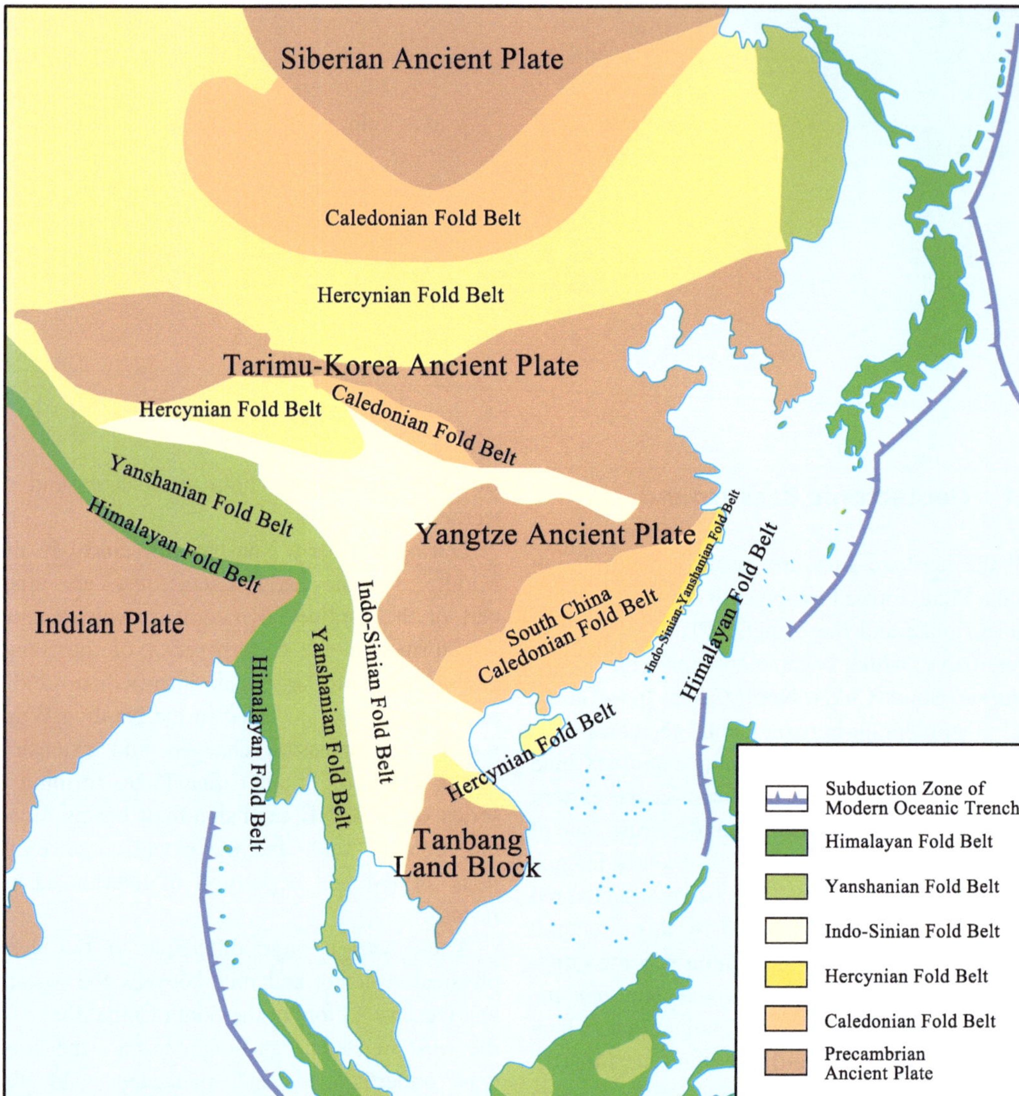

Southeast Asia Geotectonic Map

and thinning and was weakly uplifted in the Cenozoic, which opened a continental margin basin and an inland basin. The central part of the plate was a transition zone, and it was uplifted to an extent midway between that of the east and west sectors. The heritage sites essentially inherited this Mesozoic tectonic framework, with crustal uplift being the dominant force. With the onset of the Himmerian Movement in the Early Neogene, these red bed basins experienced large-scale and differential uplift, which caused erosion and incision from exogenic sources, and began the development of the modern Danxia landscapes.

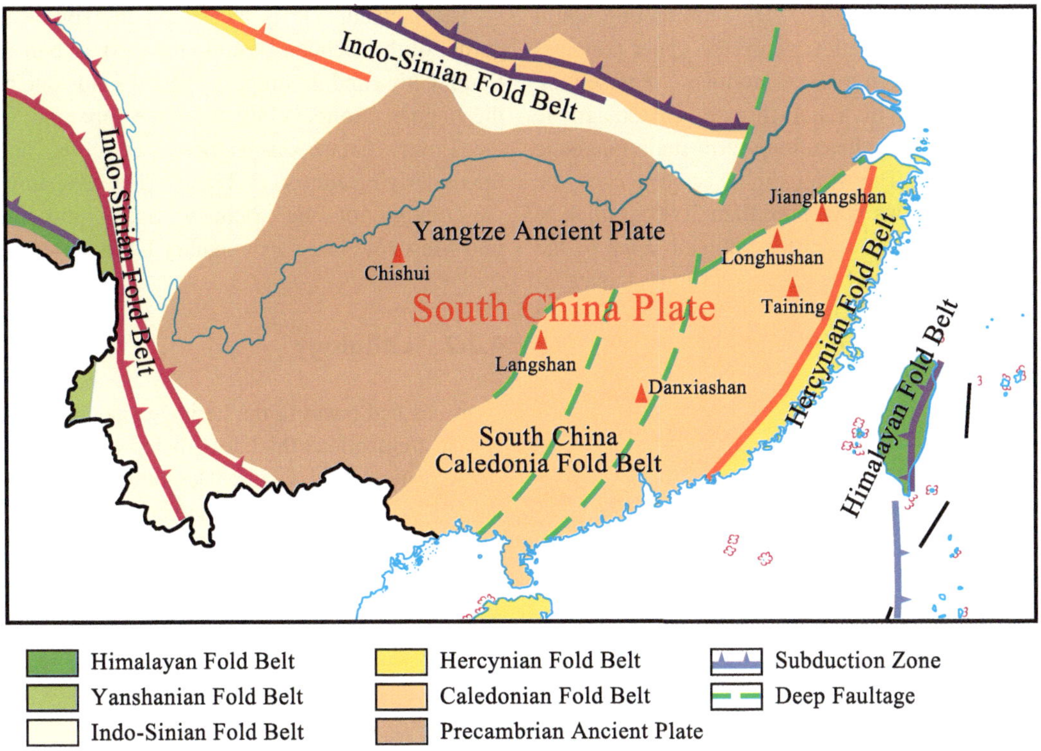

Heritage sites in relation to the South China Plate

4.2 Stratigraphic Characteristics of the Heritage Sites

4.2.1 Stratigraphy

The strata of the serial property comprise late Early Cretaceous–Late Cretaceous continental red clastic rock, with igneous rock. The specific characteristics of these strata were controlled by the structural attributes of the basin and this produced differences, as described in the following:

Xiakou Basin, Xinjiang Basin, and Taining Basin were located in the NNE magmatic activity zone in the east of South China, so they have very similar lithologic sequences, paleobiogeography, magmatic activity, and climate change characteristics. All of them were covered by a sequence of Early Cretaceous calc-alkalic rhyolitic volcanic material. The overall general situation is as follows: (1) two tectonic phases—Middle Late Cretaceous (90–80 Ma) and early Late Cretaceous–late Early Cretaceous (110–90 Ma) which each developed fault basins; the majority of sedimentary lithofacies were evolved from the bottom of slope diluvium facies, limnetic facies in the upper part of river facies, alluvial facies, etc., showing that there were two stages of faulting; at least, since 110–80 Ma, (2) the thickness of the stratigraphic sequence in the early basin within the basin center was around 2000 m, with an intercalation of bimodal volcanic rocks, indicating that the activity arose from deep within the crust. The paleobiology and fossils included Paralycoptera, Cratostracus combination, bivalve combination Trigonioides-Nippononaia-Plicatounic (TNP), Ostracoda Cypridea, and plants. In the sector dating from about 105 Ma, the lower part was gray, containing pyrite, etc., representing warm climate characteristics, while the upper part was purple, containing lime nodules, representing characteristics of a hot climate. Paleomagnetic measurements indicate long-term static magnetic polarity. (3) The stratigraphic sequence of the advanced basins was more than 2000 m thick in Jiangxin Basin, and in the collision zone of the Yangtze Plate and the ancient Chinese plate, it was about 4000 m thick. This contained a small amount of alkaline rhyolitic volcanic rock, purple in color, with gypsum salt, aeolian sand dunes, etc., showing hot dry climate characteristics. Fossils were those of dinosaurs, plants, and Euestherites. The paleomagnetic polarity of the lower strata was a continuation of the static magnetic characteristics, while the upper strata had a negative polarity. (4) The basins were covered by volcanic sequences, which mainly consisted of rhyolitic rock during the Early Cretaceous (135–120 Ma), and a volcanic sedimentary sequence containing Yanjiestheria Group sediments, Mesoclupea, etc.

Danxiashan is also located in the NNE magma activity zone of eastern South China. Within the volcanic basins in the early Early Cretaceous (135–120 Ma), a continuous continental clastic rock formation was created by a sustained depression in the early Late Cretaceous–late Middle Cretaceous (110–80 Ma), and the thickness of sediments in the basin center amounted to 3700 m. The bottom Changba Group has red beds of piedmont-limnetic facies that contained ostracods, stonewort, etc., and the upper Danxia Group has red conglomerate of alluvial-river facies containing ostracods. The main bodies of the Changba Group and the Danxia Group were in positive ultra-magnetic polarity, and the upper of Danxia Group was in negative polarity.

The Chishui site is located in the southeast corner of the Sichuan Basin in the western part of South China and is composed of red sandstone of the Jiaguan Group, belonging to the upper part of the Lower Cretaceous–Upper Cretaceous, and the underlying Jurassic red beds. They are part of the margin of the Sichuan Basin and the mid-Mesozoic large sedimentary basin series.

4.2.2 Lithology

Each red bed basin in the heritage sites is developed a tremendous thickness of red clastic sediment of terrestrial facies. The lithology of the sediments differed vertically and horizontally. For example, very thick pluvial mud and gravel ingredients often accumulated at basin margins (especially the pediment zone where the clast came from) and then changed gradually toward the center to pluvial conglomerate, gritstone, sandstone, lacustrine fine sand, siltstone, or argillaceous rock. The periphery of lake basin of fluvial outwash and piedmont pluvial accumulations were mainly made up of coarse clastic sediments like conglomerate and cemented by iron, silicon, and silt foundation, so produced a resistant rock type. The sediments of the lake basin facies consisted mostly of siltstone and argillaceous rocks, with a high content of calcium in the cement, intercalated locally with laminar argillaceous limestone, which was of lesser intensity and more easily weathered and eroded. Except for the nominated Chishui site (Bashu ancient lake basin), which was at the edge of the lake basin and consisted mostly of sandstone, and Longhushan (Xinjiang Basin) mostly of sandstone and sandstone with gravel, all other heritage sites were mainly made up of coarse clastic sediments such as conglomerate and sandy conglomerate.

Overview of the lithology and stratigraphy of the main basins in the heritage sites

District	Basin		Red Beds	Geologic Age	Thickness of Red Beds	Lithologic Characteristics	Landform Feature
Chishui	Sichuan Basin		Jiaguan Group	K_2j	942	Purple-red, brick red to gray purple thick layer of fine feldspar quartz sandstone and intercalation of multilayers of purple-red mudstone, the bottom was conglomerate	Danxia landform
Taining	Taining Basin		Chongan Formation	K_2c	1900	Fluvial and alluvial fan mauve calcareous glutenite and conglomerate, with a little argillaceous sandstone contained gravel; detritus stream and mud–sand flow facies are common	Danxia landform
			Shaxian Formation	K_2s	270	Mauve siltstone and mudstone, with sandstone and glutenite mingled; fluvial facies glutenite and sandstone alternation bottom; lacustrine facies arentilla-siltstone and mudstone, with sand conglomerate lenticular body in the small	Red bed hills
Langshan	Zixin Basin		Upper section of Langshan Formation	K_1l	50–100	Mauve pebbled sandstone, pebbled argillaceous siltstone, mingle mudstone, siltstone, and dissolubility carbonate contained varying degree	Red bed hills
			Lower section of Langshan Formation	K_1l	150–2320	Fluvial facies, pluvial facies mauve conglomerate, glutenite, sandstone, siltstone, argillaceous siltstone and dissolubility carbonate varying degree	Danxia landform and Danxia Karst
Danxiashan	Danxia Basin		Danxia Formation	K_2d	1300	Fluvial facies, pluvial facies mauve conglomerate, glutenite, sandstone and a little thin siltstone; ferroginous, sludge basal cement	Danxia landform
			Changba Formation	$K_{1-2}c$	2400	Fluvial and alluvial conglomerate and glutenite at the bottom, main in lacustrine thin argillaceous, siltstone, intercalated with cryptite partly in the middle to the upper part	Red bed hills
Longhushan	Xinjiang Basin	Guifeng Group	Lianhe Formation	K_2lh	1600	Brick red–mauve gritstone, medium-to-fine-grained sandstone and siltstone in the	Red bed hills

(continued)

District	Basin	Red Beds	Geologic Age	Thickness of Red Beds	Lithologic Characteristics	Landform Feature
					upper part; conglomerate, glutenite, medium-to-fine-grained sandstone at the bottom	
		Tangbian Formation	K_1t	462	Brick red bed–gritstone, fine sandstone contains calcium and siltstone in the upper part; brick red debris quartz sandstone, fine sandstone, and siltstone at the bottom. Dinosaur egg fossil and other fossils can be found	Danxia landform
		Hekou Formation	K_2h	687	Mauve conglomerate, glutenite, pebbled sandstone and siltstone, with algae, dinosaur egg, dinosaur skeleton, and other fossils contained	
	Ganzhou Group	Zhoutian Formation	K_2z	650	Mauve calcic fine sandstone and siltstone are an alternate layer, gypsum, calcium, and mirabilite contained. Born with plants fossils, mussel–shrimp and other fossils	Red bed hills
		Maodian Formation	K_2m	830	Brick red to mauve conglomerate, glutenite, and gravel with medium-to fine-grained sandstone and siltstone, with basalt partly. Plants and agatized wood fossils are contained	
Jianglangshan	Xiakou Basin	Fangyan Formation	K_1f	580	Purple-red, French gray conglomerate from great heavy to the massive, with intercalation of sandstone, gritstone lens	Danxia landform
		Chaochuan Formation	K_1c	1368	Purplish-red coarse sandstone with gravel, gritstone, medium-coarse sandstone, block siltstone, silty mudstone	Red bed hills

4.2.3 The Effects of Stratigraphy and Lithology on the Development of Danxia Landform

The resistance of red beds to weathering and erosion, as well as the different thickness of the sediments, played a crucial role in the development of Danxia landscapes. Research shows that the alluvial–pluvial deposits were often in thick or very thick layers; conglomerate and gritstone with a silt and sand cementation or siliceous, iron cementation were relatively hard. Danxia sediments in China's southern humid zone mostly developed on relatively thick layers of hard conglomerate, sandstone, and conglomerate. However, the siltstone, argillaceous rock deposits in the center of the basin contained much soluble material, with poor water permeability, and rich in water; thus, it was quite weak and produced the red beds. But in the arid zone, the weak argillaceous rock and siltstone can maintain large-scale cliff slopes.

The thickness of terrain was also an important controlling factor in the development of Danxia landscapes. The very thick red beds with a uniform texture often produced blocky structures, which resisted weathering and erosion well, and formed large-scale cliffs and hill blocks. But the red beds with thin layers were weak in their resistance to erosion on the whole because of the highly variable layers and formed red bed hills.

Red Beds of Different Lithology

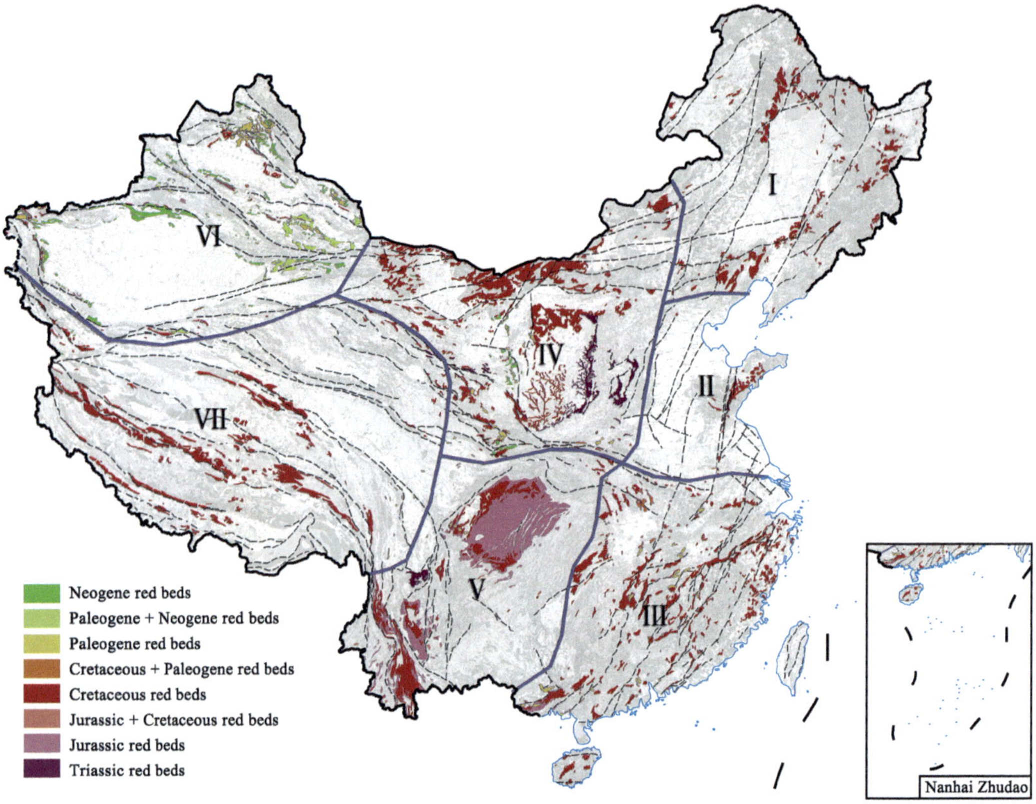

Distribution of red bed outcrops in China

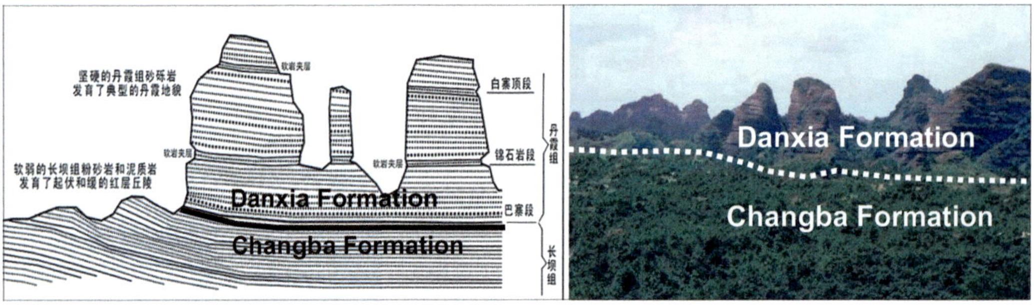

Differential Landforms in red beds with variable lithology (Danxiashan)

The Correlation of the Basins of Danxia landscapes in South China

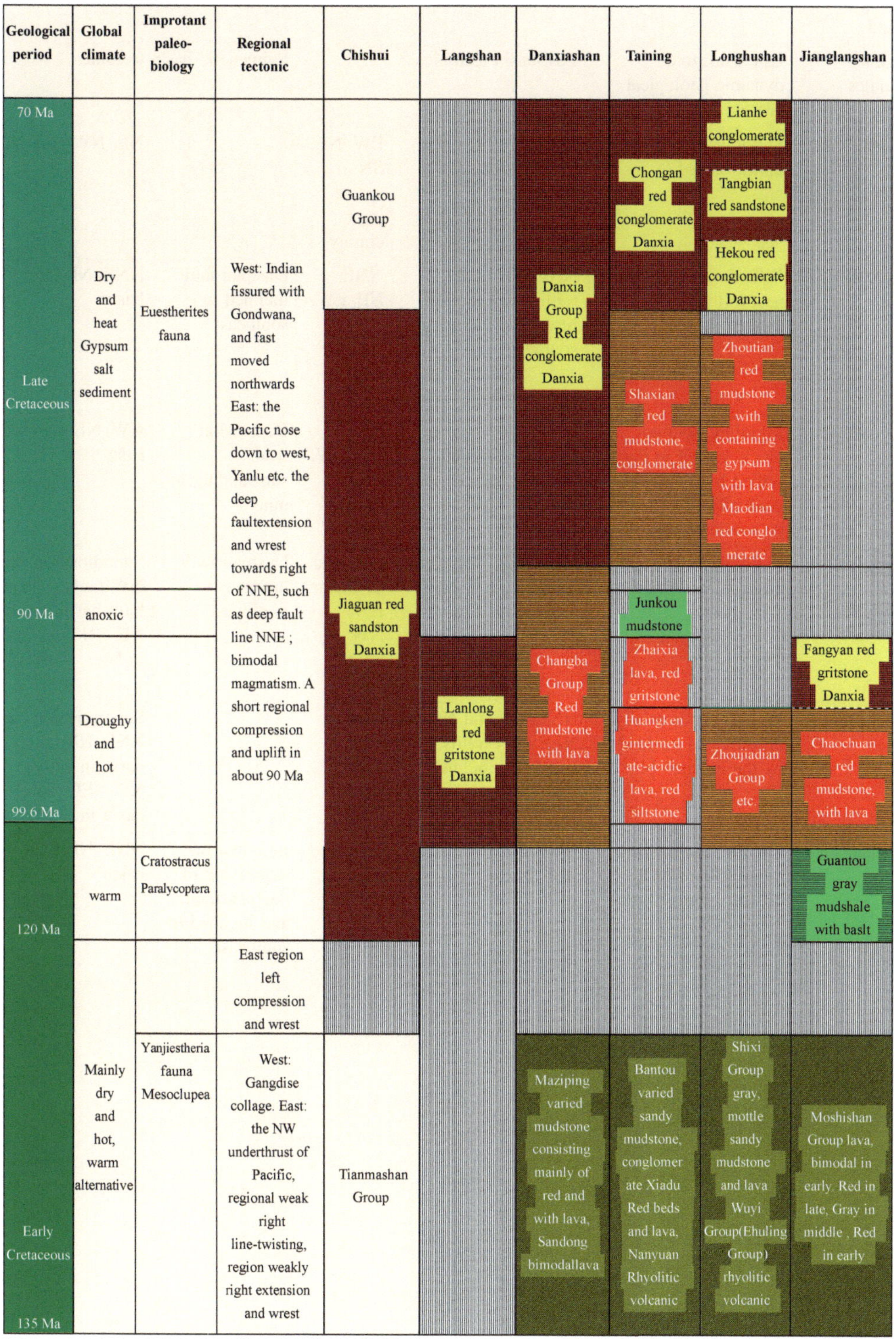

Tectonic comparison within the China Southern Danxia Basin

Geological period	Miocene	Late Cretaceous to Cenozoic		75 Ma	110 Ma
Regional tectonics	Late Cenozoic dynamic geological process	Post-basin tectonics		Syngenetic basin tectonics	
Chishui	Strongly differential uplift, closed to basin	Near level, tectonic deformation, weak fracturing	EW, NE, SN sparse joint cranny		NE, NW post-basin
Langshan	Differentially uplifted, near to the rudiment of erosional basin	NNE–NE fault trellis, tectonically lifted and inclined to southeast	NNE, NE, EW, NNW dense joint cranny	Dustpan fault inclined to southeast	NNE–NE fault basin
Danxiashan	Middling uplifted block, closed to the settled block, and gradually extended to be erosional basin	NNE left compression and depression fault trellis	NNE–NE, EW, NW joint cranny	Depression from surrounding to inner	NW–NE depression basin
Taining	Middling uplifted, no basin in surrounding	NE, NNE, NW fault trellis	NE, near SN, near EW, NW joint cranny	Dustpan fault inclined to northwest	Superposition basin, NW fault-volcanic fault basin in early, NE–SN fault basin in late
Longhushan	Weakly differentially uplifted and studded by the inclined block	NNE–NE and NE–NEE fault trellis, lifted and inclined with slowly obliquity toward north	NNE, NEE, NW joint cranny		Superposition basin, NNE–NE fault basin in early, and NE–NEE fault basin in late
Jianglangshan	Intense uplifted fault block, closed to the subsided fault block	NNE–NE fault trellis	NW joint cranny	Inherited the depression of dustpan fault, and inclined to southeast	NNE–NE fault basin

4.3 Tectonics of the Heritage Sites

4.3.1 Geological Structure

All the red bed basins of the heritage sites were NE-NNE oriented tectonic basins, which were controlled by the southeast China regional tectonics. An exception is the Chishui site on the border of Sichuan and Guizhou provinces, which was influenced by the Sichuan Basin structure, an almost SN and EW structure from outside. From the Cretaceous on (late Yanshan Movement), the main lineament was not subject to disturbance and many impacts of boundary tectonics have persisted until today. For example, the dominant faults of the Danxia Basin, like the "Shaoguan-Renhua fault," the Chongan-Shicheng NE fault on the rim of Taining Basin and the Taining-Longyan SN fault, the Dongxiang-Guangfeng fault belt on the rim of Xinjiang Basin, and Langshan NNE fault of Gongtian-Ningxiang-Xinning-Ziyuan, which used to be synsedimentary fault belts, and the main controlling fault in the regional tectonics after the basin was uplifted. The basins mentioned above experienced only minor folding or nonessential development during the Cenozoic. They developed a range of different-sized multistage faults due to regional faults, causing either integral or differential uplift and characteristic block structures. In addition, the basin developed numbers of groups of large joint systems.The distribution of tectonic linement and hill block in Mt. Danxiashan

4.3.2 Control of Faults and Joints on Hill Block Framework

On the one hand, from the study of the various units of the geological structure, the nominated region has mostly a series of sparse NNE and near NS faults, leading to the arrangement of hill blocks-oriented NNE or near NS. On the other hand, most parts of the basins in the nominated area also developed some near EW, NE, and NWW faults. These large tectonic lineaments controlled the general arrangement of hill blocks, and the small structures controlled the trend, density, and plane pattern of hill blocks. The Taining Basin was generally

The tectonic lineament and congested canyon system in the Shiwang, Taining

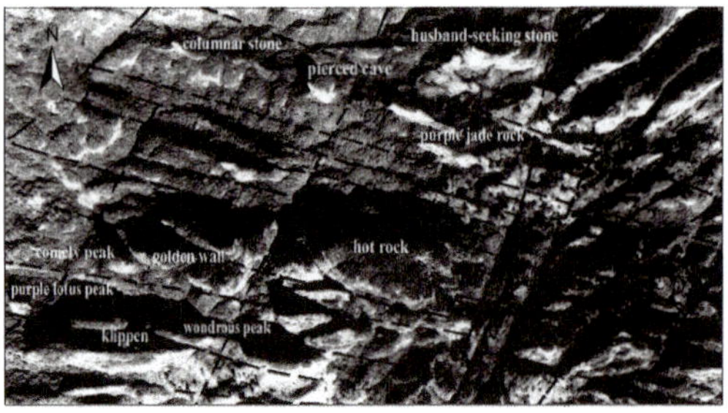

The distribution of tectonic linement and hill block in Mt. Danxiashan

Landform pattern of Mt. Longhushan controlled by grid fault structure

controlled by the Chong'an-Shicheng NE fracture and the Taining-Longyan SN fracture; therefore, SN, NE and NW faults and joints developed within the basin. The mountain block and the incisive meanders in the Shiwang District were controlled by these faults and joints. The arrangement of hill blocks in Danxiashan extended in a NNE direction along the major fault. The trend of hill blocks and the arrangement of stone pillars were mainly along the fault and large joints of the near east–west trend. The near EW and NW faults of Jiangxi Longhushan controlled the scale, trend, and contour of the Danxiashan hill blocks, but the closely spaced NNE, NEE, NW joint belts controlled the arrangement of the Danxia Village, peak clump, peak forest, stone wall, stone girder, and stone pillars. In Hunan Langshan, the big NE fault of Gongtian-Ningxiang-Xinning-Ziyuan matched with four groups of NNE, NE, near EW and NNS joints, provided a spatial framework for molding the landscape here.

4.3.3 Control of Terrain Occurrence on the Shape of Slope Surfaces

During the Neozoic crustal movement, the heritage sites experienced massive uplift. Apart from a slightly larger dip near the fault zone and basin margin, the majority of sites were located in a horizontal or near horizontal uplift. The characteristic Danxia landform produced was of a type called "flat top, steep face, and gentle piedmont." The Danxia landscapes that developed on the inclined terrain had cuesta characteristics, and the slipping summits were essentially consistent with the bedding planes.Near horizontally stratified Danxia landform (left) and the cuesta Danxia landform within the monoclinic terrain (right)

Near horizontally stratified Danxia landform (left) and the cuesta Danxia landform within the monoclinic terrain (right)

4.3.4 Influence of Neotectonic Movement on the Process of Landform Evolution

All the Danxia landscapes in the heritage sites were in a differential and intermittent uplift zone during the neotectonic movement. They formed under the influence of exogenic forces and processes such as fluvial erosion and mass movement. Where the region was uplifted in an early stage followed by a long-term stable state this allowed for the successive and gradual evolution of Danxia landscapes from infancy to old stages, such as at the Xianshuiyan scenic spot and Mazuyan scenic spot in Mount Longhushan, where the landscape is now at a relatively old age stage. However, intermittent uplift causes the development of multilayered Danxia landscapes. At Langshan Mountain, the overall slope of the terrain is from west to east, and the crustal uplift varied from the strong in the west to weak in the east. This produced three levels of platform gradually descending toward the north, with each platform separated by about 200 m altitude. At Mt. Danxiashan, Taining, Mt. Longhushan, there are five levels of platforms, with 100 m between each level.

At 400 m, there is a widespread planation surface, and above 400 m is the old age Danxia landform, which indicates that it was in a long-term steady period when the 400 m planation surface formed; after being intermittently uplifted, other planation surfaces were formed below the 400 m level.

Five Main Planation Surfaces at Mount Danxiashan

4.4 Paleogeography and Paleontology of the Heritage Sites

Most of China red beds were formed in the Cretaceous. The Cretaceous was also the time that the sudden extinction of dinosaurs occurred and that paleogene mammals flourished. So the red beds are also a principal source of dinosaur fossils. For example, the Cretaceous strata of Langshan and neighboring areas have Tyrannosauridae bone and eggs fossils. Three mammalian strata occur and divided as follows from lower into upper Paleocene of the Paleogene: Matutinia-Cocomys-Orientdophus, Baudaomys-Ampholophus, and Rhombomys-Hohomys-Heptodon. Plant pollen in strata from the Late Cretaceous includes gymnosperms 56.2%, ferns 42.4%, angiosperms 1.1%,and needle squamous conifers pine more than 20%, which indicates drought. All of the above evidence shows that the Cretaceous had a tropical or subtropical dry, hot climate. In another example, the Hekou Formation of Xinjiang Basin where Longhushan is located contains paleontologic fossils, such as dinosaur eggs, skeletons, and stonewort. Large numbers of dinosaur footprints fossils were found in the strata of the Jiaguan Formation of Chishui Basin. All this evidence confirms the characteristics of the paleoenvironment in the Cretaceous.

In general, in the Early Cretaceous, the Zhejiang-Fujian-Guangdong area was in a shoreline within an active volcanic zone. Dinosaur bone, eggs, and footprint fossils were found in isogenetic sediments, which indicate that the sedimentary environment was mainly the shore of a shallow lake.

At the beginning of the Late Cretaceous, the red beds of fluvial-lake facies, which formed the Danxia landscapes in the Zhejiang and Fujiang regions, were mainly a piedmont alluvial deposit, where fossils are limited to only a few ostracodes and stonewort. The Late Cretaceous red beds of Xinjiang Basin are composed of aeolian sand, aeolian dreikanter, gypsum, and halite, which shows that the climate became drier with a tendency to be tropical (subtropical), near-desert and saline. This situation lasted until the end of the Late Cretaceous.

Moreover, the Danxia landscape regions also contain a concentration of paleo-human cultural remains. An example is the important "Catfish Zhuan" of the Neolithic Age in the Yulinshi Cave of Mt. Danxiashan, where 80 historic relics were excavated dating from 6000 years ago. Artefacts produced 4000 years ago, such as stone axes, cutting tools, stone scrapers, stone shovels, stone balls, stone clusters, and pulverizers, were unearthed in the monument of Zhoujiashan and Baimianzhai. Neolithic Age sediments in southern Langshan contain many tools, such as bone combs and bone hairpins, as well as some ceramics such as axes, pots, bowls, and painted pottery figures.

4.5 Geological History of the Heritage Sites

All the red beds of each basin in the heritage sites are terrestrial interior basin sediments, which were formed in a very hot, dry climate during the Mesozoic time. But the red beds were shaped into Danxia landscapes only in the Late Neogene to Quaternary period. So their crustal foundation

Southeast China structural evolution

Geological time			Symbol	Age (Ma)	Tectonic cycle		Geological structure evolution	Dynamics system
Phanerozoic	Cenozoic	Quaternary	Q	2.5	Alpine Movement	Himmerian Movement	The collision of Indian and Eurasia, Marginal sea developed Modern structure landform formed	Modern Danxia landform formed
		Neogene	N	24.6				Basin returned, paleo–Danxia developed
		Paleogene	E	65.0				
	Mesozoic	Cretaceous	K	137		Yanshan Movement	The ancient Pacific Ocean and East Asia mainland collided fiercely, Intense structure–magmatic activity in southeast China	Red bed basin formed, red bed deposits
		Jurassic	J	203				
		Triassic	T	227		Indosinian Movement	The South China Plate met with North China and Indosinian Plate separately, the various plates in East Asia finally merged	South China Plate activated, Mantle fold
				251	Hercynian Movement		The structure development stage of South China Plate Hainan cracked trough closed Qinzhou geosyncline closed	South China kept stable, Covering strata deposits formed
	Paleozoic	Permian	P	295				
		Carboniferous	C	355				
		Devonian	D	408				
		Silurian	S	435	Caledonian Movement		Yangtze Plate met with China Plate South China geosyncline closed	South China Caledonian fold formed
		Ordovician	O	495				
		Cambrian	∈	543				South China ancient plate cracked, Sea trough formed
Proterozoic	Neo-Proterozoic	Sinian	Z	635	Salayier		Split between Yangtze and China Plates and then formed the South China geosyncline	
		Huananian	Pt_3	800				
		Qingbaikou-Nian		1000	Jinning		Yangtze and China ancient plate part jointed, then formed the whole South China ancient plate	South China ancient plate (Yangtze and China) crystalline basement formed
	Meso-Proterozoic	Jixianian	Pt_2	1400				
		Changcheng-Nian		1800				
	Paleo-Proterozoic		Pt_1	2500				
Archean			Ar					

was established in the Paleozoic, and the material basement was formed in the Mesozoic, and the uplift and Danxia landscape shaping occurred in the Cainozoic.

4.5.1 The Stage of Crustal Foundation

Since the Late Proterozoic, the earth's crust in the nominated area experienced three tectonic development phases, from an active block to a stable block then to an active block.

- **Active Block Phase**: All the nominated sites were in one part of an active margin belt (the South China geosynclinal system). After the division of South China system in the Late Proterozoic–Early Paleozoic, a huge thickness of flyschoid formation was deposited, lasting for a long time from the Late Proterozoic. Following the return of the Caledonian Movement folding, the South China Plate was combined with the Yangtze Plate.
- **Stable Plate Phase**: In the Neopaleozoic, the South China crust tended to be stable and formed the South China meta-platform, which was integrated with the Yangtze platform. Except for the Chishui Region, the other regions experienced deposition of carbonite and clastic rocks of shore-neritic facies from the Devonian to the Early Triassic, which indicates a stable tectonic environment and clastic rock deposition.
- **Active Plate Phase**: By Mesozoic times, new tectonic changes had occurred in the global plates. The Asian Plate was being affected by underthrust of the Pacific Plate and became active. The stable continental margin of China's southeastern coastal region became an active continental margin and produced a series of NE-NNE inland fault basins with red beds and magmatic activity from westward to the inner continental areas.

4.5.2 Red Bed Deposition Phase

As mentioned above, the South China Plate was active from the Middle–Late Triassic. The mechanism driving the plate activity was the increased movement north of the Indian Plate and collision with the Asian Plate, and the underthrusting of the Pacific Plate beneath the Asian Plate. Consequently, the Sichuan Basin, being in the convergence center of these three big plates, experienced regional submergence.

Prophase Structure: With the commencement and strengthening of the Middle–Late Jurassic Yanshanian Movement (Pacific Movement), large-scale folding, faulting, and magmatic activities occurred in the eastern part of the South China Plate. The structural foundation of the eastern heritage sites was formed in this way.

Basin Fault: The Late Jurassic to Early Cretaceous Yanshanian Movement reached a new climax and the underthrust of the Pacific Plate beneath the Eurasian Plate strengthened. NNE-oriented uplift and depression belts were formed in East Asia, initiating a new intensely active zone on the continental margin. The Sichuan Basin continued its stable submergence, and many fault basins appeared in Hunan, Guangdong, Jiangxi, Fujian, and Zhejiang. The above-mentioned basins in the east experienced large-scale magmatic activity, forming calc-alkaline lava (andesite, liparite, and basalt) and bimodal volcanic rocks (andesite, rhyolitic, and basalt), accompanied by the development of tuff pyroclastic rocks and clastic rocks.

Red Bed Deposits: The eastern arc extension fault basin experienced large-scale submergence successively from the Cretaceous, and its peripheral mountains rose relatively rapidly. Meanwhile, the climate became dry and hot, and a very thick red clastic rock series was deposited, impounded by the seacoast mountains to the southeast.

The eastern fault basins, except the Sichuan Basin, had features indicating development over a short-time, large-scale activity, fast ebbing, thick deposition, rough sedimentation with poor gradation, unstable minerals, and deposits from near the source of supply. Silt/gravel mixed composition deposits formed, such as the red pluvial-alluvial facies clastic rocks, evaporites containing gypsum salt, aeolian sedimentary rocks, dark shale, and other types of freshwater marl. The red pluvial–alluvial facies clastic rock was the main material foundation for Danxia landscape formation in southeast China.

More importantly, the Sichuan Basin, Zixin Basin, which was fault basins located inland, developed at the same time as the uplift of the Yunnan–Guizhou Plateau and Nanling, where there was no magma eruption. This is consistent with the mode of development of an inland rift—depression, unlike the basin evolution model of the eastern continental margin.

4.5.3 Basin Uplift and Danxia Landform Formation Phase

At the beginning of the Cenozoic Era, Western China was strongly squeezed by the Indian Plate, causing wide-ranging uplift in the west and crushing toward the east. The underthrusting of the Pacific Plate caused the continental margin of East Asia commence a new development process involving island and back arc marginal seas. South China, between the two tectonic zones, was uplifted. By the end of the Paleogene, all basins mentioned above finished their depositional phases. Further, uplift initiated the erosion of the red beds, and modern Danxia landscape evolution began in humid climatic conditions. Structural Plate Movement in the Heritage Sites Region

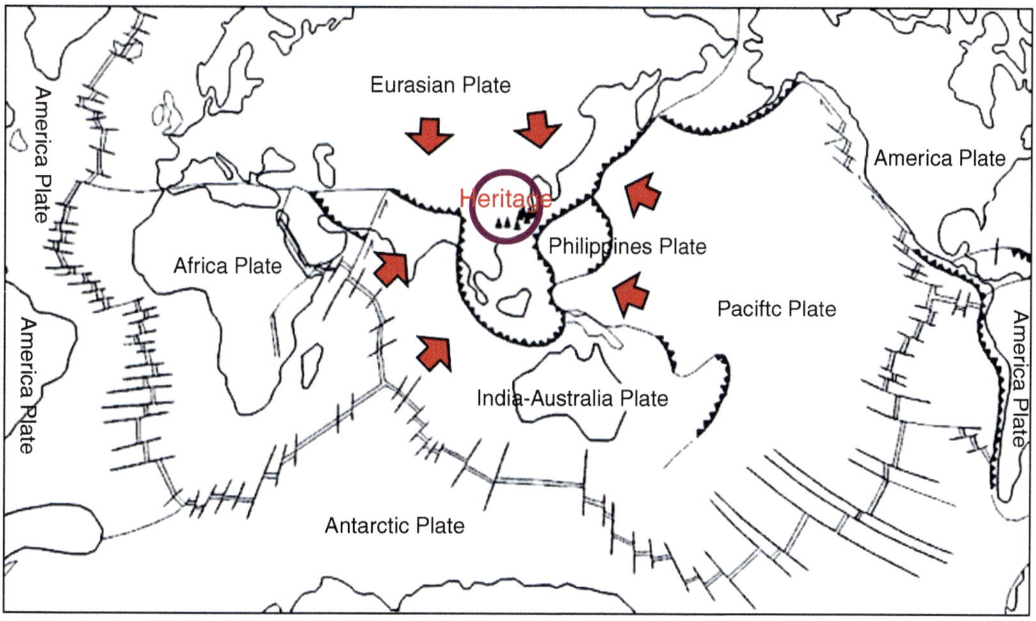

Structural Plate Movement in the Heritage Sites Region

5 Geomorphologic Features

5.1 The Concept, Types, and Distribution of Danxia Landscapes

5.1.1 The Concept of Danxia

Danxia is a specific petrographic geomorphology, named in the 1920s from Mt. Danxiashan, Guangdong Province, in China. Danxia geology and geomorphology have been studied for about 80 years in China, and the term is widely known and used both academically and in society generally.

In the geomorphic classification systems used in China, Danxia is known as a red beds landform characterized by steep cliffs. More specifically, Danxia refers to a landscape that is formed from thick continental red beds that have been regionally uplifted, intensively faulted and deeply dissected by fluvial erosion, mass movement and solutional weathering, and producing a variety of cliff bounded peaks surrounded by deep canyons. This whole process forms a spectacular landscape of great scenic beauty.

The concept above includes two essential elements—material and topography—i.e. Cretaceous coarse clastic rock and steep cliffs. It also includes two development conditions—geologic structure and exogenic forces.

Red beds are the material foundation for the development of Danxia. The red beds responsible for the development of Danxia are mainly a terrestrial red clastic rock series comprising primarily conglomerates and sandstones.

The geometrical feature of Danxia is steep cliffs. The steep cliffs of various scales and patterns are the basic elements that characterize the Danxia hilly blocks and canyon walls. Cliffs are considered to be the outstanding "thumbprint" of Danxia landscapes.

The development of Danxia is controlled by geological structure. Firstly, Danxia developed in a region where the red beds were uplifted above the regional erosion datum. Secondly, the geological structure controls the spatial pattern of the Danxia mountain blocks, and the attitude of the strata determines the geometry of the hillslopes.

Exogenic processes are the direct sculptor of Danxia landscapes. After being uplifted, the red beds experience physical and chemical weathering, mass movement, and fluvial erosion. Wind and biological weathering are other minor factors in the development of Danxia landscapes.

5.1.2 Types of Danxia Landscapes and Landforms

Based on the material foundation, geological structure, dominant exogenic forces, landform geometry, and development stages, Danxia landscapes can be categorized into the following types.

H. Peng, *China Danxia*, Springer Geography,
https://doi.org/10.1007/978-981-13-5959-0_5

Classification of Danxia Landscapes

Criteria			Type
Material	Detritus type		Conglomerate or glutenite Danxia, Sandstone Danxia, Sand shale Danxia
	Elements		Karstic Danxia, Earthy-forest type Danxia, Mud-coating type Danxia
Attitude of rocks	–		Subhorizontal Danxia; Gently dipped Danxia; Steeply dipped Danxia
Dominant force	Climate		Danxia in humid region; Danxia in arid region; Danxia in cold highland
	Formation		Water-Erosion Danxia; Weathered Danxia; Gravity collapse Danxia; Wind erosion Danxia; Sea erosion Danxia; Karstic Danxia; Artificial Danxia
Landform geometry	Single geometry	Positive	Escarpment Face; Mesa; Cuesta; Stone walls; Stone Column; Stone peak; Upland; Hill; Stone single peaks; Stone balls; Colluvial piles, and Colluvial rock block
		Negative	Danxia Valley(linear ravines, narrow gorge, canyon, incised meander, broad valley); Danxia Rock Grooves(bedding notch, bedding groove, vertical groove); Danxia Cave(bedding cave, horizontal cave, periclinal cave, superposed cave, perforated cave, honeycomb cave, niche-like perforated cave, vertical cave); Danxia Arch; Danxia pothole
	Combination structure		Plateau–canyon type Danxia; Mountainous Danxia; Peak forest-type Danxia; Single-peak type Danxia; Hilly Danxia
Developmental stages	–		Young stage; Mature stage; Old stage

5.1.3 Distribution of Danxia

Danxia landscapes are widely distributed and approximately 780 locations have been identified throughout China. They are found in tropical and subtropical humid zones, temperate humid and semi-humid zones, and semiarid–arid zones, as well as in the cold plateau of Qinghai–Tibet. Danxia landscapes also occur in areas with a wide range of altitude, from lowest in the east coast, to highest in the Tibetan Plateau where the altitude is above 4000 m. Geographically, Danxia landscapes can be divided into three regions of densest occurrence: Southeast, Southwest, and Northwest.

- **Danxia of Southeast China: Peak-hoodoo-cluster-type**. This type of Danxia landscape is found in the Jiangnan Hills Zone of the geomorphological system of China, including Zhejiang, Fujian, Jiangxi, Guangdong, and Hunan Provinces. Danxia Peak-hoodoo is characteristically developed in association with water bodies in this region, and the scenery is very attractive. Five of the six heritage sites in the serial property are located in this subarea.
- **Plateau–mountain–valley Danxia in the southwest humid region**. This type of Danxia landscape is found mainly in the transitional zone between the margin of the Sichuan Basin and its surrounding plateau, as well as in the Yunnan–Guizhou Plateau. Here, deeply-dissected, plateau–valley type Danxia and red beds-hill type Danxia developed with characteristically very irregular landforms. High red cliffs, with associated rapids and waterfalls, form a spectacular natural landscape. The candidate site of Chishui Mountain is representative of this subarea.
- **Danxia of arctic-alpine plateau and arid mountain type in the northwest**. This type of Danxia landscape is found in Gansu

Province and surrounding provinces. There are numerous cold plateau-semiarid and arid Danxia landscapes at altitudes above 3000–4000 m. In the transitional region between the Tibetean Plateau and the Loess Plateau, there is semiarid-hill type Danxia landscape formation. Along the Hexi Corridor, from the piedmont of Qilianshan Mountain to the piedmont of Tianshan Mountain, there is typical arid-hill type Danxia landscape.

Apart from China, similar landscapes and landforms have developed in red beds all around the world. These include the Colorado Plateau in western USA, the Brazil Plateau and the Andes Mountains in South America, central-southern Europe and Britain, western and northern parts of Africa, Australia, and Western Asia. Therefore, Danxia landscapes are an important landform type of global significance.

The following sections provide a brief description of the heritage sites in this serial China Danxia nomination.

5.2 Geomorphologic Features of China Danxia

5.2.1 The General Situation

The heritage sites in the serial property are all located in the subtropical humid zone of Southern China. Apart from Chishui, which is situated in the transition region between the Yunnna–Guizhou Plateau and the Sichuan Basin, the other sites are in the Jiangnan Hills Zone composed of low mountains and hills. The local relief of Danxia hilly blocks is mostly between 300 and 500 m, and the greatest is 1750 m.

From the Neogene Period, the earth crust of the red beds basin experienced differential and intermittent uplift. Uplift of the Tibetan Plateau intensified the circulation of the East Asian monsoon, and the climate in Southern China changed from arid to humid, so that fluvial processes strengthened gradually. With various geomorphic factors working together, Danxia landscapes of different development stages and topography emerged.

China Danxia is a natural complex with a unique landscape type as its foundation. China Danxia landscape forms an integrated natural system, with outstanding universal value based on its geoscience value, ecological value, extraordinary landscape, esthetic value, and associated cultural values.

5.2.2 The Features of Danxia in the Heritage Sites

- **Rich variety of geometric types of single landforms**. Because there is a different geological environment in each nominated site, with a great diversity of lithology, rock associations, distribution of faults and joints, tectonic uplift and climate, Danxia landforms have a very varied morphology. Some of the most conspicuous and spectacular landforms are cliffs, towering peaks, deep and quiet meandering valleys, and caves of various sizes and types.
- **Diverse combination of landforms**. An intricate combination of different sorts of Danxia landforms occurs. The youthful stage of geomorphic development is characterized by incised meanders, deep canyons, and narrow valleys. The mature stage of development exhibits a landscape of very strong relief with majestic peak forests. The landscapes of older age have isolated peaks with greater expanses of low land and serene rivers.
- **Integrated development stages and evolutionary processes of Danxia landscapes**. Sites representative of different geomorphological development stages of Danxia in the subtropical humid zone of Southern China, were selected to make up an integrated series of Danxia landscapes. The sequence from young to old is: Chishui, Taining, Langshan, Danxiashan, Longhushan, and Jianglangshan. These sites in combination fully display the integrated development processes of Danxia landscapes and both positive and negative landforms. The evolutionary process of positive landform development is: plateau and highland → dense peak cluster → clustered peak-hoodoo → scattered peak forest →

isolated peak, and residual hill. While the development process of negative landforms is: line ravine → lane valley → canyon→ dale.

- **A perfect combination of mountain–water–forest landscapes**. All the heritage sites have rivers or rivulets and Danxia peak clusters and peak forests were formed near water. The heritage sites have natural environments with lush forest cover and harmonious cultural and natural landscapes, which create a colorful landscape of red cliffs, blue waters, and green forests. Danxia landscapes have diversity, uniqueness, rarity, and naturalness, which makes them special in the global context.

5.3 Landform Types of China Danxia in the Heritage Sites

5.3.1 Summary of the Landform System of the Heritage Sites

- They have a common lithology—most of the nominations are composed of glutenite types of sediments, except the Chishui and Guifeng which are composed of gritstone.
- They have a common inclination of the stratigraphic beds—most of the heritage sites have subhorizontal strata, while a few or some parts have gently dipping strata.
- They have a common climate—the heritage sites are all in the subtropical humid zone.
- They are eroded by a common dominant exogenic force—all of the heritage sites were dissected chiefly by flowing water.
- They display all of the known morphological types of landform in humid-zone Danxia landscapes.
- They display all of the stages of geomorphological development of Danxia landscapes —the Shiwang district in Taining, Chishui represent the young stage, Longhushan and Jianglangshan represent the old stage, and the others are classified into the mature stage of Danxia landscape development.

5.3.2 The Main Geometric Classification of China Danxia

On the basis of their morphology, Danxia landforms can be classified as positive or negative landforms (the latter including caves), and they can be further categorized according to the type of landform, as follows:

(1) **Positive landforms**

Morphological classification of positive landforms of China Danxia

Type		Description
Danxia steep cliff		The steep cliffs with the slope $\gg 60°$ and the altitude >10 m
Danxia peak	Danxia mesa	The gentle mountain tops, steep walls on sides, take the shape of castle
	Danxia cuesta	The gentle dipped tops with 1–3 steep cliffs
	Danxia spire	The pyramidal peak composed by surrounding scarp slopes with spire tops, flat tops, and round tops
	Danxia stone wall	a kind of wall-shaped mountain block whose length is greater than twice of its width and its height > its width, the lower can be named rock beam
	Danxia column	Square or round isolated stone column, height greater than diameter; some lower one (height is smaller than diameter) called stone-mound
	Danxia isolated peak	Weathering and relict butte disseminate upon river valley plain or hillock, some lower one can be called butte or isolated stone

(continued)

Type		Description
Colluvial cone and colluvial rock		Megalith and pyramidal colluvial stones in irregular shapes distributed at the foot of steep cliffs; Rocks exist in various sizes, including huge ones of more than a hundred cubic meters
Sculpting landform		the stones or mountains with outstanding pictographic characters
Combination morphology	Peak cluster	Group of peaks, its the base has not been cut, and the height of the base is greater than 1/3 of the mountain's relative altitude
	Peak forest	Group of peaks, its the base has been cut deeply, and the height of the base is less than 1/3 of the mountain's relative altitude

(2) **Negative landform**

Morphological classification of negative landforms of China Danxia

Type		Description
Danxia valley	Linear ravine and lane valley	The ravine developed along the tectonic faults with almost parallel valley walls. Depth/width >10, the width is less than 1 m. The valley can not be passed or allowing only one man to pass, is called linear ravine; while the valley with the width between 1 m and 10 m is called lane valley
	Canyon	The depth is greater than the width; the width of the bottom >10 m, both sides of the canyon are steep; present a "V" shaped valley walls; the bottom is flat
	Mesa valley	The valley composed by surrounding curved or straight cliffs with one side open
	Incised meander	Winding river with curvature ≥ 1.5, with gorge-like valley and steep cliffs on the both sides
	Broad valley	Width of the bottom is generally between 10 m and 100 m with many peak-cluster and peak forest on both sides, with some large river passing though
Groove	Vertical groove	The groove is formed by the long term erosion of the vertical stream along the cliff
	Bedding groove	Shallow groove that is usually developed along the soft-rock strata of the cliff; the bedding groove can be continuous or discontinuous; depth > height; generally unable to pass
	Bedding rock tank	A kind of deep groove, with the depth greater than its height, is generally available to pass and developed by the rapid weathering or fluvial erosion along the soft-rock strata on the cliff
	Forehead-shaped rock tank	With the rock tank turned further deeper and higher, it forms the forehead-shaped rock tank featuring larger mouth and gentle dipped top surface
Danxia cave	Large single cave	The width of the mouth is generally greater than 10 m, and exist alone
	Niche-like cave	Distributed in a group with diverse pattern; the width of a single cave is several meters
	Honeycomb-like cave	The microcave group has the even size and dense connection, and the diameter of a single grotto is less than 30 cm, taking the shape of honeycomb
	Colluvial superposed cave	Colluvial superposed caves are formed at the piedmont due to the pilling up of giant colluvial rocks
	Danxia Karst cave	Danxia Karst cave is formed by the dissolution, suffosion, and collapse of the red beds with calcium conglomerate

(continued)

Type	Description
Danxia perforated cave	The cave that could penetrate the mountain
Stone arch and natural bridge	the perforated cave with the height greater than the thickness of the top
Pothole	On some bedrock river beds, currents carried gravel or clastic rocks for rotating movement, eroding the bedrock and forming the subcircular pothole

5.3.3 Description of the Main Morphological Landform Types of China Danxia

(1) **Danxia steep cliff**

Steep cliffs are the distinguishing feature of Danxia landscapes. The cliffs are generally formed by the collapse of overhanging strata, controlled by multi-vertical joints. Due to the effects of weathering and fluvial erosion, various cliffs types are formed, including bedding groove types, vertical groove types, and cave group types. Wavy cliff types, developed by stream downcutting and lateral erosion, are a special type of Danxia cliff.

Smooth type cliff maximum is 225 m high, and extending to 2.3 km

Smooth type cliff maximum is 225 m high

Compound type cliff

Vertical groove type cliff

Foguang rock: smooth-curved cliff

Luoxia cliff: wavy cliff

Baitai rock: niche-like cliff

Bedding groove type cliff

(2) **Danxia peaks**

Peaks are the most important landscape component of Danxia landscapes. Because of the diversity of the strata, faults, lithology, and rock combinations, there are very many and varied types of peaks. Based on the plane configuration, there are block types and strip types, and according to the solid configuration, they may be fort-like, pyramidal, mitriform, wall-like or column-shaped. Peaks constitute the uppermost components of Danxia landscapes.

Castle group (Danxia mesa)

Cuesta

Group of cuestas

Group of pyramidal peaks

Pyramidal peaks

Round-top peaks

Group of pyramidal peaks and stone columns

Danxia column

Danxia isolated peak

Butte of an old stage in the periphery of Longhushan

(3) **Danxia valley**

Valleys are the most important negative landforms in Danxia landscapes. They also provide for passage of water through the landscape. Valleys may range from extremely narrow ravines to broad valleys, depending on the stage reached in the evolutionary process of valley erosion.

Narrow shear fracture type ravine

Shear fracture type lane valley

Fluvial erosion type lane valley

Incised meander valley

Broad valley with scattered peak forest

(4) **Caves**

Caves are very common in Danxia landscapes. They are very different in origin and character from karst caves. Due to a combination of fluvial erosion, weathering and collapse, caves may take many shapes and forms.

Bedding rock tank on Lijia Rock in Taining

Forehead-like rock tank

Vertical groove

Lager horizontal cave group

A giant cave, with a width of 100 m, height of 32 m, and depth of 42 m

Niche-like cave group on the Tianqiong Rock

Honeycomb-like caves stone

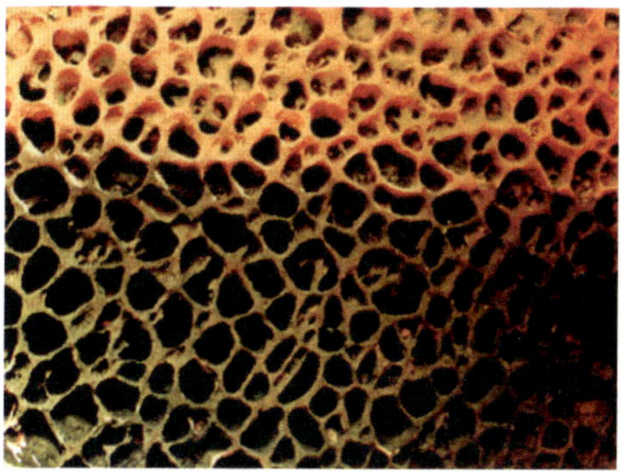

Parts of honeycomb-like caves of Longlin

Karst-like cave at Wuzhuyan

Perforation cave

Weathering cave

Tangjiaba stone arch (64 m long, 20 m high, 14 m width)

(5) **Potholes**

Commonly, on some bedrock river beds, gravel carried in the rotational flow of water grinds out a notch, which then becomes a pothole. This in turn may deepen and extend to the shores, thus forming a pool.

5.4 The Landform Development History

The red beds in the basins of the heritage sites are all constructed by red continental clastic rocks in interior basins of Cretaceous period of Mesozoic era. These red beds were developed into Danxia

Pothole group of Feihua Rivulet and Jiulong Rivulet in Danxiashan: Double-layered potholes are formed by the fluvial erosion

landscape during the late Neogene to Quaternary. It is result of the weathering, erosion, and gravitation activities after the uplift of these red beds.

5.4.1 The Form of Basin and Red Beds Deposition

During medium Triassic to late Triassic of Mesozoic, Indo-China Plate at the southwest side pushed and extruded north to South China Plate and then collided with North China Plate, the Indo-China movement which had great influence to South China happened. Crust in south part of China was activated again; the South China Plate where the heritage sites located came into inside-plate movement stage. Several widely distributed and intense fold, fault, and magmatic activities happened, thus, formed the main formation structure of South China region.

During medium Jurassic to late Jurassic, as the Yanshanian Movement (Pacific movement) began and strengthened, large-scale fold, fault, and magmatic activities happened in the east part of South China Plate. During Late Jurassic to Early Cretaceous, the pacific plate subduction to the Eurasian Plate aggravated. NNE uplift and depression belt were formed in the East Asia. A new continental margin intense activity stage began. Lots of fault basins appeared in South Plate, and thick red clastic rock series is developed. Among them there were mainly mixed sandy gravel deposition combinations, while also including gypsum salt evaporation rocks, airborne sediment rocks, freshwater marl rocks, and other sediment types that formed in drought condition, in which red pluvial-alluvial facies clastic rock type is the main material foundation of China Danxia landscape.

On the whole, red beds that composed China Danxia are all belong to Cretaceous red beds, while there are differences among the heritage sites. The Lanlong Formation which composed Langshan, the Fangyan Formation which composed Jianglangshan are both early Cretaceous red beds; while red beds in other heritage sites are all belong to Late Cretaceous red beds. In perspective of features of the formations, lithology differences among heritage sites related to the scale of red bed basin, depositional position and depositional environment, different forming time of early and Late Cretaceous red bed did not lead to lithology differences.

5.4.2 Basin Uplifts and Danxia Development

At the end of Cretaceous, the influence of the late Yanshan Movement made red beds basins of candidates uplift successively and separate form sedimentation environment. In the Paleogene, there were still some red beds basins continued sedimentation in the south and lasted as far as the end of Paleogene. In the late Paleogene, Himalayan Movement completely ended the sedimentation of red beds basins in South China and large-scale uplifts took place instead.

Large-scale uplifts are what occurred affected by Himalayan Movement since Neogene, a sort of lifting featuring large-scale and intermittence, during which red beds were eroded. In the tectonic movement, faulting deformation and fault-block movement are distinct, forming many groups of faults and large joints. During this time, the uplift of Tibetan Plateau intensified East Asian Monsoon Circulation, the transformation from arid environment to humid climate occurred in Southeast China. Meanwhile, the erosion of flows was strengthened along with gravity and biological effect. Thus, Danxia landform entered into the stage of overall development.

Since the Neogene, all red beds in which China Danxia locate have been under the condition of uplifts. According to the denudation-planation surface, there ought to be mature Danxia landform in every part of Pliocene, some have evolved into old period, such as Mt. Jianglangshan and Bazhai in Mt. Danxiashan. Since the late Pliocene, the force by which India Plate pushed Eurasian Plate northward has been strengthened and the Third Stage Terrace of China was eventually formed. Chishui area, as one of the candidates, belongs to the second stage of China terrace and distinguishes itself from Sichuan Basin by the huge lifting difference. Therefore, it actually belongs to the rapid-uplift zone inside the second stage. Though Southeast China belongs to the third stage, this area has risen fast in the recent 5 million years. Accordingly, it is the primary precondition that the crust has experienced a rapid and intermittent uplift. Finally, under the conditions of subtropical humid climate, candidates, affected by flows, are developed into beautiful and colorful modern Danxia.

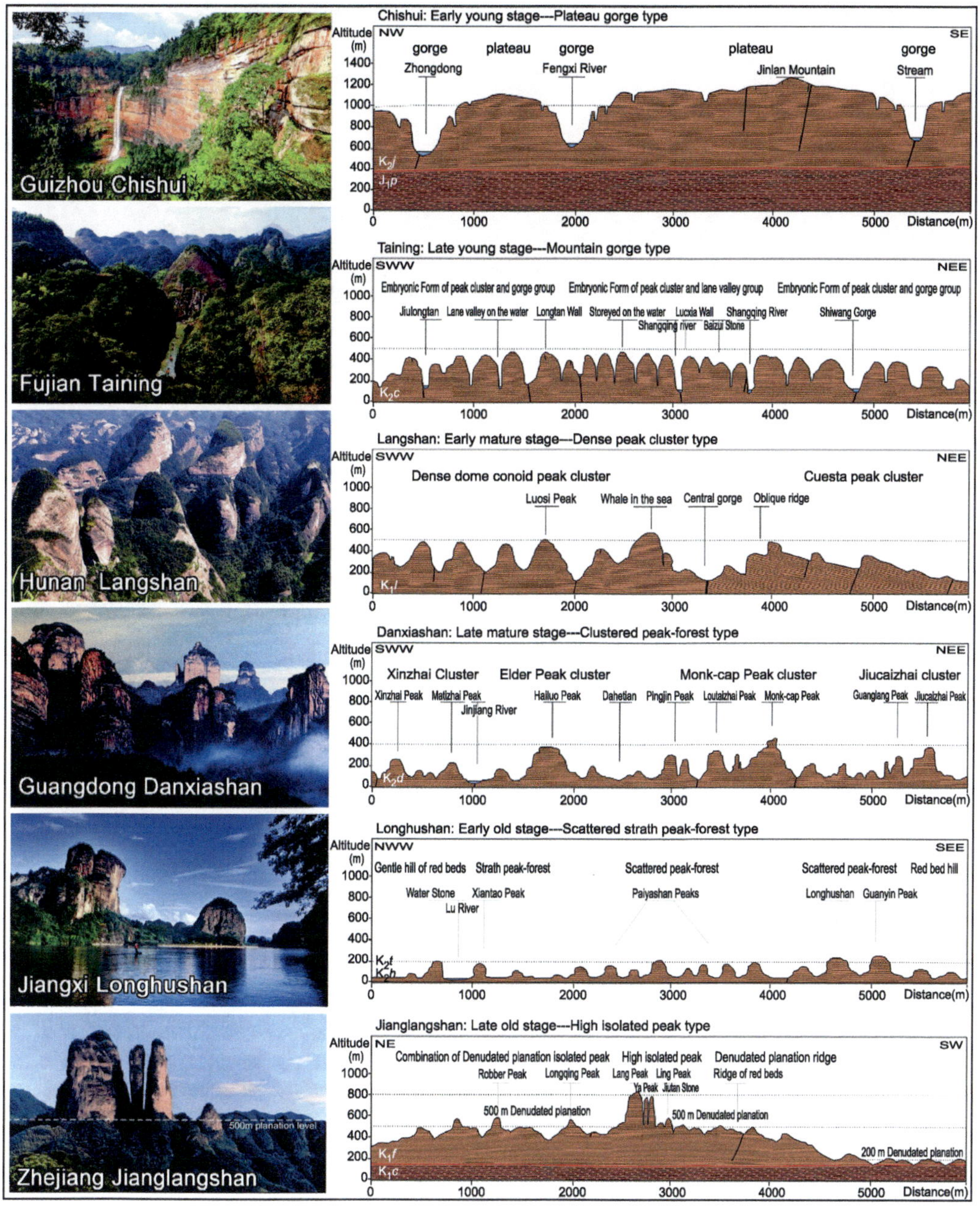

Representative landscape and topographic profile of each heritage site

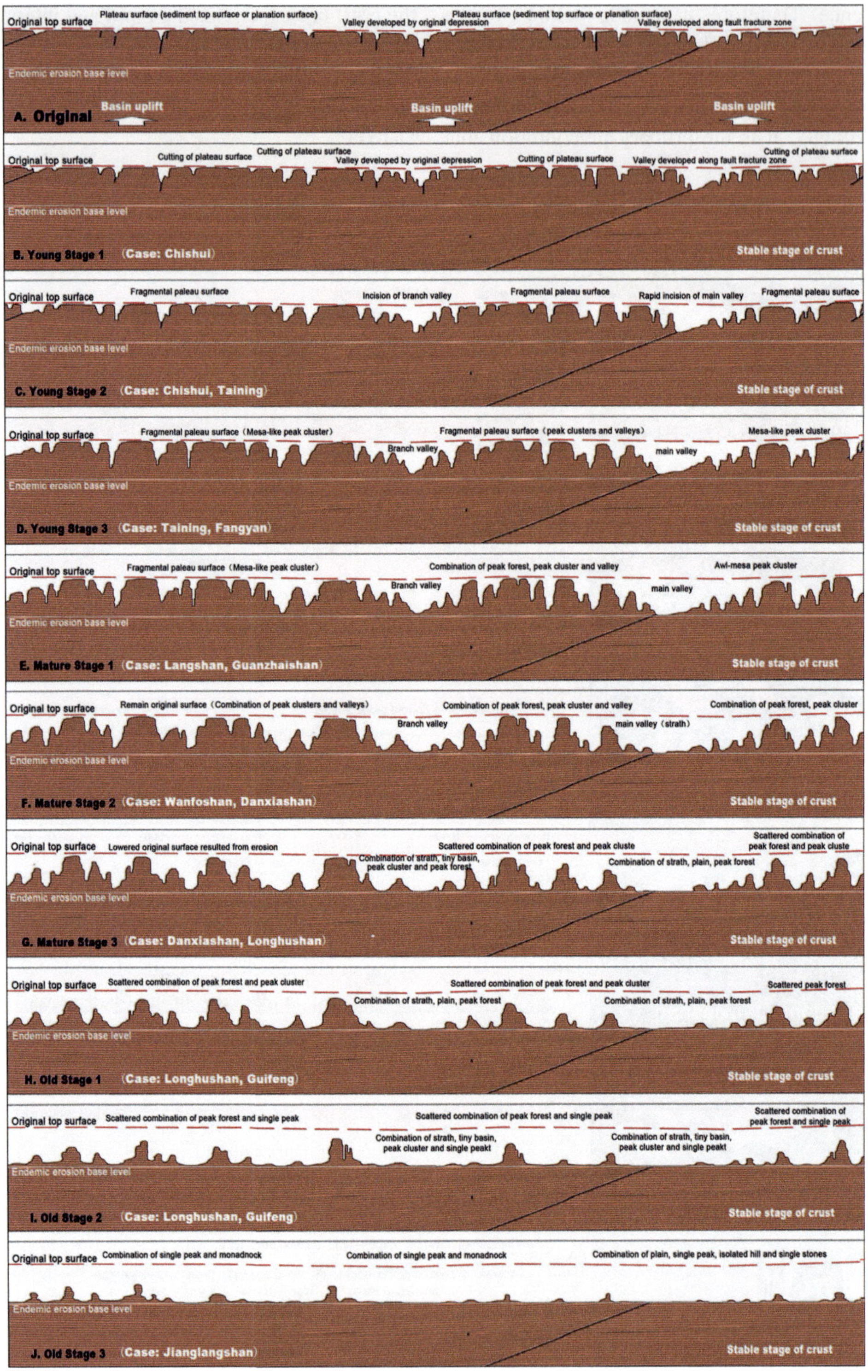

A sketch map of evolutive process of China Danxia

6 Biological Characteristic

6.1 Characteristics of Ecosystems and Plant and Animal Community Development

6.1.1 Ecological Succession of Communities

(1) **The formation and succession of evergreen broadleaf forest are derived by the east monsoon**

The evergreen broadleaf forests are developed under subtropical humid climate conditions. The most complete and extensive evergreen broadleaf forests in the world are found in the heritage sites. Distinct from prairies and deserts formed in other regions at the same latitude, the evergreen broadleaf forests exist here because of the humid climate created by the southeast monsoons. The smaller areas of typical regional evergreen broadleaf forests in each of the heritage sites combine in the overall property to create a very large expanse of natural evergreen broadleaf forest in China.

In terms of the Udvardy biogeographical system (Udvardy 1975), the heritage sites are located in the "Chinese Subtropical Forest" and "South Chinese Rainforest," Provinces within the Palaearctic and Indomalayan Realms.

H. Peng, *China Danxia*, Springer Geography,
https://doi.org/10.1007/978-981-13-5959-0_6

The evergreen broadleaf forests of the heritage sites (Arranged according to latitude)

Site	Geographical position			Main ecological factors				Flora and evergreen broadleaf forest		
				Precipitation	Temperature					
	Central latitude	Central longitude	Elevation (m)	Annual precipitation (mm)	Annual temperature (°C)	Annual radiative capacity ($kJ/cm^2/yr$)	≥ 10 °C active accumulated temperature (°C)	Tropical/temperate ratio	Regional property of evergreen broadleaf forest	Role in evergreen broadleaf forest
Danxiashan	24°57′55″N	113°42′12″E	58–625	1715	19.7	465.15	6273.0	64.8/35.2	Having tropical landscape such as buttress roots, cauliflory, and woody climber and so on. Palms are common	Transitional type of subtropical evergreen broad-leaved forest and tropical monsoon forest
Langshan	26°20′24″N	110°46′45″E	302–818	1450	15.5	406.54	5308.1	52.3/47.7	Common property of evergreen broadleaf forest	Subtropical evergreen broad-leaved forest
Taining	26°51′56″N	117°02′22″E	200–913	1778	17.1	481.50	6207.5	52.2/47.8	Common property of evergreen broadleaf forest	Subtropical evergreen broad-leaved forest
	27°00′37″N	117°13′07″E								
Longhushan	28°04′15″N	116°59′05″E	48–401	1878	18	443.65	5776.8	50.3/49.7	Common property of evergreen broadleaf forest	Subtropical evergreen broad-leaved forest
	28°19′03″N	117°25′10″E								
Chishui	28°22′11″N	105°47′39″E	240–1750	1287	18.1	361.50	5750.0	60.9/39.1	Common property of evergreen broadleaf forest	Subtropical plateau evergreen broad-leaved forest
	28°25′19″N	106°02′33″E								
Jianglangshan	28°31′44″N	118°33′43″E	644–824	1650	14	468.25	5481.0	52.9/47.1	The evergreen broadleaf forest processes property of sclerophyllous forest	Transitional type of subtropical evergreen broad-leaved forest and sclerophyllous forest

(2) **Ecological succession of biotic communities**

Danxia landscapes have experienced natural ecological succession throughout the late recent geological epoch. Original primary vegetation with no human disturbance is preserved on the denudation-planation surfaces of summits, isolated by cliffs, since the Neogene and especially the Quaternary glaciations. There are primary and secondary vegetation successions in the area.

Secondary succession: Evergreen broadleaved forest, which is the mature zonal climax vegetation, covers the ravines and foothills of the landscape. Secondary succession occurs in places disturbed by human beings. Different successional stages vary according to time and space processes. The successional series includes early pioneering forests, transitional forests, and stable forests.

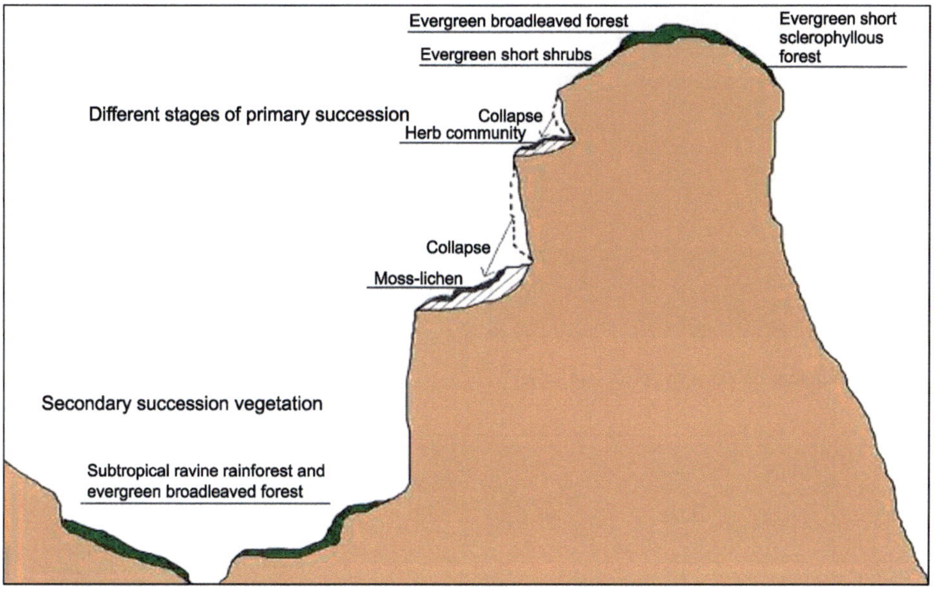

Primary and secondary vegetation succession on Danxia landscapes

Primary succession: The primary succession contains moss and herbal communities at the early stage and communities of short shrubs and trees at the middle and late stages. During the formation of Danxia landscapes, the rock surfaces are strongly corroded by seasonal and intense rainfall, as well as by fluvial erosion. Steep valley walls and deep ravines are created. The land surface is further shaped and rebuilt by weathering and landsliding, with vegetation succession commencing again on bared surfaces. Spatial patterns of vegetation succession thus emerge.

6.1.2 Ecological Effects of Danxia Landforms

(1) **Ravine effects**

Many ridgy stone peaks and narrow ravines are typical of Danxia landscape topography. The ecological factors differ between the ravines and the open areas. The particular microclimate and favorable high-moisture conditions in the ravines can provide a suitable living environment for tropical species, which prefer to hot and humid environments. The varying orientation, depth, and shape of ravines affect the availability of sunshine in the ravines. All of these aspects influence the temperature and humidity in ravines. There is abundant growth of tropical plants in most ravines. The ravines are rich in lianas, ferns, and other high-moisture tolerant plants. The presence of ravines in Danxia landscapes allows the growth of tropical plant taxa, which would otherwise not occur here at this latitude. About 10% more tropical plants are found in Danxia areas than in other areas of similar latitude. Ravines, therefore, create a latitudinal shift in plant distribution, with some species typical of south subtropics areas appearing in the mid-subtropics. These plant communities that are highly adapted to their geomorphic environment are termed geo-climax communities.

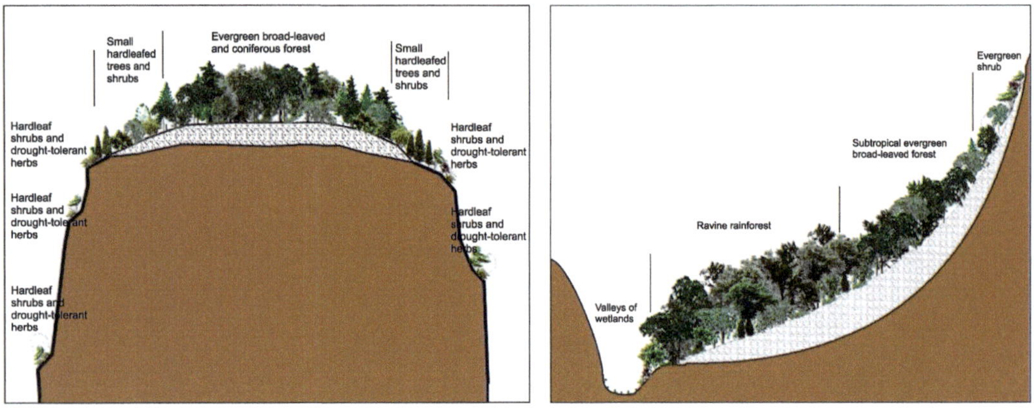

The vegetation different between the peak and valley

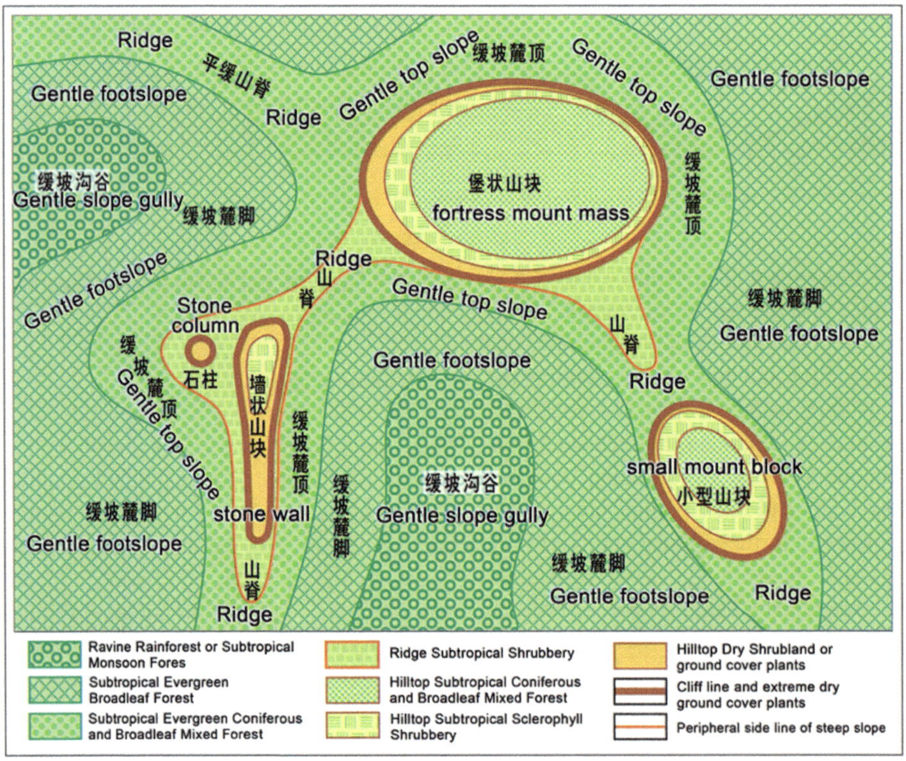

Plant differentiation caused by hilltop and ravine effects in Danxia landscapes

(2) **Hilltop effects**

The distinct topography of Danxia landscapes generates a particular hilltop ecological effect. A fundamental character of landscape morphology is "Flat hilltop, steep massif, and gentle piedmont." On cliff faces, there is essentially no soil, allowing little plant growth. This means, there is an effective separation between the vegetation on hilltops and at the base of cliffs. Hilltop vegetation is thus isolated and becomes distinctive. Also, the growing environment is very different between hilltops and cliff bases. The mean daily temperature on hilltops is higher, while the average humidity and community diversity on the hilltops are lower than those at the base of cliffs. Compared with plants at cliff bases, the hilltop plants have a smaller leaf area, lack lichen cover on the bark, are lower in height and have a smaller canopy. Hilltop effects also influence community structure and flora as well. The types of species, dominant species, and community diversity and all are markedly different between hilltops and bases of cliffs.

6.2 Biodiversity

6.2.1 Biological Species

(1) **Diversity of biological species**

In the heritage sites, there are 5772 species of higher plants, in 293 families and 1271 genera. Among these species, 5181 species are spermatophytes, belonging to 242 families and 1150 genera, and 591 species are pteridophytes, belonging to 51 families and 121 genera (detailed in the plant directories of the heritage sites). There are 836 species of vertebrates, belonging to 129 families and 37 orders (132 species of Mammalia, 364 species of Aves, 109 species of reptiles, 70 species of Amphibia, and 161 species of fish). There are 3073 species of insects in the heritage sites (detailed in the animal directories of the heritage sites).

Higher plants found in Danxia landscapes

Classifications	Families	Percentage of families in China (%)	Genera	Percentage of genera in China (%)	Species	Percentage of species in China (%)
Pteridophyta	51	–	121	–	591	–
Gymnospermae	10	100	26	76.47	54	27.98
Angiospermae	231	79.38	1124	38.15	5127	21.05

Animals and insects in Danxia landform

Classifications	Orders	Families	Genera	Species
Mammalia	8	27	76	132
Reptilia	3	13	55	109
Amphibia	2	10	26	70
Aves	18	60	191	364
Pisces	6	19	102	161
Insecta	25	329	1914	3075
Total	64	520	2447	3679

(2) **Rare and endangered species**

- **Rare and endangered plants**: There are 214 species of rare and endangered plants in the heritage sites. Of these, 34 species belonging to 23 families and 29 genera are on the IUCN Red Data List. Among them, two are classed as CR, seven are EN, 13 are VU, one is LR/cd, and 11 are LR/nt. Some 104 species belonging to seven families and 48 genera are listed by CITES (101 in Appendix, three in Appendix III). Some 49 species belonging to 30 families and 41 genera have State-level Protection, among which six are in the first class and 43 are in the second class. Some 180 species belonging to 40 families and 103 genera are on the China Species Red List (four are CR, 23 are EN, 83 are VU, and 70 are NT) (detailed in the lists of rare species of plants for each of the heritage sites).

Apart from their large number, the rare species in the heritage sites are also unique. Some species are rare to found elsewhere. For example, the large area of natural undisturbed Danxia wetland means this is the only habitat for the *Isoetes sinensis* community.

- **Rare and endangered animals**: There are 189 species of rare and endangered animals in the heritage sites. They belong to six classes, 26 orders, and 54 families. Among these species, 45 species are on the IUCN Red List (two are CR, 11 are EN, 9 are VU, and 23 are LR). Some 66 species are on the list of CITES (19 are in AppendixI, 47 are in Appendix II, and 10 are in appendix III). Some 80 species have State-level Protection (12 are in the first class, 68 are in the second class). Some 145 species are on the China Species Red List (46 are NT, 76 are VU, 16 are EN, and seven are CR) (detailed in the lists of rare species of animals for the heritage sites). A prominent example is *Mergus squamatus,* which is called "the living fossil that can fly" and has survived since the Quaternary Ice Age. More than 150 of these birds winter in the heritage sites each year, which is the largest wintering group of these birds anywhere in the world.

Isoetes sinensis (first-class national protected plant) and its habitat-Danxia Wetland

Mergus squamatus in Luxi River of Mt. Longhushan

Summary of rare plants and animals in the heritage sites

	Plant (number)				Animal (number)			
	First-class national protected	Second-class national protected	IUCN	CITES	First-class national protected	Second-class national protected	IUCN	CITES
Species	6	43	34	104	11	67	21	63
Families	30		23	7	6 class, 26 orders, 52 families			
Genera	41		29	48				
Total	118				288			

6.2.2 Endemic Species

Because of the special landscape character, there are many endemic species in the heritage sites. The heritage sites have about 600 species endemic to China and about 40 species endemic to the heritage sites. For example, based on the investigations over several years, it is known that *Firmiana danxiaensis*, *Ranunculus xinningcnsis,* and *Chirita langshanica* are found only in the heritage sites. *Ranunculus xinningcnsis* and *Chirita langshanica* are found only on the stone walls of Langshan Mountain.

Firmiana danxiaensis widely distributed in ravines

Ranunculus xinningcnsis

Chirita langshanica

In the heritage sites, there are 11 vertebrate species endemic to China, among which eight are mammalian, six are avian, four are amphibian, 16 are fish, and 10 are belong to other classes. There are also many endemic insect species in the heritage sites, such as *Toacris yaoshanensis*, *Flatocerus nankunshanensis,* and *Carabus melli.*

6.2.3 Ancient Trees

Because there has been relatively little artificial disturbance of these areas due to the complex topography and protection from the government, there are many ancient trees in the heritage sites. According to incomplete statistics, there are 4843 trees older than 100 years, among which 3455 are aged from 100 to 299 years, 904 are aged from 300–499 years, 484 are aged above 500 years, 25 are aged above 1000 years. A *Juniperus formosana* in Jianglang Mountain is 3600 years old. The large number of ancient trees here indicates the original, unmodified condition of plants and communities in the heritage sites.

communities and nine formations and nine associations of rock vegetation. The aquatic plant community is also widely distributed, including 20 formations and 31 associations, and with many river-wetland plant communities.

- **Vertical differentiation of plant communities**: Differentiation of plant communities according to altitude is very obvious. This is different from the vertical differentiation of climatic conditions. It occurs mainly because of differences in sunlight, temperature, moisture, and soils—the patterns of which differ according to the stage of geomorphic development of the landscape. For example, macrophytes generally do not grow on the steep cliffs lacking water and soil. Summit surfaces are generally very limited in extent, and the soils are thin with poor water holding capacity, so the summit vegetation often has the drought resistance characteristics. In deep gorges, sunlight is generally insufficient and river erosion is vigorous, so the trees are often low in height and brush communities are common. The lower slopes and hills are the best-watered environments, with better soil development and more moderate temperatures, which favor the growth of subtropical evergreen broadleaf forest. There are some modified vegetation communities in the broader, flatter river valleys.

Ancient trees in the heritage sites

	Age above 500 (number)	Age 300–499 (number)	Age 100–299 (number)	Total
Chishui	241	517	1450	2208
Taining	17	38	310	365
Langshan	122	245	729	1096
Danxiashan	23	13	760	796
Longhushan	61	70	147	278
Jianglangshan	20	21	59	100
Total	484	904	3455	4843

6.2.4 Biotic Communities

(1) **Plant communities**

- **Plant community types**: Vegetation type (high-rank unit), formation (median unit), and association (basic unit) are used for classifying vegetation in China. There are 23 vegetation types, 261 formations, and 424 associations within the heritage sites (as shown in appendix: Summary sheet of vegetation types in the heritage sites). Among the vegetation types are evergreen broadleaf forest, evergreen-deciduous broadleaf mixed forest, deciduous broadleaf forest, coniferous forest, and sclerophyllous forest. Evergreen broadleaf forest is the typical zonal vegetation of Danxia landscapes. It includes 70 formations and 102 associations. There are rich and diverse communities of lianas and attached rock vegetation types, with 19 formations and 21 associations of the attached lianas

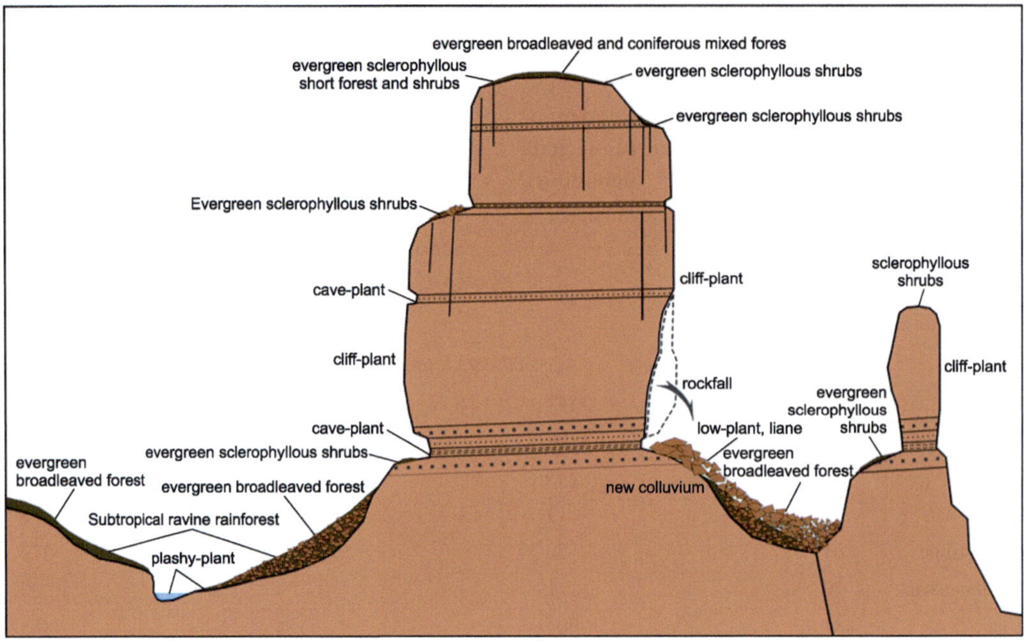

Schematic diagram of the vertical distribution of vegetation in Danxia landscapes

- **Humid evergreen broad-leaved forests in southeast China**: Large areas of original humid evergreen broad-leaved forest of the mid-subtropics are well-protected within the heritage sites. For example, in Guifeng area of Longhushan, natural evergreen broad-leaved forests at low altitude (48–100 m) include many families such as *Fagaceae, Lauraceae, Theaceae, Elaeocarpaceae, Symplocaceae,* and *Magnoliaceae*. Similarly, there is good protection of low-altitude evergreen broad-leaved forest in Taining, including *Phoebe bournei* forest, *Altingia gralilipes* forest, and *Castanopsis sclerophylla* forest. Mt. Langshan has *Castanopsis eyrei* forest, *Castanopsis fargesii* forest, *Cinnamomum camphora* forest, *Phoebe* forest, and *Lindera megaphylla* + *Phoebe* forest in the valleys; *Castanopsis carlesii* forest, Castanopsis *eyrei* forest, *Castanopsis eyrei* + *Schina superb* forest, *Schima argentea pritz* forest, and *Illicium laceolatum* forest on the foothills; *Lithocarpus hancei* forest, *Cyclobalanopsis chungii* + *Lithocarpus hancei* forest, *Nothopanax davidii* forest, *Ternstroemia gymnanthera* + *Symplocos crassilimba* forest, *Symplocos crassilimba* forest, and *Symplocos crassilimba* + *Quercus phillyraeoides* forest on the hilltops.

Comprehensive characteristics of physical geography in the heritage site with subtropical evergreen broad-leaved forest (Chishui)

Evergreen broad-leaved forest (Danxiashan)

Subtropical ravine rainforest (Danxiashan)

- **Endemic communities**: The special and isolated habitats created within the compel topography of Danxia landscapes favor the growth of endemic communities. The heritage sites have three main types of endemic communities: i) xeric plant communities on summits and crags, such as *Selaginella tamariscina, Quercus phillyraeoides*, *Quercus oxyphylla,* and *Acer oblongum.* ii) mesic communities ravines, such as *Torreya jackii*, *Phoebe bournei*, *Isoetes siensis,* and *Calamus hoplites*. iii) communities of rare endemic species, such as *Ranunculus xinningcnsis* + *Pogonatherum crinitum*, *Chirita langshanica* (Mt.Langshan), and *Firmiana danxiaensis.* Communities of rare species endemic to Danxia landscapes include *Taxus chinensis*, *Cibotium barometz* (Mt.Danxiashan), *Cotinus coggygria*, and *Exochorda racemosa*.

(2) **Animal communities**

The great diversity of ecosystems and complex natural environments provide many and varied habitats for the wildlife. There are habitats for cave-dwelling animals, ground-living animals, and tree-living animals. This means, there is a high diversity of animal communities. They can be subdivided on the basis of habitats, examples of which are forest animals, Phyllostachys edulis forest animals, shrub animals, farmland animals, cultivated zone animals, aquatic animals, flood plain animals, and cave animals.

6.2.5 Ecosystems

(1) **Diversity of ecosystems**

According to the global habitats classification system of IUCN/SSC, there are eight first-level types of habitat in the heritage sites, which account for 61.5% of the total number of first-level habitat types in the world. These habitats mainly include forest, bush, wetlands, bare rock, and caves, forming various terrestrial, aquatic, and wetland ecosystems. There are 47 second-level ecosystem types in the nominated site, including 14 natural ecosystems, 15 artificial ecosystems, and 18 compound ecosystems.

Occurrence of First-level IUCN/SSC Habitat Types in Danxia landscapes

First-level IUCN/SSC Habitat	Danxia landform
1. Forest	✓
2. Savannah	
3. Shrubland	✓
4. Grassland	✓
5. Wetlands (inland)	✓
6. Rocky barren areas	✓
7. Caves and subterranean	✓
8. Desert	
9. Sea	
10. Coastline/Intertidal	
11. Artificial–Terrestrial	
12. Artificial–Aquatic	✓
13. Introduced vegetation	✓

(2) **Special ecosystems**

Danxia landscapes have a complex topography with a network of valleys and ravines. Many special and unusual ecosystems are created, including cave ecosystems, cliff ecosystems, and other complex ecosystems.

- **Cave Ecosystems**: Caves are prominent in Danxia landscapes. They have many different shapes and sizes. There are water erosion caves, potholes, collapsed caves, notch caves, flat caves, arched caves, frontal caves, domal caves, and natural bridges, among others. These caves are not only of scenic interest, they are also the basis for development of a cave culture. Also, they provide good natural refuges for various kinds of birds and cave animals.
- **Cliff Ecosystems**: Lianas (vines) grow on soil-covered hill slopes and also extend up cliffs and other bare surfaces. Their growth also favors the establishment of mosses and herbaceous plants, which can promote plant succession in the formation of cliff ecosystems in Danxia landscapes.
- **Complex Ecosystems**: There are many different habitats and ecological niches in the complex topography of Danxia landscapes. Many of these ecosystems are of very small scale.

(3) **Integrity of the ecosystem**

- **Integrity of ecological succession**: The heritage sites have many complete series of plant successions, both primary and secondary ones. Primary succession has mosses and herbaceous plant communities at the early stage, communities of short shrubs at the middle stage, and tree communities at the late stage. The development of Danxia landscapes involves chemical weathering and fluvial erosion. Steep valley walls and ravines result. Continued weathering and mass movement shape the landforms. Plant succession begins anew on new and bared surfaces. Shrub vegetation is succeeded by forests, and pioneer forests (coniferous forest and broad-leaved forest) may develop into transitional forests (secondary evergreen-deciduous broad-leaved forest) and eventually stable forests (evergreen-deciduous broad-leaved forest) develop. Human disturbance or a high degree of natural disturbance may destroy ecosystems and form secondary bare areas. Secondary succession begins when the disturbance has concluded. Vegetation varies in its successional state and processes both in time and in space.
- **Primary plants and integrity of ecosystems**: Especially in the core areas of the heritage sites, there is little human disturbance of vegetation and primary plants are abundant. There are 25 trees older than 1000 years, 484 trees older than 500 years, even a 3600 years old *Juniperus formosana* occurs in Jianglang Mountain. These reveal the ancient state of some forests and their preservation over a long period. Much of the heritage sites have intact forest ecosystems. All the heritage sites in the heritage sites have typical evergreen broad-leaved forests of different types.

6.2.6 Biota

(1) **Biogeographical realms**

Based on a Udvardy's biogeographical system (Udvardy 1975), the heritage sites occur within the Chinese Subtropical Forest and South Chinese Rainforest Biogeographical Provinces, which in turn are within the Palaearctic and Indomalayan Realms.

Classes of Udvardy's Biogeographical Provinces occurring in Danxia landscapes

No.	Realms	Provinces	Biome types
2.1.2	The Palaearctic Realm	Chinese Subtropical Forest	Subtropical and temperate rain forests or woodlands
4.6.1	The Indomalayan Realm	South Chinese Rainforest	Tropical humid forests

(2) **Floral characteristics**

The flora of Danxia landform is part of the East China plant region and the South China plant region of the Chinese–Japanese forest plants subregion, which in turn are part of the broad North Polar plant region. The flora is principally tropical—subtropical but also has a component of the East Asian region. The massive sedimentary rocks, which stand in peaks, and the ravines and rivers in the Danxia landscapes have a strong influence on the flora. The many microclimates in the complex topography enable the growth of a greater proportion of tropical flora here, compared to areas of similar latitude. Tropical flora occupies more than 50% of the total genera of vascular plants in Danxia landscapes and temperate flora take second place.

The areal types of genera of spermatophytes in Danxia landform

Areal types	Families	Genera	Percentage of total genera (%)
1. Cosmopolitan	14	86	7.48
2. Pantropic	15	122	10.6
3. Trop. Asia to Trop. America	6	42	3.65
4. Old World Tropics	16	60	5.22
5. Trop. Asia to Trop. Australasia	13	60	5.22
6. Trop. Asia to Trop. Africa	17	58	5.04
7. Trop. Asia	23	112	9.74
8. North Temperate	25	139	12.09
9. Eastern Asia to North America	23	136	11.83
10. Old World Temperate	28	108	9.39
11. Temperate Asia	12	42	3.65
12. Mediterranean W. Asia to C. Asia	11	20	1.74
13. C. Asia	5	8	0.7
14. E. Asia	24	124	10.78
15. Endemic to China	9	33	2.87
Total	241	1150	100

(continued)

(3) **Faunal characteristics**

Fauna within the nominated region belongs to the eastern Central China Region and northern fringe of South China Region of Indo-Chinese Subrealm, Oriental Realm. And there is obviously mutual permeation phenomenon between The Palaearctic Realm and Oriental Realm. Faunal composition has typically subtropical and tropical characteristics, The North and South type has at the same time, but by south type primarily.

(4) **Ancient biota**

The biota in the heritage sites contains many ancient components that are relics from the Cretaceous and Tertiary. For example, Lycopodiaceae originated in the Paleozoic, Osmundaceae and Gleicheniaceae oringinated in the Triassic, Dicksoniaceae originated in the Jurassic, Lauraceae, Fagaceae, Palmae, and Aquifoliaceae originated in the Cretaceous, *Pinus* in the Tertiary, and *Acer*, *Betula,* and *Fagus* in the late Cretaceous. Other examples of true plant relics are *Ginlcgo bilobo*, *Cunninghamia lanaolata*, *Eucommia ulmoide*, and *Camptotheca acuminate*.

7 Natural Landscapes

7.1 Natural Landscapes of China Danxia

The natural landscape of China Danxia can be summarized as being a rich and colorful landscape of monotypical geology. It is a unique order in a variety of groups of mountain landscapes. It is also a colorful landscape of red rock, clear waters, green vegetation, blue skies, and white clouds, all of which give the landscape a great aesthetic value. The landscape in the heritage sites is the best and most beautiful representation of Danxia landscapes in China and in the world.

7.1.1 Geomorphological Landscape

Danxia landscape is considered to be a special scenic landscape, formed mainly by cliffs, stony peaks, peak groups, caves, valleys, and other distinctive landforms. The heritage sites represent Danxia landscapes in their young stage, mature stage, and old stage, within the humid area of Southeast China.

(1) **Landscape of a monotypical rock type**

The Danxia landscape type is diversified and distinctive. There are many varied positive rock landforms, such as red walls and cliffs, abruptly rising stony peaks, stone fortresses, stone pillars, and supernaturally shaped landforms. There are also many different negative rock landscapes, such as canyons and lane valleys and dales, and many types of caves and stone arches.

H. Peng, *China Danxia*, Springer Geography,
https://doi.org/10.1007/978-981-13-5959-0_7

Danxia escarpment

Danxia escarpment landscape

Danxia escarpment landscape

Danxia pyramidal peaks group landscape

Danxia stone columns group landscape

Bunch group-type Danxia peaks forest Danxia peaks clump

Intensive Danxia peaks clump

Danxia butte peaks group

Bunch group-type Danxia peaks clump

(2) **Landscape**

Groups of Danxia: Danxia landscapes are a macroscale example of mountain types in China. Danxia landscapes are discrete, orderly, multi-storied, and diversified. The mountain blocks have high towering peaks; the valleys are deep; and the peak groups are forest covered and diverse. The whole landscape presents a colorful three-dimensional space with beautiful configuration. For example, Danxiashan has orderly peak clusters and peak forests of the cluster-group type, combining sparseness of peaks in places with a high density of peaks in others. Langshan has supernatural landforms and splendid peak clusters and peak forests of great density; Taining has dense canyons and peak forests with an intimate combination of mountains and water; Jianglangshan has abruptly rising single peaks and high pedestals; Chishui has majestic plateau, canyons, and waterfalls in close association.

There are imposing red walls and red cliffs in Danxia landscapes, with numerous caves. The gorges are deep, peaceful, and quiet. The rocks and mountain take on very many and varied shapes. The spectacle of the landscape is no less than that of granite landscapes, and the beauty of the landscape is no less than that of karst landscapes. The differences in the lithology of red beds and the surrounding non-Danxia environments make Danxia landscapes that much more rich, colorful, and distinctive. China's Danxia landscapes are the most distinguished representative of the world's red bed landscapes, demonstrating supreme natural beauty and a unique personality.

Chishui is located in the most extensive area of Danxia landscapes, in Sichuan and Guizhou provinces. The relief here is the strongest found in Danxia landscapes. This site also has the most typical multi-stepped valleys and the most spectacular waterfalls. These are characteristic of the plateau–canyon Danxia landscape in an early–young stage of geomorphic development.

The Shiwang Scenic Spot in the Taining candidate site is in a young stage of landform development. It displays a peak planation surface at an altitude about 400 m high with canyons more than 400 m deep. Overall, it has a distinctive reticulate valley floor system and red mountain blocks, with the main Shangqingxi River, retained within a canyon-like valley.

Plateau and canyon waterfall combination landscape in Chishui

Crowded peaks clump and canyons group combination landscape in Taining

Langshan represents the dense peak hoodoo type of landscape. During intermittent and differential uplift of the crust, the red beds were eroded and developed many colorful landscape types. Fuyi River which passes through the east of the region has beautiful scenery.

The Danxiashan candidate site is 168 km^2 in area with more than 1000 peaks above 20 m high. In the core area of the Danxia landscape, the

Crowded peaks clump in Langshan

density of spires, mesas, stone walls, and stone columns maybe 20 per km^2; the entire landscape has the appearance of clustered peak hoodoo.

The Longhushan heritage site is located in the southwest of the peneplane in Xinjiang Basin. The Luxi River, connecting the clusters of peaks

Bunch group-type peaks forest landscape in Danxiashan

on both sides of the site, looks like a beautiful jade belt. The landscape is distinguished by scattered peaks with rounded summits. It is in a late–mature stage to early–old stage of development.

The Guifeng area landscape has scattered forested peaks and is in the late–mature stage of development. The site derives its name Guifeng, meaning Turtle Peak, from the turtle-shaped hill forms. The landscape is renowned for its spectacular and colorful scenery.

In the Jianglangshan heritage site, the red beds of Fangyan Formation cover an area of 11.86 km^2, and the scenic spot is famous for its three isolated peaks called "sanpanshi" These

Dispersal wide valley peaks forest landscape in Longhushan

Corroded remaining peak forest landscape in Guifeng

formed during a second phase of crustal uplift after the landscape reached the old stage of development. The "Sanpan Stone" rise 350 m above the ancient planation surface which is at an altitude of about 500 m.

(3) **Modeling landscape**

Landscapes in Danxia can display many dramatic shapes and patterns which imitate such things as structures, people, birds, and animals.

Corroded remaining butte peaks landscape in Jianglangshan

The outside star person (Guifeng)

The old man stone (Longhushan)

The General stone (Langshan)

Elephant Nose Rock (Longhushan)

Spindle-shaped stone peak (Langshan) and Danxia stone eggs (Chishui)

The camel peak (Langshan)

Double breast stone and the Yinyuanshi (Female Stone) (Danxiashan)

The Capsicum Stone (Langshan)

Yangyuanshi (Male Stone) (Danxiashan)

7.1.2 Water Landscapes

Each of the heritage sites is still experiencing uplift, so there are many streams and rivers flowing through and actively eroding the landscapes. The combination of red blocks and basins, with red mountains and clear water alongside each other, creates a beautiful appearance. The many ravines in the mountains, with rapids, waterfalls, and deep pools give added splendor. Eight of the nine heritage sites are of the low-altitude peak forest and peak cluster type of landscape. Streams are suitable for paddling small boats and rafts. They are livable and accessible. However, Chishui in the western part of the Danxia realm has a landscape of high mountains, deep valleys, and many waterfalls. It is an excellent wilderness.

Aquatic Guifeng landscape

Luxi River valley in Longhushan

Great waterfall of Shizhangdong in Chishui

Red mountain blue water in Danxiashan

Canyon scenery landscape in Taining

Aquatic Danxia in Taining

7.1.3 Ecosystems Landscapes

The varied topography of Danxia landscapes means they are among the best places for preserving subtropical evergreen broad-leaved forests at low altitudes in southern China. The red mountains are just like many rubies in a green ocean. Furthermore, the landscape is ornamented by the vegetation of great variety at small scales. This variety increases with landscape scale and gives the whole landscape a very colorful and attractive appearance. Also, each site preserves famous and old trees which have ornamental value, ancient towering trees in ravines and on isolated mountains, and many songbirds in the forests giving the natural environments great life and vitality. There are many picturesque elements among the wildlife, such as *Macaca mulatta*, *Viverricula indica*, Sambar, Otter, *Tragopan caboti*, *Lophura nycthemera*, *Francolinus pintadeanus, Bamboo partridge*, egret, and a large number of birds of prey. Hence, the heritage sites are a good place for ecotourism, ecological education, and the popularization of science.

Bamboo's sea

Spinulosa plant community in Chishui

Summit cover forest on Danxiashan

The old tree and the plant strangle phenomenon in Danxiashan

7.1.4 Best Combination of Biology and Ecology

All the zonal vegetation of the heritage sites is subtropical evergreen broad-leaved forests with abundant subtropical species and communities. In the Quaternary Ice Age, the heritage sites acted as refuges with a relatively warm climate and thus preserved a great number of rare plants. Therefore, the heritage sites can claim to have the most abundant species and the best combination of physical and biological landscapes among all similar geomorphological regions in the world. In China, it is colloquially said that the "flat top, steep face, and gentle piedmonts," the most typical geometrical features of Danxia landscapes, are ornamented by the vegetation and turned into an aboriginal "handsome boys and pretty girls" with "straw hats, tattoos, and hula skirts."

7.2 The Aesthetic Importance of Danxia Landscapes

The heritage sites are representative of Danxia landscapes in the humid zones of China. It has many different landscape features, and landscapes at different developmental stages, which are discrete, orderly, multi-storied, and diversified. It is the best-preserved and most beautiful Danxia landscape with subtropical evergreen broad-leaved forest and associated ecosystems in China and in the world.

From the perspective of the aesthetics of shape and form, Danxia landscapes possess rich and colorful landforms of a single rock type. The mountain groups combine sparseness of distribution with denseness of concentration, and their spatial patterns have an orderly manner or a tuneful rhythm. Red mountains, clear water, green trees, blue sky, and white clouds add to the beauty and color. In terms of aesthetics, the height and steepness of the red walls and cliffs, the supernatural landform shapes and the beauty and elegance of the landscape gardens, the quiet and serenity of valleys and forests, and the mystery and fantasy of clouds and fogs, all give Danxia landscapes extraordinary beauty and superb aesthetic quality.

The beauty of Danxia landscapes has promoted the development of China's aesthetic culture and given birth to series of exclusive terms of Danxia aesthetics. For example, Danxia itself is a term of great aesthetic perception. Danxiashan is named for its colorful appearance akin to that of rose-colored clouds. The Chinese phrase "*Chi Bi Dan Ya*" is used to describe the red walls and red cliffs; "*Wan Gu Jin Cheng*" (the ancient castle built by gold) is used to describe Danxia mountain blocks; "*Zi Fu Dong Tian*" is used to describe Danxia caves (*Zi Fu* is the palace for a king or emperor; *Dong Tian* is the ground where the gods live); and "*Dan Shan Bi Shui*" is used to describe the composition of the landscape (red mount and blue water).

The purple hues of Danxia landscapes give people a sense of heavenliness and are associated with authority, wealth, and good fortune in traditional Chinese culture. Purple is also the main color associated with China's religions. The castle shapes of Danxia landforms are called "*Wan Gu Jin Cheng*" or "*Zi Fu Dong Tian,*" having an ideal state for religious people. People built temples in Danxia areas or in Danxia caves, and this strengthens the authority and mystery of the religious places and their environmental associations. Many Danxia areas are religious sites, such as the Taoist civil court in Longhushan; the Buddhist grotto temples in Danxiashan and Taining; Nanyanshan Mountain in Guifeng. Danxia landscapes are also become a rest areas for writers. In addition, there are cultural phenomena, such as funerals in Danxia caves made by ancients to ensure passage to heaven that is closely related to the unique shapes and individuality of Danxia landforms. The homogeneous structure, density, and solidity of Danxia sediments allow the rocks to be carved easily, thus leaving large numbers of stone inscription, statues, and murals on the cliff walls and in rock caves, such as at Nanyanshan Mountain in Guifeng and the carved stone groups in Danxiashan.

This close relationship between landscape and people creates a particular special cultural association in Danxia areas that can be called a "Danxia culture," which is well recognized in China.

8 Human Activities

8.1 History of Human Activities in the Heritage Sites

8.1.1 Prehistoric Human Activities

The human activities associated with the use of natural resources in Danxia areas can be traced back to prehistoric times. There are prehistoric cultural sites in most of the heritage sites. Most of these are associated with hunting, fishing, gathering, and farming in valleys or the building of dwellings in natural caves.

Archaeological research has shown that there were human activities in most of the sites as early as the Neolithic. For example, there are important ancient human cultural sites of "catfish turn" in the Neolithic Age (6000 years ago). As many as 80 types of artifacts such as stone tools and stoneware have been excavated.

In 5000 year old Neolithic sites in Zhoujiashan Mountain and Baimianzhai, in the southern part of Langshan, artifacts, such as stone axes, cutting tools, scraping stones, stone chisels, stone spades, stone balls, stone clusters, and grinding stones, have been excavated. There are also some bone artifacts such as needles, hairpins, and crockery, including axes, cans, pots, and other pottery pieces. Research has shown that hunting and fishing, supplemented by gathering, were the main human activities at that time in Langshan.

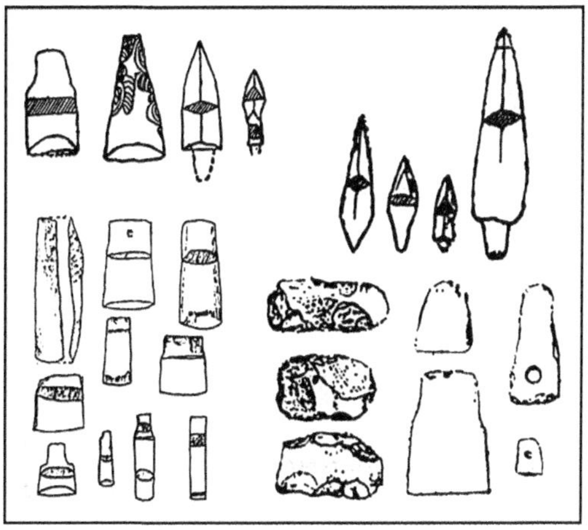

Stoneware found in Yulinshi Cave In Danxiashan

H. Peng, *China Danxia*, Springer Geography,
https://doi.org/10.1007/978-981-13-5959-0_8

Stoneware and porcelain excavated in Langshan

A large number of stone agricultural tools, such as cutters, spades, and axes, and a large number of stone hunting tools, such as spears, arrowheads, and shooting stars, have been excavated in cultural sites dating from the Neolithic era (4500 years ago) in Longshan, and also in Longhushan. Archaeological research also shows a transition to an era of original rice farming and hunting. Thus, in excavations of sites from the late Neolithic Age (4000 years ago) in Taining, stoneware such as swords, arrowheads, and axes, have been found together with clay pottery such as tripod legs, and spinning wheels.

8.1.2 Ancient Human Activities

The valleys and plains on the margins of the heritage sites entered into an era of more developed agricultural civilizations during the Xiashangzhou Period (BC 2070–BC 256). This saw the emergence of stable residential areas including cemeteries, pottery workshops and kilns, as revealed by the grains, animal bones, ceramic containers and bronzes that have been excavated.

Later, from conceiving a harmony between humanity and the earth, people began to appreciate the aesthetic value of Danxia landscapes and landforms, and they gave it a mysterious aura. People then began building dwellings in Danxia caves, or used them as cemeteries. In the so-called Spring and Autumn Period (BC 770–BC 217), people began to bury their ancestors in cliff caves in Longhushan. The caves were used because they were places that people and animals found difficult to reach, while at the same time they provided ease of ascent to heaven, This epitomises the close association between people and nature.

Harmony between humanity and heaven on the cliffs (Swan Lake in Longhushan)

Ancient cliff tombs in Longhushan in the Spring and Autumn Period

Cave dwellings in Danxiashan

The color, morphology, shape, and sense of holiness of Danxia landscapes were first appreciated by the masters of religion, who began to build Taoist Courts and Grotto Temples in Danxia areas, especially in caves. In the East Han dynasty period, the founder of Taoism, Zhang Daoling (34–156 AD) educated himself according to religious doctrine and did alchemy in Longhushan. He then founded "Tianshi Taoism" in 142 AD, making Longhushan the birthplace of China's Taoism and Taoist's civil courts. There were also people who studied and lived according to Taoist doctrine in natural caves in Taining and Danxiashan since the Han Dynasty. In the Tang Dynasty, at Lingnan in Danxiashan, there was a Buddhist resort, known as the birthplace of "Yunmen Zen" in the southern Zen. After the Tang Dynasty, came the emergence of Buddhism in the Song Dynasty and especially in the Ming and Qing Dynasties. The custom of establishing schools sprang up in southeast of China from the time of the Song Dynasty, with many schools and colleges being created in famous mountain areas. Looking at the uses that peoples have made of Danxia areas, we can detect a gradual transformation from original farming settlements to present day scenic spots. They are places which became associated with local folk beliefs, immortal cultures, religions, and college cultures. These uses and activities stem from recognition of, and wonder at, the beauty and aesthetic qualities of Danxia landscapes, and realization of the close harmony between humanity and nature.

8.2 Main Human Activities in the Heritage Sites

8.2.1 Farming Activities

Archaeological research has revealed that there were farming activities in the valley plains of heritage sites beginning thousands of years ago, but this was limited largely to traditional farming, aquaculture, fishing, and hunting. Because of the difficulties of settlement in the mountainous Danxia landscape areas, there were only small populations. The undulating terrain prevented the expansion of arable land, and most of the area remained unpopulated. Therefore, although agricultural civilization and traditional farming activities have existed here for thousands of years, their influence on the natural environments of the Danxia landscape has been minimal.

The residential population in the neighboring valleys also remains low. There are no industrial activities in the nominated area at present nor any large-scale human disturbance. The heritage sites are, therefore, largely preserved in their natural state. Over a very long period concepts of harmony in Chinese traditional culture have had a strong influence in the protection of environmental resources in the heritage sites. For the most part, local residents have maintained a tradition of living in harmony with their natural surroundings and respecting nature.

8.2.2 Religious Activities

The most influential indigenous religion in China is Taoism. The highest God Taoist belief is Laozi, and the highest code is "Dao De Jing" (Morality Lection). The core theory of "Dao De Jing" is humanity takes his law from the earth; the earth takes its law from heaven; heaven takes its law from the Tao. The law of the Tao is "everything being what it is." This means that people are restricted by the earth, the earth is restricted by heaven, heaven is restricted by the Tao, and the Tao is restricted by the nature. In other words, the law of nature is the highest law. Also, the Tao in its regular course does nothing (for the sake of doing it), and so there is nothing which it does not do. This means if you do not do things that are contrary to natural law, and do not do things to alter nature by deliberate action, you can achieve great accomplishments. This is a central philosophy of "Tao." So, Taoism seeks harmony between humans and heaven through the process of respecting and cherishing nature, promoting the religion beliefs of heavenly laws, undertaking good works and treating evil with punitive justice. From the perspective of the modern nature conservation, Taoist thought is positive a positive force for establishing harmony between people and nature. Such beliefs have had a very positive impact on nature protection in the heritage sites.

Additionally, Longhushan, Danxiashan, Taining, Jianglangshan, and Langshan are Buddhist centers, or local Buddhist resorts, of considerable importance and influence. Natural grotto temples are a common manifestation of Buddhist religion in the landscape. The number of grotto temples found in Taining is more than 80 today, 30 of which are still thriving. Buddhist buildings are also located in mountains, where they use the terrain and environment skillfully for siting and construction. Thus, the religious atmosphere of the landscapes and the sense of harmony between nature and culture, heaven and earth, are strengthened.

Other religious beliefs in some regions always have their ideological roots within either Taoism or Buddhism, and they also play a positive role on the protection of nature and public education in conservation. For example, the central religious belief of the Dong Minority is to worship ancestors and nature. They believe that their ancestors and everything on earth have intelligence, and practice pantheistic totem worship of many kinds. They follow the teachings of their ancestors strictly, and thus protect the natural environment strictly.

8.2.3 College and Academic Activities

"Gentle people like mountains and wiser people like water." China's literati know that a quiet and peaceful environment can purify the soul, and that beautiful mountains and rivers can shape a person's temperament. Therefore, since the times of the Tang Dynasty, some cultural celebrities have sought beautiful scenic places for establishing their schools, and this gradually became common practice. The term "College" refers a place to write, print and collect books, lecture and sacrifice in the ways of the ancient Chinese. At the same time, a college has academic activities. Colleges are also a concourse for Confucianism for discussing doctrine, such as the Xiangshan College near Longhushan, which was one of Four Colleges in the Southern Song Dynasty. Additionally, Jianglang College of Jianglangshan, Dieshan College near the Nanyan area of Guifeng, the Schooling-Rock of Jinhu in Taining are all colleges stemming from ancient Chinese religious culture. The architecture of these colleges has traditionally been skillfully in harmony with the shapes and forms of the Danxia landscapes. Many of the colleges use quiet and peaceful natural caves to absorb the atmosphere and other elements of heaven and earth, and thus realize the integration of nature and humanity.

8.2.4 Tourism Activities

Many of the nominated Danxia sites have been important scenic spots for a long period in Chinese history. For example, more than 4000 years ago, when the *Emperor Shun* traveled south and saw the Danxia mountain groups, he climbed a mountain and played the music of "Shao," naming he area "Shao Shi Shan" (in the southern part of Danxiashan), which became a scenic resort in the Sui and Tang Dynasties. Longhushan became a famous tourist attraction before the Tang Dynasty, as seen by the beautiful poems and inscription on Danxia cliffs. The fantastic landscape of Jianglangshan caught people's attention long ago and became a common inspiration for writers from the times of the Tang Dynasty. The development of religion in the Danxia areas helped them to become famous scenic attractions and tourist destinations in very early times. On the other hand, tourism development in places like Chishui, which are located in more remote western areas, is more recent. As a whole, tourism in the heritage sites is not intensively developed, so most of the Danxia areas remain unaffected by any detrimental impacts of tourism and are essentially in their natural state. Limitations on transportation and access, and a lack of tourism promotion, have helped to reduce the environmental impacts of tourism.

8.2.5 Scientific Expeditions and Scientific Activities

The Danxia landscapes in the eastern region of China have captured the attention of Chinese geographers since ancient times. Xu Xiake (1586–1641), an ancient Chinese geographer, conducted a scientific expedition in Longhushan and Jianglangshan, which was reported in detail in the very comprehensive "Xu Xiake Travel Notes."

Expeditions of geological, geomorphological, and geographical experts in more recent times have focused on mainly on geological history, landform features, the natural environment, and the value of these geological places as special tourism resources. In 1928, the academician Feng Jinglan termed the red beds in the Danxiashan region as "Danxia Formation" when describing the landform features there. In 1939, the academician Chen Guoda named the landscape here as "Danxia Topography." Expeditions to Danxia areas have continued over a long period. Large-scale and intensive studies on Danxia landscapes began nation-wide from the late 1980s, when Professor Huang Jin made national inspections and surveys. In 1991, the "National Symposium on Danxia Landform and Tourism Development" was held. By the end of 2009, Professor Huang Jin had investigated more than 900 Danxia landscape sites throughout China and promoted further research and survey of Danxia geology. He also encouraged the development of the research on special subjects within the Danxia landscapes, such as biological resources and biological landscapes, water resources and water environments, land resources, tourism resources and development, natural disasters, historical culture, and socio-economic development.

To date, domestic Chinese scholars have published more than 600 articles or books on Danxia landscapes and landforms, and related research on associated environmental resources. Some well-known domestic and international experts have inspected the nine heritage sites of the Danxia serial property area, resulting in publication of a considerable amount of research information. Research has also promoted the development of scientific tourism and ecotourism in these Danxia landscape areas. Consequently, they have become important bases for teaching, research, and popular science education, used by many institutions of higher learning and scientific research.

8.2.6 Size and Nationality of the Population in the Heritage Sites

The total area of core zone in China Danxia nominated properties is 821.51 km^2 with the population of 34,026 persons, and the population density is 41 persons/km^2. The total area of buffer zone is 1362.06 km^2 with the population of 100,259 persons and the population density is 74 persons/km^2. The area in sum is 2183.57 km^2 with the population of 134,285 and the population density is 62 persons/km^2. Among them, there is no population living in the core zone of Jianglangshan; the population density of Taining core zone is only 6 persons/km^2; Danxiashan is 9 persons/km^2.

People of the heritage sites all belong to the Han nationality. There are also some individual ethnic minority villages in Langshan and Chishui.

9 Value of the Heritage

9.1 Criteria Under Which the Inscription Is Proposed

China Danxia meets the requirements for a serial nomination, according to the *Operational Guidelines*, Paragraph 137: "provided it is the series as a whole, and not necessarily the individual parts of it which are of outstanding universal value."

The China Danxia serial heritage site is proposed under the World Heritage criteria: (vii), (viii), (ix), (x).

Each nominated site of the serial property contributes to more than two of these criteria.

9.1.1 Criterion (vii): Contain Superlative Natural Phenomena or Areas of Exceptional Natural Beauty and Esthetic Importance

China Danxia has a scenically superlative landscape and landforms, displaying huge vertical red walls and cliffs, combined with various natural color elements and natural environment elements such as peaks, rocks, forests, and water. It has exceptional natural esthetic value.

China Danxia is a natural landscape type characterized by a fantastic combination of natural colors such as red (in the rocks), green (in the forests), blue (in the waters and sky), and white (in the waterfalls and clouds). It has an array of natural elements such as mountains, waters, and forests. It has a great diversity of landform types and many rare landscape characteristics. It is a rich and colorful mountainous landscape, with orderly mountain masses, and an integrated landscape of red mountains, blue waters, green forests and vegetation, blue sky and white cloud, giving the whole landscape a unique personality. The property is the outstanding representative of China Danxia, including sites that are the most typical and beautiful Danxia landscapes in the world.

The heritage site is serial one with individual sites representing the young stage, mature stage, and old age stage of geomorphic development in the humid zone of south China. The deep gorges and valleys of plateau-like mountains of young age, the strong high-relief peak clusters and peak forests of the mature stage, the fewer and lower mountains, and gently flowing rivers in broad valleys of old age are well displayed. They surprise people with their beauty and have a huge visual impact. The sublime and steep red walls and red cliffs, the marvelous and delicate pictograph landscape, the elegant and quiet mountains and waters, the deep and serene ravine forests, and the profound and wonderful clouds and mists, all give China Danxia exceptional natural beauty.

These beautiful landscapes have a special place in Chinese culture. The mountains like

H. Peng, *China Danxia*, Springer Geography,
https://doi.org/10.1007/978-981-13-5959-0_9

"ancient gold castles" give a sense of solemnity and sacredness. At the same time, they symbolize authority, richness, and honor. The landscape color has become associated with traditional religions. Thus, the Danxia landscapes are holy lands in religious terms, and perfect places for refined scholars to rest, and become immortal.

9.1.2 Criterion (viii): Be Outstanding Examples Representing Major Stages of Earth's History, Including the Record of Life, Significant Ongoing Geological Processes in the Development of Landforms, or Significant Geomorphic or Physiographic Features

China Danxia is an outstanding example of a vitally important step in the process of evolution of the earth's crust, including the geological origin of landscapes, the ongoing landform evolution and geomorphic processes, the important record of biological life, and the development of distinctive landscapes and landforms. In sum, Danxia landscapes have outstanding universal geological value.

- **China Danxia is a product of regional crustal dynamics**. The Chinese red beds underwent large-scale development in the late Mesozoic. They contain important geological and ancient geographical information. The red beds were uplifted in the Cenozoic and progressive developed Danxia landscapes on a relatively stable crustal plate. Danxia landscapes show the growing maturity of geomorphic change in the regional crust in China. They show how an intact regional crust develops through a process of "movement-stabilization-activation" in sedimentary red beds, over a period of more than 60 million years of geological evolution.
- **China Danxia shows both past and modern ongoing evolutionary geological processes**. Each of the candidate sites of China Danxia has a distinct array of Danxia landforms, showing different stages in the overall development of the landscapes. They are natural laboratories and museums for studying geological and geomorphological phenomena.
- **The "Danxia phenomena" in China reveals global change of earth terrestrial system from the late Mesozoic to the Cenozoic**. They provide an ideal place in which to understand global change including climate change.

9.1.3 Criterion (ix): Be Outstanding Examples Representing Significant Ongoing Ecological and Biological Processes in the Evolution and Development of Terrestrial, Freshwater, Coastal and Marine Ecosystems, and Communities of Plants and Animals

The serial Danxia property belongs Udvardy (1975) biogeographically to the Palaearctic Realm and Indo-Malayan Realms. It also belongs to the "Chinese Subtropical Forest" and "Chinese Southern Rainforest" Biogeographical Provinces. The most complete and largest evergreen broadleaf forests in the world are found here, giving them global significance. Distinct from the prairies and deserts formed in other regions at the same latitude, the evergreen broadleaf forests occur because of the influence of the southeast monsoon. There forests in the different sites all combine to create a very large expanse of natural evergreen broadleaf forest.

The sites also provide a complete series of both primary and secondary vegetation succession indicating the processes of vegetation community evolution. The primary succession begins from moss and herbaceous communities at the early stages, to the communities of short shrubs and trees at the middle and late stages. The red sediments are strongly corroded by seasonal and torrential rainfall and experience fluvial erosion as the dominant landforming process—creating steep valley walls and deep ravines. Weathering and landsliding add to the erosional processes of streams. The succession of ecosystem restarts on the bared surfaces of collapsed cliffs and other rock and soil surfaces. Succession varies in time and space. Secondary succession occurs in places where the vegetation is disturbed by natural or human influences. Secondary succession proceeds from shrub or pioneer forests (masson pine forest and pioneer broadleaf forest) to transitional forests (secondary broadleaf forest) and stable forests (evergreen broadleaf forest).

The topography of Danxia landscapes produces different ravine and hilltop effects in the vegetation. The ravine effect replaces spatial variation with vertical variation. Southern subtropical and even tropical plants can thrive in the nominate property which occurs generally in the mid-subtropical zone. The climax communities, controlled so much by topography and geology, are termed "geo-climax" communities. Therefore, Danxia landscapes have some special biological characteristics and unusual ecological processes.

9.1.4 Criterion (x): Contain the Most Important and Significant Natural Habitats for In-Situ Conservation of Biological Diversity, Including Those Containing Threatened Species of Outstanding Universal Value from the Point of View of Science or Conservation

According to the "Global 200 Eco-regions for Saving Life on Earth" the property belongs to the "Indo-Malayan" region and "Southeast China-Hainan Moist Forest" areas. The protected status of the ecological zones is critical or endangered (CE). In terms of the Udvardy biogeographical scheme, the heritage site is the ecotope with the highest biodiversity located in the "Chinese Subtropical Forest" and "South Chinese Rainforest" biomes. The property has vitally important natural habitats for in-situ conservation of biological diversity.

The heritage site has a high level of ecosystem diversity and species diversity. There are eight first-level types of habitats recognized by IUCN/Species Survival Commission in the heritage site. There are various terrestrial ecosystems, aquatic ecosystems, and wetland ecosystems. There are 47 types of second-level habitats ecosystems, and of them, there are 14 natural ecosystems, 15 artificial ecosystems, and 18 compound ecosystems. There are 23 vegetation types (including artificial vegetation types), 261 formations, and 424 associations. Ecosystem diversity is the basis of the development of species diversity. The heritage site protects a total of 5772 higher plant species, 836 vertebrates, and 3073 types of insects. There are 400 rare and endangered biotic species in the nominated region, and 55 plants and animals are listed in the IUCN Red Data List, and 167 types are listed under CITES. There are more than 600 species endemic to China and more than 40 species which are regionally endemic. *Firmiana danxiaensis*, *Ranunculus xinningcnsis,* and *Chirita langshanica* are three species found only in the nominated region. The property protects three endemic bird zones of the Endemic Bird Areas of the world: Priorities for Biodiversity Conservation, of which four species are limited to the EBAs. A "living fossil" *Mergus squamatus* is found here. The world's largest known wintering flock of this bird species is protected here. In addition, the heritage site is one of the four fragmented habitats critically important for the survival of the first-class national protected plant *Isoetes sinensis*.

It can be concluded that the nominated region meets the criterion as having the most important natural habitats for in-situ conservation of biological diversity.

9.2 Accepted Criteria for Inclusion by the World Heritage Committee

That was believed on the *Decision: 34 COM 8B.1* by the World Heritage Committee: Inscribes China Danxia, China, on the World Heritage List under criteria (vii) and (viii). Takes note of the following provisional Statement of Outstanding Universal Value:

(1) **Brief synthesis**

China Danxia is a serial property comprising six areas found in the subtropical zone of southern China. The term "China Danxia" describes the physical landscape developed from continental (terrestrial) reddish conglomerate and sandstone, also known as "red beds," in a warm, humid monsoon climate. These landscapes developed on continental red terrigenous sedimentary beds influenced by endogenous forces (including uplift) and exogenous forces (including weathering and erosion). The process of development is characterized by a particular rock sequence, tectonic background, warm and humid climatic conditions, and resulting erosion processes and landforms.

China Danxia is unrivaled in its rich warm and humid climate red beds sandstone landform geomorphologic features characterized by spectacular red cliffs and a range of erosional landforms, including dramatic natural pillars, towers, ravines, valleys, and waterfalls. China Danxia is also noted as a natural esthetic landscape comprising red rocks, green vegetation, blue water, and white clouds.

The rugged landscapes in the heritage site have helped to conserve subtropical broad-leaved evergreen forests and these forests are found within all six serial sites. A range of important microhabitats is also found. The nature of the Danxia landforms leads to intensive fragmentation and isolation of ravine and mountain top habitats. The natural habitats host many species of flora and fauna including endemic, endangered, and threatened species of conservation significance.

(2) **Criterion (vii)**

China Danxia is an impressive and unique landscape of great natural beauty. The reddish conglomerate and sandstone that form this landscape of exceptional natural beauty have been shaped into spectacular peaks, pillars, cliffs, and imposing gorges. Together with exuberant forest, winding rivers, and majestic waterfalls, China Danxia presents a resplendent natural picture. The sharp contrast of red rock against green forests and blue rivers is a striking feature of China Danxia and renders great scenic appeal. China Danxia sites have long been appreciated by both the general public and the academic world and further celebrated by artists. It is one of the most important scenic identities of China and has even attained significance as religious shrines. Its significance is further elaborated by the countless paintings, poems, and articles eulogizing these unusual beautiful sites since ancient time.

(3) **Criterion (viii)**

Compared with other similar areas, China Danxia is an outstanding example of warm and humid climate red beds landform in the world. As a result of favorable geological, hydrological, and climatic conditions since at least late Mesozoic period, China Danxia areas preserve and display much richer geomorphological, ecological, biological, and scenic features for warm and humid climate red beds landform than any place in the world. The component parts represent the best examples of "least eroded" to "most eroded" Danxia landforms, displaying a clear landform sequence from "young" through "mature" to "old age," and with each component site displaying characteristic geomorphologic features of a given stage. China Danxia contains a wide variety of well-developed red beds landforms such as peaks, towers, mesas, cuestas, cliffs, valleys, caves, and arches. Being shaped by both endogenous forces (including uplift) and exogenous forces (including weathering and erosion), China Danxia provides a range of different aspects of the phenomenon of physical landscape developed from continental (terrestrial) reddish conglomerate and sandstone in a warm, humid monsoon climate, illustrating both the range of landforms in relation to the forces and processes that formed them.

9.3 The Statement of Outstanding Universal Value

9.3.1 Comprehensive Statement

Danxia, meaning red cloud glow, is an absolutely wonderful name for a geological phenomenon. About 80 years ago, Chinese geologists used its name a landscape formed by red rocks in Danxiashan, Guangdong, China. The term later gained scientific and popular acceptance.

"Red beds" are distributed widely throughout the world, including red beds from the Precambrian, the Paleozoic, and the Meso-Cenozoic geological periods. Of all places in the world, China is the one where red beds are best displayed. The sediments are unique for their striking red color and various distinctive landform shapes.

The term Danxia is a synthesis of geology, landforms, and related natural landscapes formed from continental red bed deposits. It has been of interest since ancient times. The scientific study of Danxia is already 80 years old.

This heritage is for a natural serial property called China Danxia, and it is included in the China State Party Tentative List of World Heritage properties. The property overall includes six geographically separate sites, each of which has unique characteristics that, collectively, give the property outstanding universal value. The sites are the best representatives of China Danxia. They are representative of the Danxia landscapes in the southern humid zone of China. Collectively, they display geological diversity, landform diversity, biological diversity, ecology uniqueness, and landscape rareness. The outstanding universal value is based on the following qualities:

- A record of the geological history of a continental crustal fault basin in the South China area of the Eurasian Plate since the Cretaceous. This includes the record of geological and climatic changes in the Mesozoic and since that time. Evidence of important geological events is included such as gypsum-salt deposition, eolian accumulations, dinosaur life and extinction, basin uplift, continental volcanism, and crustal deformation—all occurring in a region of hot-arid climate.
- The development of landscapes and landforms with a distinct morphology, shape, color, and appearance.
- The presence of various typical and outstanding, rare and endangered biological species and habitats.
- A history of people adapted to their natural environment, demonstrating the harmonious coexistence between people and nature.
- A series of sites very well protected and managed by national and regional authorities.

The natural phenomena, natural evolutionary process, and natural beauty, all compressed into one serial property, are of global significance. Thus, China Danxia makes an irreplaceable contribution to world natural heritage.

9.3.2 China Danxia Is the Outstanding and Most Beautiful Representative of Red Bed Landscapes in the World

This serial property has an outstanding natural character and beauty. It is composed of natural elements such as red rocks of symbolic significance, a great variety of landforms, green vegetation, and blue waters—all constituting a place of great natural beauty. Red bed landscapes are widely distributed in the world, but the China Danxia is the best example of this phenomenon.

(1) **Landscape components**

The geomorphological, hydrological, and biological landscapes are still in their original natural state and have very high esthetic value.

China Danxia is a special form of landscape, constructed of many distinctive elements such as red cliffs at a variety of sizes and scales. The scenic elements consist of red walls and red cliffs, peaks, sculptured landforms, caves, ravines, and valleys all of great variety.

There are rivers and streams flowing through each candidate site have a close association physically and esthetically with the land. The many ravines have tremendous torrents of water, waterfalls, deep ponds, and potholes. In the eastern low-altitude landscapes of peak forests and peak clusters, the surrounding rivers tend to be more gently flowing. In Chishui, in the west, the mountain masses are higher with deep canyons, torrential river flows, and spectacular waterfalls.

The heritage site has the best-preserved low-altitude subtropical evergreen broad-leaved forest in south China. The red mountain groups look like rubies in green sea. More color is added through the presence of forests. There are also many ancient and famous trees. The property has good places for ecological tourism and popular science education.

(2) **Natural beauty**

The landscapes in the property are not only beautiful, but they also have a unique personality. The individual sites in the property represent the full range of landscape development from youth through maturity to old age within the humid region of Southern China. The landscapes display elements of dispersion, order, multi-layering, and diversity. Landforms range from high mountain peaks to deep ravines and valleys. The landscapes have a rich and colorful three-dimensional space dimension, with landforms sculpted into many strange and beautiful shapes. China Danxia is the most outstanding and beautiful representative among the red bed landscapes of the world.

Youthful Danxia landscapes have incised meanders, high gorges, and deep glens. Mature landscapes are dominated by peak clusters and peak forests. Old age landscapes have a lower altitude and more gently sloping surfaces and broad river valleys, with isolated upstanding peaks in the more resistant rock types.

(3) **Esthetic values of nature**

First, there is the perspective of formal esthetics. China Danxia has outstanding formal beauty of a colorful mountain-stone landscape, structured beauty of layered and orderly mountain groups, rhythmic beauty of staggering mountains, color beauty of red mountains, blue waters, green forests, blue sky, and white clouds. Second, there is the perspective of artistic composition. The sublime and steep red walls and red cliffs, the marvelous and delicate morphological landforms, the elegant and quiet mountains and waters, the deep and serene ravine forests, and the profound and wonderful clouds and mist, combine to capture the imagination of the artist and of all people alike.

(4) **The cultural associations and values of China Danxia**

Chinese esthetic culture recognizes such elements as "chi bi dan ya (red walls and red cliffs)," "wan gu jin cheng (ancient gold castles)," "zi fu dong tian (Zi Fu is the palace for the emperor, Dong Tian is the ground where the gods live)," "dan shan bi shui (red mountains and blue water)."

Chinese traditional religious culture including Taoist culture (Longhushan heritage site) is strongly associated with the Danxia landscapes. Landscape components often have religious significance. The red color of Danxia landforms gives a sense of solemnity and sacredness, authority, richness, and honor. It is also the traditional color for advocating the doctrines and beliefs of religious thought. The red mountains mostly are castle shaped, thus are called "wan gu jin cheng (ancient gold castles)", "zi fu dong tian" (Zi Fu is the palace for welkin or the emperor, Dong Tian is the Paradise of the ground where the gods live). Thus the Danxia landscape areas are an ideal realm for the conduct of religious observances and the building of temples in places such as caves. Danxia landscapes have, therefore, great religious significance in the Chinese culture. Therefore, there kept lots of cliffside carvings, cliffside statues and cliffside drawings, and so on. In Longhushan, there is a cliff tomb group which is 2600 years old based on the caves in cliffs along the Luxi River. A former ancient Yue Nationality culture was based here also. Religious traditions were thus closely associated with landscapes and landforms. The landscapes take on the significance of an immortal realm of sublime natural beauty and great esthetic quality.

9.3.3 China Danxia Reveals the History of Earth's Geological Processes Since the Mesozoic

The serial heritage sites are developed from a unique tectonic background and are a special regional geomorphological complex. It is mainly composed of late Cretaceous continental red beds and associated landforms and other natural environmental elements. It is the natural product of evolution of the continental crust from the late Mesozoic to the Cenozoic, containing outstanding geological and geographic features. The sites in the property are all located in the southeast part of the Eurasian Plate within the South China structural area. They experienced influences of the west Pacific Ocean structural area and the Tethys Ocean structural area, including the controlling influence of the Yangtze Plate and the Cathaysia Plate and the rupture zone between them. They bear testimony to important geological and tectonic events, such as collision between the Indian Plate and the Eurasian Plate, uplift of the Tibetan Plateau, western subduction of the Pacific Plate and NNE-strike slip shearing.

All candidate sites in the property are located in the humid climate zone of the southeastern part of Eurasia. Because of the unique geological structural background and natural geographic location, China Danxia contains outstanding earth science value. The property displays the earth's history since the late Mesozoic, showing geological and landform process of the continental crust, and is a model site for the study of the evolution of earth's continental fault basins and ongoing geological process. It also is an ideal place to study change in the global climate over long geological time periods.

(1) **China Danxia displays geological evolutionary processes**

Danxia landscapes are a product of a special phase in the development of the regional continental crust in China. The symbol of the landscapes is the "red beds." These developed during the reactivation of plates that formed enclosed intermontane basins, and it is only when these red beds were uplifted that the Danxia landscapes were formed.

The differences between the appearance and distribution of continental red beds in the world reflect the differences of regional crustal evolution. The earliest large-scale red beds appeared in the late Proterozoic era on the southern Gondwana supercontinent: the red beds in the North American–Russia ancient continent appeared in the early Paleozoic era, and the red beds in China formed in the Mesozoic era. All of these reflect the diversity in the evolution of ancient crustal plates around the world.

The process of development of Danxia landform in the nominated properties reflects the special history of regional crustal evolution: The red beds basin in China all developed on land blocks within the continental crust, reflecting an uninterrupted cycle of active zone—stable zone—reactivated zone. The basins in which the candidate sites are located experienced the whole process of Caledonian southern China geosyncline series of events—Upper Proterozoic to Lower Paleozoic geosyncline sedimentation—Caledonian fold inversion—late Paleozoic era meta-platform development— Upper Paleozoic Capping beds—Mesozoic activation (active continental margin rifting basin-red beds sediment)—late age uplift (Himalayan uplift—shaping of Danxia landscapes), all of which are distinctive of the evolution of the South China crust.

(2) **China Danxia landscapes are the most important red bed landscapes on earth**

The Danxia landscape is characterized by great variety in its configuration, unique shapes, varied and vivid colors, and a special combination of red mountains-green vegetation-blue waters. The serial heritage is designed to be representative of the full range of Danxia landscapes and landforms, with associated biotic elements. It contains the best such landscapes in the humid areas of the world from Mesozoic to Cenozoic times.The Danxia landscape is characterized by great variety in its configuration, unique shapes, varied and vivid colors, and a special combination of red mountains-green vegetation-blue waters. The serial heritage is designed to be representative of the full range of Danxia landscapes and landforms, with associated biotic elements. It contains the best such landscapes in the humid areas of the world from Mesozoic to Cenozoic times.

Six sites are chosen for inclusion in the serial property to demonstrate the full range of landform evolution from youth to maturity to old age.

The landscapes display geological and landform relics of different stages since the late Mesozoic, including fault and joint structures; relics of differential uplift of the crust and different stages of development of denudation planation surfaces; relics of river terrace, relics of corrosion of soluble elements; and relics of colluvial deposits and flake denudation deposits. Taining shows geological process of dissection and fluvial erosion dictated by structure. Danxiashan shows geological process erosion and crustal uplift at different stages. Longhushan shows geological process of ongoing water erosion and river erosion, and Langshan shows geological processes associated with rapid uplift of the crust. Chishui shows geological process of intense down-cutting due to rapid uplift and vigorous fluvial erosion processes.

The landscapes display a full suite of typical Danxia landforms of twelve types:

- "Danxia peak forests": including compact type peak forest (Langshan), cluster flat-top peak forest (Danxiashan), round scattered type peak forest, and scattered erosive remaining peak forest (Longhushan).
- "Peak cluster": including, conical cluster type peak cluster (Danxiashan, Langshan), round peak cluster (Paiya peak of Longhushan).
- "Cliffs and walls": found in each nominated site, these are the most outstanding landscape element in Danxia landscapes.
- "Mesa": represented by Danxiashan and characterized by flat tops, steep cliffs, and gentle lower slopes.
- "Cuesta": found at different stages of development in each site, typical cases such as the Danxiashan, Langshan, and Guifeng of Longhushan.
- "Peak walls": the big peak walls group in Herd of Elephants peaks of Danxiashan is the most typical.
- "Stone columns": the most typical is the scattered type stone column group in Danxiashan.
- "Isolated peaks": the most outstanding are high isolated peaks in Jianglangshan and Danxiashan; Longhushan contains lots of old stage isolated peaks in the surrounding area.
- "Valleys": the most typical are the dense deep valley-meander group in Taining and the big ladder-type valley group in Chuishui.
- "Caves": the most typical are the cliff cave group in Taining and along the bank of the Luxi River in Longhushan.
- "Holes and stone arches": represented well in Danxiashan and Langshan.
- "Shaped rocks": the most typical and unique are the various rock shapes in Guifeng of Longhushan, Danxiashan, and Langshan.

Red bed is a widely distributed phenomenon of the earth's terrigenous crust, with special earth science values. Thus, Danxia landscapes are special geological places of global significance.

China has the widest distribution of Danxia landscapes in the world. The heritage is confined to Danxia landscapes in the southern humid region of China. They exhibit not only geology and landforms but also the typical ecologic systems and biodiversity of this geotype.

China Danxia has been studied for 80 years and makes a significant contribution to geology and landform science worldwide.

(3) **China Danxia represents important landform evolutionary and ongoing processes**

Each of the candidate sites is in a crustal uplift zone. Each is in a different stage of geomorphic development. Therefore, the heritage site is a natural laboratory and for studying modern landforming processes. To date, each site has established places for observing geologic processes and landform development at work, mainly for observing weathering and erosion of soft-rock sediments, stresses in slope generation, gravitational movement of blocks, geological hazards, fluvial erosion processes and volumes and retreat of valley and cliff walls.

9.3.4 Study of Ecological Processes of Red Beds Areas

9.3.4.1 Ecological Values of the Heritage Site

(1) **Rich community and ecosystem diversity**

The heritage has many and varied terrestrial, aquatic, and wetland ecosystems. There are eight first-level types of habitats recognized by IUCN/SSC, such as forests, bush, wetland, bare rock, and caves. These cover 61.5% of total number of first-level habitat types in the world. There are 47 second-level ecosystem types and higher diversification of ecosystems at smaller scales. These reflect the complex array of physical habitats in Danxia landscapes. There are special ecosystems also, as the result of special landforms. The community diversity is rich, containing 23 vegetation types (including artificial ones), 261 formations, and 424 associations, making the property extremely important for in-situ biodiversity conservation.

(2) **Representative of evergreen broadleaf forest in the east monsoon zone**

The evergreen broadleaf forest extends over a vast area of the subtropical realm in China. It is one of the most extensive tracts of this forest type in the world. Subtropical evergreen broadleaf forest is also the typical vegetation type in the heritage. Normally, there are grasslands and deserts at this latitude elsewhere in the world. The forest exists here because of the influence of the east monsoons, and this makes the forest of particular scientific and conservation value globally.

(3) **Intact ecological succession processes**

Changes in vegetation communities occur with changes in landscape development. Primary succession starts from collapsed surfaces with mosses and lichen communities. Secondary succession develops from pioneer forests to transitional forests and finally to essentially stable mature forest.

(4) **Evolution of climax communities in a special Landscape**

The influence of a ravine topography creates special ecological effects, contributing to latitudinal shift of tropical plant distribution. Some south subtropical species now occur in the mid-subtropics. The zonal vegetation is typical mature subtropical evergreen broadleaf forest in the ravines. This can be called a geo-climax community as it is controlled primarily by topography. While the Danxia landscapes here are similar to those elsewhere, they differ in having different climax vegetation types, and this makes them special in the world.

9.3.4.2 Biodiversity of the Heritage

(1) **Rich biodiversity**: The heritage belongs to the "Chinese Subtropical Forest Biogeographical Province" and the "Chinese Southern Rain Forest Biogeographical Province" in the Udvardy scheme. It is a key biodiversity center in the world. There are 5772 species of higher plants, comprising 293 families 1271 genera; 836 species of vertebrate animals, with 37 orders 129 families, and 3073 species of insects. Compared to other World Heritage sites in China, the number of species of plants and animals is greater than Wuyi Mountain in Fujian Province and Sanqingshan in Jiangxi Province. This emphasizes the relative importance of the biodiversity in China Danxia.

Comparison of the numbers of higher plants in World Heritage areas

Heritage site	Danxia regions	Sanqingshan	Huangshan	Taishan	Japanese Yakushima	Yosemite National Park of USA
The number of species	5772	2373	1304	1056	1900	2003 (Vascular plants 1900)
%[a]	100.0	41.11	22.59	18.29	32.92	34.70

[a]The percentage of the number of species in nominated sites with its affiliated regions

The biodiversity comparison between the nominated sites and adjacent domestic World Heritage sites

Name		Plants	Animal	Vegetation and ecosystem	Unique species
The nominated sites	China Danxia	5772 kinds of higher plants. Of 5181 kinds of seed plants, and 591 kinds of ferns	132 kinds of mammals, 364 kinds of Aves, 109 kinds of reptiles, 70 amphibians, 161 kinds of fish; 3073 kinds of insects	Vegetation type in a subtropical evergreen broad-leaved forest, mixed broad-coniferous-leaved forest, coniferous forest and other 23 vegetation types, 261 vegetation formations, 424 associations. Mainly in subtropical evergreen broad-leaved forest. Biomes with ancient and integrity	Local endemic species: *Firmiana danxiaensis*, *Isoetes sinensis*, *Ranunculus xinningcnsis*, *Chirita langshanica*, *Synotis lanshanensis*, *Cynodontium gracilescens*, *Hypericum hengshanense*, *Phlegmariurus mingchegensis*, *Asplenium juxtapositum*, *Dryopteris jiangshanensis*, *Shibataea chiangshanensis*, and other more than 40 species
The existing World Heritage Sites	Wuyi Mountain in Fujian province	282 kinds of ferns; 1976 kinds of seed plants	71 kinds of mammals, 256 kinds of birds, 73 kinds of reptiles, 35 amphibians, fish 40; insects 28 orders 341 Sect. 4557 kinds	The area in a typical area of vegetation in the subtropical evergreen broad-leaved forest with rich species composition in flora complex, unique and, residual species distribution. However, longer exploitation and development in Wuyi Mountain, less subtropical ecosystem integrity	*Berberis wuyiensis*, *Maddenia fujianensis*, *Prunus fukienensis*, *Photinia wuyishanensis*, *Rubus caudifolius*, *R. fujianensis*, *R. jiangxiensis*, *R. rosaefolius var. wuyishanensis*, *R. yanshanensis*, *Sorbus amabilis var. wuyishanensis*, *Rhododendron wuyishanicum*, *Salvia chunganensis*, *Sinosenecio wuyiense*, *P. wuyishanensis*, *Y. wuyishanensis* etc. 22 kinds
	Three Parallel Rivers Flow of Yuannan	Plants more than 210 branches, more than 1200 genus, and more than 6000 kinds	173 kinds of mammals, 417 kinds of birds, 59 kinds of reptiles, 36 species of amphibians and 76 kinds of fish, 31 species of Papilionidae category insects	Gathered all of types including the northern hemisphere, South subtropic, the North subtropical, temperate, cold temperate and boreal, and other types of climate and biological communities. 10 vegetation types, the 23 vegetation subtypes, more than 90 formations. Not covered by mainland glacier during Quaternary Ice Age, many ancient original vegetation left, a residue development behind during the Pleistocene	5 species: *Dipentodon sinicus*, *Cephalotaxus lonceolata*, *Magnolia rostrata*, *Davidia involucrata var. vilmoriniana*, *Taiwania flousiana*

(2) **Rare, endangered and unique species**

The nominated region is located in the "Global 200 Eco-regions for Saving Life on Earth", and the "Tropical and Subtropical Moist Broadleaf Forest" region of the "Indo-Malayan" realm. Protection of ecosystems and endangered species is therefore given special emphasis in the heritage.

In the heritage, there is a very high level of endemic species diversity. There are 394 species of rare and endangered plants and animals. Among, there are 214 species of rare and endangered plants, including 34 species identified by IUCN for protection, 104 species under CITES protection, 49 species of plants belonging to China national key protection, and 180 species on the Red Species List of China. There are 189 species of rare and endangered animals, of which 45 species are IUCN-listed, 66 are listed under CITES, 88 are China national key protected species, and the Red Species List of China contains 145 species.

Isoetes sinenss, an ancient and relict plant, is found in Danxishan. According to records, *Isoetes sinenss* is also found in Xiuning in Anhui Province, and Songyang and Jiande in Zhejiang Province, but the widespread occurrence in valley wetlands of Danxiashan is the most important occurrence of this plant. Langshan is a special habitat region, revealing co-evolution between basal groups of the angiospermae and insects. For example, *Kadsura japonica* has a highly specialized pollination mechanism which requires insects. So, the presence of such "living fossils," among both plants and animals demonstrates tenacity and continuity of biota evolution in the nominated area.

The world's largest wintering population of the Chinese merganser *Anatidae mergus*, which is extremely endangered, is found on the Luxi River and in the Qinghu section in Yiyang of the Xin River in Longhushan. About 150 birds have wintered here over many years, accounting for approximately 4% of the total global population. The sites have the status of Wetlands of International Importance, under criteria ii and vi of the International Wetlands Convention. This is an extremely significant population of this species for scientific study and conservation.

According to the description and classification of "Endemic Bird Areas of the world: Priorities for Biodiversity Conservation," the heritage is in three endemic regions,as follows: Southeast Chinese Mountains, Chinese subtropical forest and Central Sichuan Mountains. The property is mostly in the Southeast China Mountain region, and there are four endemic species—*Arborophila gingica, Tragopan caboti, Syrmaticus ellioti,* and *Latoucheornis siemsseni* with limited distribution.

There are more than 600 species of endemic plants, including regional endemics such as *Firmiana danxiaensis, Ranunculus xinningcnsis,* and *Chirita langshanica*, limited to the heritage. *Ranunculus xinningcnsis* and *Chirita langshanica* grow only on cliffs in Langshan, revealing its very limited ecological distribution.

The rich biodiversity and the presence of rare, endangered, and endemic species emphasize the global significance of the heritage for biological conservation.

(3) **Ancient flora and fauna and pristine communities**

The heritage was not directly affected by Quaternary glaciers, and it become a refuge for many species of biota. The flora has ancient origins and includes many relict species from the Cretaceous and the Paleogene. Of special importance are the 5444 trees more than 100 years old, 25 trees more than 1000 years old, such as *Taxus chinensis*, *Cephalotaxus sinensis*, and *Sabina chinensis*. They reflect the antiquity of habitats and the pristine state of biological communities here and thus their importance in the world.

9.3.4.3 A Model Area for Biological Research of the World's Red Sedimentary Landscapes

(1) **Research values for succession theory**

The intact primary and secondary succession in communities in the property, especially the primary succession series, which are generated by special physiognomy, means that this is an ideal place to study the theories and principles of biotic succession processes and patterns. The intact primary and secondary succession series here can be used as reference models in the process of vegetation restoration.

(2) **Geomorphological effects of species distribution**

Species distribution regions vary in shape and size. Species are constantly shifting and expanding their range. Study of the relationship between species distribution and environment is a key element for understanding biogeography. Climate, landforms, soil, and organisms all affect species distribution. Climate is the most important factor affecting species (especially plant) distribution.

In the heritage, the ecological influences of the ravine topography make Danxia landscapes different from all others. The differences mean that tropical plant flora is relatively more important here than in other places at the same latitude. Ravine topography causes a latitudinal shift of plant distribution, which means some typical southern subtropical species appear in the mid-subtropics. The ravines of Danxia landscapes are, therefore, excellent places to study the interrelationships of landforms and plant communities.

(3) **Research values of conservation biology**

- **Natural laboratory for island biogeography theory**: Species abundance in islands is controlled by two processes—import of new species and species extinction. When the rates of introduction and extinction are equal, the island species numbers are in dynamic balance. While the species number is relatively stable, the species composition is continually changing and rejuvenating. This is the core of island biogeographical theory.

The hilltop island effect in Danxia landscapes demonstrates these principles. The summit areas of Danxia landscapes are very isolated. Biological groups on the summits are simple and population numbers are much less than those at the base of the mountains. Therefore, Danxia landscapes are important places for research into the theory of island biogeography.

- **Research values of rare and endangered species in conservation theory**: Biological conservation, especially rare species conservation, has an international focus today. Mechanisms by which species become endangered are a core interest in conservation biology and rare species protection.

In the heritage, there are many rare species (including birds, reptiles, and amphibians). There is also an abundance of rare habitat types—large marshes and water areas, including three regions identified in the "Endemic Bird Areas of the world: Priorities for Biodiversity Conservation," and five species of limited distribution—*Arborophila* gingica, *Tragopan caboti, Syrmaticus ellioti, Gorsachius magnificus,* and *Latoucheornis siemsseni*—identified in EBAs. The property is, therefore, an ideal site for research on mechanisms causing rarity among species and for pursuing rare species conservation.

- **Research values in relationships of special species**: The correlation between species development and their habitats is a crucial issue for biological and ecological research. Especially of interest are some special species, which exist in special local environments and have a closely dependent relationship with their habitat.

Langshan has special habitats revealing co-evolution of angiosperm communities and insects. Additionally, in the heritage sites, there are many other special species, such as *Firmiana danxiaensis*, *Ranunculus xinningcnsis*, *Chirita Langshanica*, and *Sedum fangyanense*. Species studied widely over many years are *Firmiana danxiaensis*, *Ranunculus xinningcnsis*, and *Chirita Langshanica*. *Ranunculus xinningcnsis* and *Chirita Langshanica*, for example, grow only on cliffs in Langshan.

9.3.5 China Danxia Is an Outstanding Global Example of Harmonious Coexistence between Humanity and Nature

In addition to the natural values of the heritage site, there are also very significant historical and cultural values. In particular, Taoist, Buddhist, and Confucianist cultures are well known in the nominated sites from the abundance of archeological artifacts and other relics. These relics extend back over 2600 years, providing a very long-term picture of the associations between religious beliefs and natural landscapes. The cultural history also reveals very strikingly the close harmonious relationships that have developed between people and nature, and which continues today. The property also includes places that gave birth to the Taoist religion and associated culture. The core Taoist principle of harmony between people's lives and their natural environment is fundamental to the protection of these Danxia landscapes over many generations and is a key reason why the landscapes and their biota are still intact and in a natural state.

10 Comparative Analysis

10.1 The Basis of Comparative Analysis

10.1.1 General Analysis

"Danxia" defines an exogenetic landscape type developed on continental red terrigenous sedimentary beds. It is characterized by a particular rock sequence, tectonic background, climatic conditions, erosional processes, and suite of landforms. While there are some other locations in the world that bear similar visual, geological, and geomorphological properties with Danxia, none are as well-defined, complete, or extensive as the Danxia landscapes of China. Indeed, the landscape type was first recognized in China, and it is fitting that the name "Danxia" is derived from the classical landscape of the Danxiashan National Park in Guangdong province. One problem in undertaking this comparative analysis is that while the Danxia landscape type exists in other parts of the world, its geomorphology has not been as extensively studied as in China, and good scientific information—the knowledge base—on which to undertake a meaningful comparison is scarce.

The heritage site stands distinct in the world from other inscribed and non-inscribed sandstone landscapes because through its six sites in Southeast (SE) China, it represents a complete geomorphic sequence, or system. It is in fact a series of landscapes, each site having some elements different from the other, that collectively demonstrates the complete evolutionary development of Danxia in a warm, humid environment, from youthful to late mature stages. Any other similar site in the world represents only an individual component in the Danxia evolutionary sequence, or it may show forms more typical of a drier morphoclimatic environment ("dry Danxia", to form a later second stage of this serial heritage). Even then, only a handful of comparable sites are as well developed, as scientifically important, or a visually impressive, as the China Danxia.

A further feature of the Danxia in SE China is that it is characterized by a distinctive suite of landforms. Thus, continuing structural uplift of Danxia sedimentary blocks, and dense networks of faults and joints, guide rapid fluvial incision to create deep clefts and narrow valleys, while also leaving unstable slopes which collapse to form precipitous cliffs. At later stages, valleys become wider and plateau remnants become isolated table mountains, mesas, buttes, and columns. Still later, mountain-tops become rounded by physical and possibly some chemical weathering activity. By the final stages, only isolated, low, and rounded monadnocks remain.

H. Peng, *China Danxia*, Springer Geography,
https://doi.org/10.1007/978-981-13-5959-0_10

10.1.2 Basis for Comparative Analysis

It is important that any comparative analysis should compare like with like. It is clear that Danxia sites can be compared on the basis of geology (continental red bed lithology), although the geomorphology is rather more problematic. The Danxia landscape is a product of dissection by fluvial action of greatly fractured, usually near-horizontally bedded (sometimes with low dip) terrigenous sediments (not solely sandstones, but also conglomerates, siltstones, and some evaporites). Logically, this process would lead to the production firstly of plateaus dissected by river-cut canyons and gorges, then to valley widening and valley-side collapse, to produce more isolated table lands, and ultimately to a more open landscape with mesa and buttes representing the remnants of the former plateau. Theoretically, such a process would lead to angular or rectilinear landscape forms, particularly in arid and semi-arid climates, but in the warm humid climate of SE China, many Danxia sites have rounded and fluted slopes, with groups of bell-shaped hills, indicating a very active surface weathering regime. Thus, it appears that the process of tectonically controlled fluvial dissection of plateaus described above is supplemented or overprinted by significant surface weathering, in some sites producing a landscape visually analogous to cone karst (but it is not correct to call Danxia sandstone karst or "psuedokarst").

A fundamental geomorphological question is the nature of the weathering that causes slopes to become rounded and fluted. The Danxia sediments are relatively young (Cretaceous), mainly sandstones, siltstones, and conglomerates. While some quartz dissolution of the sandstones may occur, some of the Danxia rocks may have a carbonate cement, which is more susceptible to dissolution than quartz. However, dissolution is not thought to be the most important weathering process, and the rounding and fluting of slopes are more likely to be an effect of weakening of the shear strength of the clayey sedimentary rocks due to poor mineral overgrowths and saturation with water.

The particular values of this serial heritage based upon which it should be compared with other similar sites, therefore, are:

- Danxia geomorphology is more extensively developed and diverse in China than anywhere else in the world
- The serial heritage illustrates a complete geomorphic system and evolutionary sequence
- The serial heritage embraces a unique assemblage of landforms, as a group not replicated with the same quality or as complete in any other world location
- The China Danxia's red bed geology signifies a particular geological climatic period and environment not well represented in the World Heritage List
- The Danxia landscapes of SE China have formed in one particular group of rocks, under a particular morphoclimate regime, and under particular tectonic conditions
- The China Danxia harbors largely untouched forest, of high biodiversity and with important protected species
- The Danxia landscapes of SE China have unrivaled, majestic, and colorful scenic beauty. Each landscape is a combination of imposing vistas of table mountains and peak forest, abundant exposed red rock in cliffs and ravines, extensive forest cover, superb and highly visible wildlife, and abundant watercourses (as large rivers, mountain streams, and lakes)
- The serial heritage claims Outstanding Universal Value under scenic, geomorphological, and biological criteria (vii, viii, ix, x).

Technical note on the difficulties of comparative analysis and on the titles of the columns in the following tables.

Danxia-type landscapes are present throughout the world, but very little research has been undertaken on any of them, in particular of their biological and geomorphological values. However, both of these values are heavily influenced by climate. Climate, therefore, is an important comparator.

Nevertheless, while biology responds relatively rapidly to climate change, modification of landscape form (geomorphology) is much slower, and it is suspected that some Danxia-type landscapes throughout the world may still exhibit forms that were developed in a previous climatic regime. There is also some evidence that the length of time that Danxia-type rocks have been exposed to the atmosphere (i.e., weathering and erosion) may have influence on the landforms we see today. Thus, when considering the evolution of any Danxia-type landscape in any part of the world, it is essential to know the length of time the rocks have been exposed to exogenetic processes, and how these processes might have changed (e.g., from humid to arid-type processes, or vice versa) over that period of exposure. What we observe in any physical landscape may, therefore, be an illusion of the truth. It is, therefore, necessary to ask the question: To what extend are the landscapes we see today fossil landscapes, fossil landscapes undergoing modification in the contemporary climate, or wholely the product of the contemporary climate? The answers to this question are not known in any Danxia-type landscape in the world. It is, therefore, not possible with the current state of knowledge to compare Danxia geomorphological landscapes on the basis of their morphogenesis, which means that comparison is only possible on the basis of the visual appearance, or form, of the landscape.

A simple grouping of the different types of landscape is, therefore, necessary. As noted above, some Danxia landscapes contain rounded slopes, while others are angular or rectilinear, and still others have combinations of these land forms. While we may speculate on possible causes for these different landscape forms (e.g., angular forms from mechanical weathering in arid or cold climatic conditions, rounded forms caused by mixtures of physical and chemical weathering in warm, and humid climatic conditions), the determining processes are not known definitively, and so any classification must be solely descriptive. Even here, the "types" of Danxia landscapes are not clear-cut because some (as in this serial heritage) have formed discrete mountainous area, while others occur because sedimentary red beds have become exposed and eroded through the widening of major river valleys (e.g., the Grand Canyon, or Three Parallel Rivers) or previously glaciated valleys (e.g., Pyrenees–Mont Perdu, or Canadian Rocky Mountains).

Another problem in identifying like for like is the degree of completeness, or integrity, of the geomorphological landscapes. As a first base to identifying Danxia world-wide, it is possible to locate important outcrops of continental red sedimentary beds, but these do not always form complete, discrete (closed?) geomorphic landscapes. The places where such landscapes are known in the world are very few and in any case on the basis of our present scientific knowledge are difficult to identify.

In summary, Danxis can only be compared around the world by description of form and not by origin. Furthermore, while climate is an important controlling factor, climatic information is only of sufficient to be able to use it at the macro (world) scale, rather than at the local or site level. Geology, geomorphology, and climate are also controls on biology, but again a lack of information from sites across the world precludes any attempt to produce a deep, meaningful comparison, other than in world climatic zones.

10.1.3 Red Beds Landscapes on the World Heritage List

In August 2008, the World Heritage List contained 878 properties in 145 countries. These properties included 174 inscribed in the natural heritage category, 25 in the mixed category, and 679 in the cultural heritage category. In the 199 properties in the natural and mixed categories, 16 have red bed geology (including one natural and two mixed category sites in China), while of the 679 cultural properties, there are 16 properties with red bed geology (including three properties in China) (32 properties in total, 6 being found in China). Table 10.1 shows that these properties are distributed across all continents, except Antarctica.

Table 10.1 World Heritage Sites that contain continental red beds geology and Danxia-type landscapes

Name of property	Country	Date of inscrip-tion	Criteria under which inscribed	Age of beds	Climatic zone	Udvardy Biological zone	Dominant landscape form				Notes	Photo
							Fold mtns and valley sides	Recti-linear (table mtn)	Rounded	Peak Forest		
Grand Canyon National Park	USA	1979	(vii) (viii) (ix) (x)	Pre-Cambrian to Paleogene	Subtropical Semiarid/arid	Rocky mountains; Mixed mountains/highland systems with complex zonation	✓	✓			Valley exposing sequence with Mesozoic continental beds forming upper part. Landforms are mainly large valleys; the layered cliffs along the two banks are majestic and spectacular and cut into block mountains. This kind of red beds landforms is widely distributed in Colorado plateau	
Dinosaur Provincial Park	Canada	1979	(vii) (viii)	Cretaceous	Temperate Semi-arid	Grasslands Temperate grasslands			✓		There are mainly Cretaceous gray-brown, sandy beige red bed and colorful rocks. They are red bed accumulation 75 million years ago, contains steep cliffs of tens of meters. There were several intense glaciations in Quaternary ice age	
Canadian Rocky Mountain Parks	Canada	1984, 1990	(vii) (viii)	Pre-Cambrian to Cretaceous	Temperate Highlands	Rocky Mountains; Mixed mountain systems	✓				Fold mountains with range of strata, among which in Banff Nat. Park is Cretaceous gray-brown sed. rock. Suffered intense glaciations	
Purnululu National Park	Australia	2003	(vii) (viii)	Devonian	Subtropical Semi-arid	Savanna; Tropical grasslands/savannas			✓	✓	Dark brown-mixed quartz sandstones eroded into tower and cone-shaped hills separated by linear valleys/canyons, eroded over 20 million years. Cliffs range up to 100 m high	

(continued)

Table 10.1 (continued)

Name of property	Country	Date of inscrip-tion	Criteria under which inscribed	Age of beds	Climatic zone	Udvardy Biological zone	Dominant landscape form				Notes	Photo
							Fold mtns and valley sides	Recti-linear (table mtn)	Rounded	Peak Forest		
The Greater Blue Mountains Area	Australia	2000	(ix) (x)	Trias	Subtropical subhumid/ Highlands	Eastern Sclerophyll; Evergreen sclerophyllous forests, scrubs or woodlands		✓	✓		Sandstone plains, cliffs, and valleys. Nearly horizontal, medium-thin bedding, cut by canyons into blocks and columns	
Canaima National Park	Vene- zuela	1994	(vii) (viii) (ix) (x)	Creta-Ceous	Tropical wet	Guyanan; Tropical humid forest		✓			Cretaceous quartzites and red-brown sandstones rest on Pre-Cambrian rocks. Strata have been eroded for millions of years to create flat-topped plataeus and table mountains cliffs over 1000 m	
Three Parallel Rivers of Yunnan Protected Areas	China	2003	(vii) (viii) (ix) (x)	Tertiary	Humid subtropic	Mixed mountains/ highland systems with complex zonation	✓		✓		Deep, parallel gorges of the Jinsha Jiang, Lancang Jiang, and Nu Jiang. Large area containing tertiary red beds. Typical Danxia mesas, peak clusters, red cliffs, and caves in Lao Jun Mountain. Turtle-like weathered convex hills are unique	
Uluru-Kata Tjuta National Park	Australia	1987	(v) (vi) (vii) (ix)	Cambrian/ Ordo-vician	Subtropical Arid/Semi-arid	Central Desert; Warm deserts/ semi-deserts			✓		After long-term weathering and erosion, the red beds of Cambrian/Ordovician have become rounded hills. The remnant Ayers Rock big stone and a group of smooth domes of the Olgas make up spectacular beautiful scene	

(continued)

Table 10.1 (continued)

Name of property	Country	Date of inscrip-tion	Criteria under which inscribed	Age of beds	Climatic zone	Udvardy Biological zone	Dominant landscape form				Notes	Photo
							Fold mtns and valley sides	Recti-linear (table mtn)	Rounded	Peak Forest		
Kakadu National Park	Australia	1981	(i) (vi) (vii) (ix) (x)	Creta-ceous/ Protero-zoic	Tropical Subhumid/ Sub-dry	Northern Coastal; Tropical dry or deciduous forests		✓	✓		The Proterozoic chocolate brown middle-thick layer sandstones and sandy conglomerate are covered by the Cretaceous rock group and the continental Petrel rock group. Suffered severe weathering and erosion, it formed the following landscapes: residual mesas and isolated hills lying in the wide canyon, flat plateau surface, cliffs up to 100 m at the edge of the plateau and smooth rock edges	
Drakensberg Park	South Africa	2000	(i) (iii) (vii) (x)	Trias	Moderate Marine	South African Highlands; Mixed mountains/ highland systems with complex zonation		✓			Composed of nearly horizontal red sandstones, shale, and basalts of early Trias. Because of the deep cut of valleys, the landform is table-shape mountain with flat top as a whole, also with surrounding cliff slope tens of meters high	
Tassili n'Ajjer	Algeria	1982	(i) (iii) (vii) (viii)	Pre-cambrian to early Paleozoic	Arid	Sahara; Warm deserts/semi-deserts				✓	The base of Pre-Cambrian rocks, overlain by an accumulation of early Paleozoic brown and beige colored sandstones and conglomerates of, near-horizontal occurrence. Various red bed stone forests and peak clusters, with lots of rock peak scattering in big desert, with relative height of tens of meters	

(continued)

Table 10.1 (continued)

Name of property	Country	Date of inscription	Criteria under which inscribed	Age of beds	Climatic zone	Udvardy Biological zone	Dominant landscape form				Notes	Photo
							Fold mtns and valley sides	Recti-linear (table mtn)	Rounded	Peak Forest		
Cliff of Bandiagara (Land of the Dogons)	Mali	1989	(v) (vii)	Late Precam-brian	Semiarid /arid	West African Woodland/ savanna; Tropical dry or deciduous forests or woodlands		✓			The littoral–neritic facies red beds of late Pre-Cambrian comprise of red quartz sandstone and gravelly sandstone. There are cliffs on the edge of the red bed plateau and table-shape mountain. Some rounding of slopes	
Meteora	Greece	1988	(i) (ii) (iii) (iv); (vii)	Oligocene	Mediterra nean	Balkan Highlands; Mixed mountains/ highland systems with complex zonation			✓		Gently inclined red beds accumulated in a piggyback basin during the Oligocene. The rock peaks and steep cliffs developed on the red beds are of tens to hundred of meters' high, with smooth edges and corners	
Pyrénées–Mont Perdu	France/ Spain	1997, 1999	(iii) (iv) (v); (vii) (viii)	Cretac-eous-Paleogene	Coastal Mediterra nean climate	Iberian Highlands; Evergreen sclerophyllous forests, scrubs or woodlands	✓	✓			Large scale of steep cliffs and mesas were developed along the joints. The calcareous massif of the Mount Perdu displays classic geological land forms, including deep canyons and spectacular cirque cliffs	
Mount Wuyi	China	1999	(iii) (iv); (vii) (x)	Upper Cretacic	Subtropical humid monsoonal climate	Chinese subtropical forest; Evergreen broad- leaved forest or sclerophyllous forests			✓	✓	Made up by mauve and sandy conglomerate of upper Cretaceous series, it developed Danxia peak cluster, massive mountains stone columns, and etc.	

(continued)

Table 10.1 (continued)

Name of property	Country	Date of inscrip-tion	Criteria under which inscribed	Age of beds	Climatic zone	Udvardy Biological zone	Dominant landscape form				Notes	Photo
							Fold mtns and valley sides	Recti-linear (table mtn)	Rounded	Peak Forest		
Mount Emei Scenic Area, including Leshan Giant Buddha Scenic Area	China	1996	(iv) (vi); (x)	Cretaceous	Subtropical monsoonal climate	Oriental Deciduous Forest; Evergreen sclerophyllous forests, scrubs or woodlands	✓	✓			Vertical joints are not well developed here. The red cliffs along the riverside are typical types and the famous Giant Buddha was carved on the cliff on the bank	
Mesa Verde	USA	1978	C(iii)	Cretaceous	Semiarid	Rocky Mountains; Mixed mountains/ highland systems with complex zonation		✓			Colorado plateau, constituted by Cretaceous red sandstone, formed various sized canyons and cliffs through river force. The caves developed on red bed sandstone became dwelling for Ancient Indians	
Rock-Art Sites of Tadrart Acacus	Libyan Arab Jamahiriya	1985	C(iii)	lower Cambrian	Arid	Sahara; Warm deserts/ semi-deserts		✓	✓		The middle section of lower Cambrian is red bed graywacke, which is widely distributed in this area and belongs to the sedimentation of intertidal zone and coastal environment. The uppermost and lowermost are similar, but lack volcanic rocks and in macroscopic view, feature in erosion tableland and monadnock in red bed of the arid area	
Capivara National Park	Brazil	1991	C(iii)	Silurian-Devonian	Tropical semiarid	Caatinga; Tropical dry or deciduous forests or woodlands			✓	✓	Consist of Silurian- Devonian brown and beige colored sandstone deposited in fluvial, delta, coastal and neritic environments. The area comprises red sandstone peak forest and peak cluster of tens of meters high	

(continued)

Table 10.1 (continued)

Name of property	Country	Date of inscrip-tion	Criteria under which inscribed	Age of beds	Climatic zone	Udvardy Biological zone	Dominant landscape form				Notes	Photo
							Fold mtns and valley sides	Recti-linear (table mtn)	Rounded	Peak Forest		
Petra	Jordan	1985	C(i), (iii), (iv)	Cretaceous	Arid	Arabian Desert; Warm deserts/ semi-deserts		✓			Cretaceous vermeil, brown and filemot sedimentary rock with various colors and gentle occurrence. Some beds form flat hills, tens of hundreds of meters across. The narrow canyon in which the archeological site is located is 1.5 km long, and 70–100 m high	
Bamiyan Valley	Afghanistan	2003	C(i) (ii) (iii) (iv) (vi)	Neogene	Semiarid	Hindu Kush Highlands; Warm deserts/ semi-deserts		✓			Neogene continental red bed, in the bottom with lower Cambrian, Ordovician, upper Carboniferous, Permian, lower Cretaceous and late Tertiary marine red conglomerate, sandstone and shale of hundred meters thick. Scarp slope of 50-100 meters is developed along the valley	
Ouadi Qadisha and the Forest of the Cedars of God	Lebanon	1998	C(iii), (iv)	Cretaceous	Mediterra-nean	Mediter-ranean Sclerophyll; Warm deserts/ semi-deserts				✓	Consist of Cretaceous ocher and sandy beige conglomerate and sandy conglomerate, with nearly horizontal occurrence. Numerous red bed massifs, canyons and scarp slope	
Rock Shelters of Bhimbetka	India	2003	C(iii) (v)	Proterozoic	Tropic monsoon	Bengalian Rainforest; Tropical humid forest	✓	✓			Proterozoic brown and ocher sandstone are widely distributed in the middle of India, with nearly horizontal occurrence. Numerous caves and shelters are developed in the rock and scarp slope of tens of meters high developed along the ravine	

(continued)

Table 10.1 (continued)

Name of property	Country	Date of inscrip-tion	Criteria under which inscribed	Age of beds	Climatic zone	Udvardy Biological zone	Dominant landscape form				Notes	Photo
							Fold mtns and valley sides	Recti-linear (table mtn)	Rounded	Peak Forest		
Old Walled City of Shibam	Yemen	1982	C(iii) (iv) (v)	Cretaceous	Tropical arid	Arabian Desert; Warm deserts/ semi-deserts		✓			The Cretaceous reddish-brown quartz sandstone, siltstone and shale of littoral neritic facies have horizontal altitude. The mountain has a flat top, steep face, and gentle piedmont, rising 200 meters above the Masila valley	
Nubian Monuments from Abu Simbel to Philae	Egypt	1979	C(i) (iii) (vi)	Carboni-ferous to Cretaceous	Tropical desert climate	Sahara; Warm deserts/ semi-deserts		✓	✓		It belongs to the thick-bedded Nubian Sandstone of the Carboniferous-Cretaceous, from filemot to ocherous color. Outside the reservoir, there are residual platforms with height of few meters, also with steep cliff slope in part	
Twyfelfon-tein	Namibia	2007	C(iii) (v)	Neoprote-rozoic to Cretaceous	Subtropical arid and semi-arid	Namib; Warm deserts/ semi-deserts		✓	✓		The macro-landform is mainly residual red sandstone platform	
Mapun-gubwe Cultural Landscape	South Africa	2003	C(ii) (iii) (iv) (v)	Cretace-ous-Paleogene	Tropical savanna climate	Malagasy Rain Forest; Tropical dry or deciduous forests or woodlands		✓	✓		It belongs to the ocher-red sandstone and sediment rock of Cretaceous-Paleogene, with subhorizontal altitude. There are steep cliff slops with a height of several meters on the red residual hill and platform which is of old stage	

(continued)

Table 10.1 (continued)

Name of property	Country	Date of inscrip-tion	Criteria under which inscribed	Age of beds	Climatic zone	Udvardy Biological zone	Dominant landscape form				Notes	Photo
							Fold mtns and valley sides	Recti-linear (table mtn)	Rounded	Peak Forest		
Ksar of Ait-Ben-Haddou	Morocco	1987	C(iv) (v)	Triassic–Jurassic	Semi-arid tropical desert	Mediterranean Sclerophyll Warm deserts/semi-deserts		✓	✓	✓	The red sandstone, shale and basalt of Triassic–Jurassic have subhorizontal altitude. It developed the residual hill and golden sandstone stone fortress with subhorizontal top or perfect circle peak	
Al-Hijr Archaeo-logical Site (Madâin Sâlih)	Saudi Arabia	2008	C(ii) (iii)	Cretaceous	Arid tropical desert	Arabian Desert; Warm deserts/semi-deserts			✓		It is marked by a number of sandstone outcrops of various sizes and heights	
Mountain Resort and its Outlying Temples, Chengde	China	1994	C(ii) (iv)	Cretaceous	Humid temperate continental monsoon	Oriental Deciduous Forest; Deciduous forests, scrubs or woodlands			✓		This isolated mountain is formed by conglomerate, sandstone, and shale. It develops peak and queer stone with typical wall shape, pillar shape, and block shape	

(continued)

Table 10.1 (continued)

Name of property	Country	Date of inscrip-tion	Criteria under which inscribed	Age of beds	Climatic zone	Udvardy Biological zone	Dominant landscape form				Notes	Photo
							Fold mtns and valley sides	Recti-linear (table mtn)	Rounded	Peak Forest		
Dazu Rock Carvings	China	1999	C(i) (ii) (iii)	Lower Cretaceous	Humid subtropical monsoon	Oriental Deciduous Forest; Evergreen sclerophyllous forests, scrubs or woodlands	✓			✓	The light pink and mauve massive sandstone of lower Cretaceous developed red hills, with typical red cliffs and red walls in part. They were often developed into grottoes	
Mount Qingcheng and the Dujiangyan Irrigation System	China	2000	C(ii) (iv) (vi)	Cretaceous–Paleogene	Subtropical humid-wet monsoon	Oriental Deciduous Forest; Evergreen sclerophyllous forests, scrubs or woodlands	✓			✓	Mainly formed by amaranth conglomerate from Cretaceous to old lower Tertiary with the development of vertical joint, and then it formed red cliffs and red walls after erosion and corrosion	

General points arising from analysis of the table are as follows:

(1) Only eight of the 32 properties with red bed geology were inscribed under criterion (viii) in recognition of their important geological values, and just three of these eight properties —Canaima (Venezuela), Tassili n'Ajjer (Mali) and Purnululu (Australia)—were inscribed specifically in recognition of their distinctive geomorphic features. The other 24 properties do contain red beds, but the extent of any Danxia-type geomorphic landscape is variable. Furthermore, in these latter properties, red bed geology is only a subordinate part of their general geology, and the presence of these beds is incidental to their inscription.
(2) Of the 32 properties in the table, just nine can be identified as representing an integral, discrete Danxia-type landscape system in the model of the sites forming this nomination. These sites are Grand Canyon, Dinosaur Prov. Park, Purnululu, and Canaima, Three Parallel Rivers, Uluru-Kata Tjuta, Tassili n'Ajjer, Meteora, Capivara, and Bamiyan Valley.
(3) The six properties identified above each represent a single, isolated landscape occurrence, and none forms part of a series that demonstrates a sequence of evolutionary stages, as the China Danxia serial sites do.
(4) The geological basis of the China Danxia is red beds from the Mesozoic era, reflecting a particular period of continental environmental conditions poorly represented in the sites listed in the table.
(5) Rock lithology and structures in the China Danxia are unique and different from those exhibited in all other red bed properties. Especially important to landscape formation in the Danxia of SE China, poorly represented in other sites in the world, are the very dense networks of faults and master joints.
(6) Of the six properties identified in two above, only three falls in the same climatic regime as the Danxia of SE China. The others occur in arid/semi-arid or Mediterranean type climates.
(7) Despite the fact that 15 of the 32 red bed World Heritage properties are inscribed under esthetic criterion (vii), just seven are specially recognized for their distinctive red bed geomorphological scenery (Purnululu, Uluru-Kata Tjuta, Canaima, Tassili n'Ajjer, Cliffs of Bandiagara, Meteora, and Mt Wuyi), and usually, this element is subordinate to other values (i.e., merely one component of the landscape, as in the Grand Canyon, Three Parallel Rivers of Yunnan, Mt Purdu, Drakenburg Mountains, Dinosaur Provincial Park, Drakenberg Park, Kakadu NP, Canadian Rocky Mountains Parks, etc.).
(8) In all red bed World Heritage properties, the association between geology and culture is utilitarian (i.e., cultures has been related to the economic, defensive, or religious value in the rock structures), whereas cultural associations with the China Danxia are much broader, emphasizing the spiritual harmony between people and nature.

10.1.4 Danxia in Existing World Heritage Sites in China

With over 700 separate areas of red beds landscapes, China is the country in which Danxia geology and geomorphology is most prominently displayed and recognized. It is probably also the only country which contains a complete developmental series of both warm-humid and arid climate landscapes, as well as a complete range of Danxia landform types. It is, therefore, not surprising that many protected areas in China contain Danxia, although as the table shows, only six of China's 37 existing World Heritage Sites have notable red beds geology and geomorphology.

Although present, Danxia geology and geomorphology is not well developed in any of the six World Heritage Sites, with the exception of Mt Wuyi. The Danxia landscape of Mt Wuyi is representative of the mature stage cuesta-type peak cluster. The red cliffs are well developed, especially in the area of the Nine Bend River, forming a beautiful landscape of red cliffs and green river, while the best landscape is to be seen in the Tianyou district. While some features,

Table 10.2 Existing World Heritage Sites in China that contain Danxia geology and geomorphology (these sites are also listed in Table 10.1

General situation		Geologic structure	Geomorphologic feature	Physical environment	Photograph	Geomorphology value
Name of the property	Three Parallel Rivers of Yunnan Protected Areas	The property is of outstanding value for displaying the geological history of the last 50 million years associated with the collision of the Indian Plate with the Eurasian Plate, the closure of the ancient Tethys Sea, and the uplifting of the Himalaya Range and the Tibetan Plateau. These were major geological events in the evolution of the land surface of Asia and they are ongoing. The diverse rock types within the site record this history The purple conglomerate and calcareous siltstone interbed in the Baoxiang Temple formation (E_2b) with the upper portion of light gray conglomerate; the sandstone with middle-thick laminiplantation, gentle dipped occurence and well-developed vertical joints. Danxia was chiefly formed on the red beds of Baoxiang Temple formation	The landforms are predominantly typical Danxia mesas, peak cluster, steep cliffs, and caves in Laojun Mountain of Lijiang. The near-surface caves called "color waterfalls" are common in this area. In particular, the turtle-back shaped weathering convex hull (ramous convexity) has unique characteristic and formed the "Qianguishan(a thousand turtles)" Incomplete Danxia cliffs were developed, the height of which is greater than 80 m	The natural phenomena and natural beauty of the parallel gorges of the Jinsha, Lancang, and Nu Jiang are outstanding and irreplaceable. Due to the huge rises and falls of the terrain, this site may be the most biologically diverse temperate region on earth, and is of outstanding universal value	 A scenery of Laojunshan of Lijiang  The mountain type Danxia of Laojunshan	Danxia of Lijiang in the site, is one of the many geological elements, belongs to the plateau- mountain type Danxia of humid zone. The turtle-shaped weathering convex hull has unique characteristic. The overall scenery is monotonous relating to the Danxia of the eastern area
State Party	China					
Date of Inscription	2003					
Type	N					
Criteria	N(vii) (viii) (ix) (x)					
Name of the property	Mount Wuyi	From late Paleozoic to mid-Triassic, Wuyi Mountain region was in a state of uplift. From late Triassic, it formed a series of fault basin, in which were deposited extremely thick sandstone and volcanic rock. Some small-scale fault basins developed in the east, and in which were deposited Cretaceous purple sandstone and glutenite The red beds in this area are mainly represented by the purple thick layer of sandstone of the Cretaceous Chishi group. The Chishi group belongs to the pluvial phase and fluvial facies, with 707–2246 m thick. The red beds generally face toward the NW, the inclination is 20°–30°	Most of the shape of peak plane is arc type (I do not understand), while area present conical and cuesta forms.. The maximum altitude of the steep cliff is 200–300 m, whose geometry has a stepladder shape There are some high cliffs near the ravine and river. Some caves along bedding planes are distributed on the cliffs. The geomorphologic features include Danxia cuesta, stone peak, caves, lane valley, stone wall, and stone castle	Mount Wuyi is the most outstanding area for biodiversity conservation in south-east China and a refuge for a large number of ancient, relict species The dramatic gorges of the Nine Bend River have serene beauty, with numerous temples and monasteries	 The scenery along the Nine Bend River  The Shaibu Cliff beside the Nine Bend River	The red beds of Cretaceous sandstone and glutenite,(?) developed the landscape of Danxia peak cluster, block mountain and stone column. The scenery is the most concentrated along the Nine Bend River. However, the scenery away from the Nine Bend River is quite simple
State Party	China					
Date of Inscription	1999					
Type	C/N					
Criteria	C (iii) (vi) N (vii) (x)					

(continued)

Table 10.2 (continued)

General situation		Geologic structure	Geomorphologic feature	Physical environment	Photograph	Geomorphology value
Name of the property	Mount Emei Scenic Area including Leshan Giant Buddha Scenic Area	Leshan is located in the southwest of Sichuan Basin, and the area is noted for its massive red sandstone; The red beds are of the Cretaceous Jiaguan formation (K_2j) chiefly belonging to the brick red, massively bedded feldspar quartz sandstone, which is 300–900 m thick and subhorizontal in occurrence. There are also purple, brick red mudstone, and sandstone of Guankou formation (K_2g)	The landscape is of scattered Danxia hills, the side near water was cut into to form cliffs that reach 30-80 m in height The Giant Buddha was curved out of the cliff near water.	The Giant Buddha of Leshan was carved out of a hillside and look is down on the confluence of three rivers. The Giant Buddha is the most remarkable and the largest Buddha in the world	 The Giant Buddha was curved out of the red beds	Leshan, formed by the Cretaceous brick red sandstone intercalated by mudstone and shale. It is a typical Danxai cliffs along the river
State Party	China					
Date of Inscription	1996					
Type	C/N					
Criteria	C(iv) (vi); N(x)					
Name of the property	Mountain Resort and its Outlying Temples, Chengde	The outer mountains are composed of the Cretaceous Chengde conglomerate, Pingquan layer sandstone, shale and conglomerate	The landform is mainly composed of cutting hills and other well-developed Danxia landform such as Danxia cliff, mesa, stone peak, and column. One of the most typical landscapes is Qingchui peak (wooden club mountain)	The property belongs to the subhumid temperate climate, and the predominant vegetation is deciduous broadleaved forest. This is an outstanding example of harmonious existence of architecture and natural	 Qinchui peak: Danxia column	The mountains are composed of the Cretaceous Chengde conglomerate, layer, and shale, where the typical Danxia was developed well
State Party	China					
Date of Inscription	1994					
Type	C					
Criteria	C(ii) (iv)					
Name of the property	Dazu Rock Carvings	Dazu district is mainly composed of sandstone and shale of upper Jurassic Penglaizhen formation (J_3p), with subhorizontal occurrence. The average altitude of the hills is 500 m, and the relative cutting depth is 50–100 m. The steep hillsides of the Dazu area contain an exceptional series of rock carvings dating from the ninth to the thirteenth century	The Dazu carvings were carved on the hilly red beds area, as the sand shale is relatively easy to be weathered and eroded, so the steep cliffs are not very typical. The famous Dazu carvings were carved on some steep cliffs	East of the grottoes is the Pingqiu watershed with a NNE-SW distribution. The top of the Grotto Budda Gulf is flat. The vegetation belongs to the subtropical evergreen broad-leaved forest	 The Danxia cliff with grottoes	The lower Cretaceous light pink, purple massive sandstone, developed into the red bed hills. Some Danxia cliffs are developed as grottoes
State Party	China					
Date of Inscription	1999					
Type	C					
Criteria	C(i) (ii) (iii)					
Name of the property	Mount Qingcheng and the Dujiangyan Irrigation System	Qingcheng Mountain is composed of the Mesozoic red beds, including red conglomerate and sandstone dating from the Cretaceous to the Palaeogene, which is a typical coarse-grained molasse construction. The rock occurrence is gentle, and vertical joints are well developed	Mt. Qingcheng is a low-hill or middle-hill landscape area, and it belongs to the strong uplifting and deep cutting mountainous Danxia landform area. A single red steep cliff is not high	It belongs to the subtropical monsoon climate zone, the lush forest. Qingchengshan is a famous Taoist Mountain in China, known as the "lucky place" and "worldly paradise"	 The cliff on the upper part of Mt. Qingcheng	The purple Cretaceous—Paleogene conglomerate was eroded to form the mountainous type Danxia. The caves are often used as venues for religious activities
State Party	China					
Date of Inscription	2000					
Type	C					
Criteria	C(ii) (iv) (vi)					

such as the peak forest are well developed, individual landforms are better represented in one or other of the six component sites of the serial property. Moreover, the area of Wuyi Mountain is smaller than any of the serial sites of the nomination, while in the area of the Wuyi Danxia very little of the original forest remains.

The Table 10.2 supplements the information on the China Danxia sites to be found in Table 10.1.

Conclusion

Compared with the sites of this serial heritage, the Danxia landscapes in existing World Heritage properties in China are unremarkable. The best representative of Danxia is in the Mt. Wuyi property, which has the best landforms and has also benefitted from some research. The Danxia of the other five sites is deemed to be ordinary and lacking any significant research. The scientific and landscape values of the Danxia landforms of these sites are, therefore, relatively low.

Compared with the six existing World Heritage Sites, the sites forming this serial heritage have outstanding geological and geomorphological qualities as follow:

(1) The Danxia sites of the serial heritage have been chosen in a systematic way for their unique but related qualities. These sites form a complete geomorphological story, which has great integrity. They possess outstanding esthetic value and many of have the highest biological values.
(2) The geological structures, rocks, and landforms of the Danxia sites forming the serial heritage contain invaluable information on the geological, environmental and biological evolution, and ongoing geological processes, of Southern China. This information is all the more important because the sites represent different, scattered sedimentary basins, and therefore, the information they contain is more extensive and comprehensive of the history of the earth's crust since Mesozoic period in Southern China than if they were all found in a single, large basin.. The Danxia of the existing China World Heritage Sites does not remotely achieve the same value.
(3) The nomination is able to reflect the ecosystem and species diversity of the China subtropical evergreen broad-leaved bio-geographic region. The heritage sites also represent the ecological succession of the unique Danxia ecosystem in the recent geological past and have preserved prominent species diversity. Therefore, as a collection of sites, the nomination has greater importance than the existing World Heritage Sites with Danxia landscapes in meeting criterion (x).

10.2 Comparison with Other Danxia-Type Sites from Across the World not Inscribed on the World Heritage List

Table 10.3 lists other red bed sites that are located throughout the world. This list is the product of a literature search, and because of limited available detailed scientific information on the geology and geomorphology of individual sites, it does not claim to be definitive.

The research was helped particularly by publications by Wray in 1997, and the proceedings of two conferences on Sandstone Landscapes in Europe. In all, the world-wide search revealed a total of 49 red bed sites. The table shows that while red beds are well-represented throughout the world; their rocks date from widely different periods. Also notable is that, while many of the red bed landscapes listed in the table are located in arid or semi-arid, morphoclimatic zones, this contrasts with the sites forming this nomination which are all located in a warm, semi-monsoonal climate. It is unfortunate that because of the dearth of scientific information, the extent and integrity of the geomorphological system of the principal sites in the table cannot be ascertained. Nevertheless, the most important conclusion to be gained from reviewing this table is that no-where in the world is Danxia as well represented as in the sites forming the subject of this nomination in SE China.

The table does include some iconic red bed landscapes, particularly in the USA and Australia. Some of these sites bear some visual resemblance to the Danxia of SE China, but frequently, the scale is much smaller and karstic processes (e.g., quartzite dissolution) may be more dominant than in the Chinese sites.

Table 10.3 Other sites from around the world with red bed geology and Danxia-type landscapes

Country	Number	Geographic location	Stratum and structure	Danxia landform	Memo
Canada	1	Fundy National Park, New Brunswick in the southeast	Lower Paleozoic mauve sandy conglomerate with the dip 30^{σ}	Cliffs, rock columns, and stone arches along coast, the angularity of rocks is not smooth	Figure 1
	2	Prince Edward National Park in the southeast	Paleozoic nearly horizontal red sedimentary rocks with soft lithology	Sea cliffs are more than 10 m in height with smooth angularity	Figure 2
USA	3	Monument Valley is located in the intersection of Utah and Arizona, and to the west of Masienvaught	Red nearly horizontal sandstone formed by 250 MaBP	Rock peaks with the height less than 100 m are scattered hundreds of meters apart from each other in red desert. These rock peaks feature stone wall, column, and needle, resembling many monuments	Figure 3
	4	Rainbow Bridege is located nearly 100 miles to the west of Monument Valley	Pink and lavender gentle-inclined sedimentary rocks	A natural bridge which goes across a valley is 94 m long, 10 m wide, and 88 m deep. The angularity of rocks is smooth	Figure 4
	5	Bryce Canyon park is located in the east of Pausaugunt Plateau	The red beds are formed by 60 MaBP with nearly horizontal occurrence. The color is transitioned from orange in the upper to the red in the lower	Cliff slope is broken by vertical joints into dense almost linked rock columns, which are usually as high as tens meters with smooth angularity	Figure 5
	6	Capitol Reef National Park, Utah	Red beds formed by 65 MaBP, the colors of red and red-yellow dominate, and the occurrence is nearly horizontal	The cliff slope with hilly tip top is more than 100 m high. The summit-slopes of hills around display in circular-arc form	Figure 6
	7	Natural Bridge National Park, 50 m to the southeast of Hiatt, Utah	Jurassic continental red sandstone with nearly horizontal occurrence	Huge number of red sandstones, penetrated by caves and natural bridges. There are more than 1,000 natural bridges, among which Landscape Arch is 89 m long and 32 m high	Figure 7
	8	Natural Stone Arch, Missouri	Red-yellow nearly horizontal sedimentary rocks	The arch gate spans tens of meters, the cliff slopes around are tens meters high	Figure 8
	9	Sedona Talc Park in the south of Flagstaff	Nearly horizontal brown sedimentary rocks, the upper of some rock peaks is white	The height of Danxia rock peak cliff slope is from tens meters to one hundred meters	Figure 9
	10	Gold Valley, Death Valley National Park, South California	Early Cambrian red-brown and mauve sandy conglomerate with gentle-inclined occurrence	Plenty of cliffs with height ranging from tens meters to one hundred	Figure 10

(continued)

Table 10.3 (continued)

Country	Number	Geographic location	Stratum and structure	Danxia landform	Memo
	11	Around Hoover Dam, Dark Valley, Colorado	Jurassic-Cretaceous mauve sedimentary rocks with nearly horizontal occurrence	Cliff valley, the slopes are over 100 m with blunt rock angularity	
Guyana	12	In the west of the country adjacent the Canaima National Park in the neighbor country, Guyana Plateau	Early Cambrian red-brown sandstone, including quartzite and conglomerate with horizontal or gentle-inclined occurrence	Danxia-type Mesa, many waterfalls, cliffs around are hundreds of meters high	
Brazil	13	In the north of Roraima, adjacent the Canaima National Park, belonging to Guyana Plateau	Early Cambrian red-brown sandstone, including quartzite and conglomerate with horizontal or gentle-inclined occurrence	The Mt.Roraima with the altitude of 2810 m is the boundary mountain of Venezuela, Guyana, and Brazil; Danxia-type mesas are developed	
Australia	14	Kings Canyon, which belongs to northern region, is on the way from Alice Springs to Aias	Thin yellow horizontal bedding red sedimentary rocks included	Rock peaks display in circular-arc form, the side slopes display in convex cliffs with height from tens to hundred meters. The rock angular is smooth	Figure 11
	15	Funke Gorge National Park, to the west by north of Funke Town in the north	Red-yellow sedimentary rock, thin-layer and nearly horizontal	Rock peaks display in circular-arc form, steep valley. The height is from tens to hundred meters. The rock angularity is smooth	Figure 12
	16	Victoria River Valley in the Northwest of northern region	Red-yellow sedimentary rock with gentle-inclined occurrence	The height of valley cliff is over 10 m	Figure 13
	17	Carr Boyd Ranges in the northern part of West Australia, to the southeast of Wyndham	Early Cambrian brown sedimentary rocks with nearly horizontal occurrence	The rock peaks feature round top, partly flat top with the height of 300 m. The height of cliffs is over 100 m	
	18	Kimberley, in the Kimberley Plateau	Early Cambrian brown nearly horizontal sedimentary rocks with yellow thin beddings	300 m high hills, in Geikie Canyon Park, Canyon	
	19	Dales Gorge, Hamersle Range, West Australia	Red-brown sedimentary rocks with nearly horizontal occurrence	The height of valley cliffs is over 10 m	Figure 14
	20	Murrav Valley, northeast of Adelaide, West Australia	Early Cambrian red sedimentary rocks with nearly horizontal occurrence	There are 30 m high cliff slopes along the river bank	
France	21	Swiss-Sachsen, in the southern part of France Central Massif and the northern part of the Vosges	Cretaceous red sandstone, including thin-layer shale in which vertical joints developed	Rock peak, rock column and rock needle are all developed on red beds tectonic platform. The cliff of rock peak is shown in forehead-like cliff	
Germany	22	Bohemia Plateau in South Germany and the area of this plateau from which the Elbe flows	Cretaceous red sandstone with vertical joints and horizontal occurrence	Rock peak, rock column, and rock needle are all developed on red beds tectonic platform. Danxia natural bridges and basking-cloth rock landform exist also	

(continued)

Table 10.3 (continued)

Country	Number	Geographic location	Stratum and structure	Danxia landform	Memo
U.K.	23	Orkney Islands, northeast Scotland	Devonian red sandstone containing conglomerate and siltstone, belonging to continental sediments with nearly horizontal occurrence	Plenty of sea cliffs, Hoy Island has sea columns with the height of tens meters	Figure 15
	24	Duncansby Head, northeast Scotland	Devonian red sandstone containing light-colored rock with nearly horizontal occurrence	Sea cliffs are developed with tens meters height	Figure 16
	25	Edinburgh, on the southern side of False Bay, Scotland	Devonian red sedimentary rocks	Brown-red cliffs with the height of tens meters	
	26	Brecon Beacons National Park, South Wales	Devonian red sandstone containing conglomerate and siltstone with gentle-inclined occurrence	Over 10 rock peaks, uni-facial hills with cliffs, Cliffs can be seen in valleys	
	27	Gower Peninsula and South Glamorgan, South Wales	Devonian red sandstone containing conglomerate and siltstone	See cliffs and valley-slope cliffs	
	28	The coast of South Devon and Dorset England	Red sandstone, feldspar breccias, weathering sandy-colliculus rock mainly formed in Permian	Developed sea cliffs	
	29	Norfolk, Southeast England	Late Cretaceous red sedimentary rocks	Developed sea cliffs, some of whose upper parts are white sedimentary rocks	
Greece	30	Mt. Olympus, west of the west side of Aegean Sea	Oligocene gentle-inclined red beds which are developed in tectonic window	Plenty rock peak cliffs with the height of tens to one hundred meters, the rock angularity is relatively smooth	
	31	Kastraki Village, Thessalia	Red-brown sandstone with clear beddings, gentle-inclined	The cliffs are over 200 m high with smooth angularity	Figure 17
	32	Corinth Bay, Peloponnesian Peninsula	Cretaceous mauve nearly horizontal sedimentary rocks	The cliffs are tens or more meters high	
	33	Sounion Cape, Peloponnesian Peninsula	Cretaceous red-brown sedimentary rocks	The sea cliffs are 10 or more meters high	
	34	Thera Island, the furthest south of Cyclades Archipelago, Aegean Sea	Brown nearly horizontal rocks	The sea cliffs are tens or more meters high	Figure 18
	35	Amorgos Island, near Syra Island	Brown and drab sedimentary rocks	The sea cliffs are tens meters high, the angularity of rocks is not smooth	
Jordan	36	Moon Valley, 56 km to the northeast of Agaba, 50 m to the south of Petra Ancient City	Cretaceous red-brown sedimentary rocks with gentle-inclined occurrence	Canyon, steep cliffs are tens meters high with blunt angularity of rocks	

(continued)

Table 10.3 (continued)

Country	Number	Geographic location	Stratum and structure	Danxia landform	Memo
Saudi Arabia	37	Northwest Saudi Arabia	Cambrian-Ordovician sandstone, the color ranging from brown to black, thick-layered, from medium to coarse grain, partial cross-stratum with hard lithology and 600 m in thickness	Group of cliffs	
	38	Middle Saudi Arabia	Permian V-sandstone whose lower part is brown, thick-layered, cross-stratum, quartz sandstone from medium to coarse grain, the thickness is 150 m	Mesas, located in red desert of this country with little population where red beds formed stone columns, rock needles can be seen. The height is tens meters	
Egypt	39	Luxor, on the both sides of Nile, 675 km to the south of Cairo	Tertiary little drab and brown mudstone and sandstone with nearly horizontal occurrence	Hilly cliff slopes are tens meters high. There locates many historic relics and artificial landforms, such as Thebes City, Valley of the King and ancient temples	Figure 19
	40	Mt. Sinai, Sinai Peninsula	Oligocene brown and red sedimentary rocks	Group of cliff slopes are hundreds of meters high with tip top and blunt rock angularity	
Algeria	41	Constantine, Constantine Province	Cretaceous light red-brown containing light color sedimentary rocks with nearly horizontal occurrence	The city is built on top slope with valleys around, the height of cliff slope is tens meters	
Mali	42	Manding Plateau, to the west of the capital of Bamako which is in the southwest of Mali	Late early Cambrian littoral facies red graywacke with great thickness and horizontal or gentle-inclined occurrence	Mesas, cuestas, T and Mande (two grand cliffs) with the height of hundreds of meters, rock columns and mushrooms included	Figure 20
	43	Gangdamiya—Hongbo hills, in the east of Dayton, the south of Grand Meander of Niger River	Late early Cambrian littoral facies red sandstone and conglomerate with horizontal or gentle-inclined occurrence	The highest altitude of 1155 m is a mesa; relative height is hundreds of meters. It is very majestic	
	44	Timeteine Plateau, Taoudenni	Cretaceous-Tertiary red sedimentary rocks with horizontal or gentle-inclined occurrence	Cuesta, the relative height is from tens to over 100 meters. It is the largest area in four of Danxia-type landscape in Mali	
Guinea	45	Mande Plateau (also in Mali)	Late early Cambrian littoral facies red sandstone	Mesa, cuesta	
Burkina Faso	46	Dogon Plateau (also in Mali),	Late early Cambrian littoral facies red sandstone	Mesa, similar to Dogon Plateau, Mali	
Tanzania	47	The Olduvai Valley, Lake Natron	Gray-brown volcanic clastic rocks with nearly horizontal occurrence	Cliffs with the height of over 10 m	
South Africa	48	The Barren Coast of India Ocean from Edvard Harbor, south of Durban to Morgan Bay	Triassic or earlier drab rock series with nearly horizontal occurrence	Sea cliffs and arches	
	49	Shedlberg, Western Cape	Red rock series with nearly horizontal occurrence	Duo-layered stone arch	

Annexes: Images of Some International Red Bed Sites

Fig. 1 Canada Fundy National Park By (CAN) Michael Lvory Canada. 2002

Fig.2 Canada Prince Edward National Park By (CAN) Michael Lvory Canada. 2002

Fig.3 America Monumant Valley By Yuan Kejuan From the Arctic to the Antarctic .2003

Fig.4 America Monument Valley By Shi Yingtian Nation Ethnic.1999

Fig.5 America Rainbow Bridge By Rechard Phelport 100 Natural Wonders .1999

Fig.6 America Bryce Canyon By Fang Hao. America .1977

Fig.7 America Arches National Park

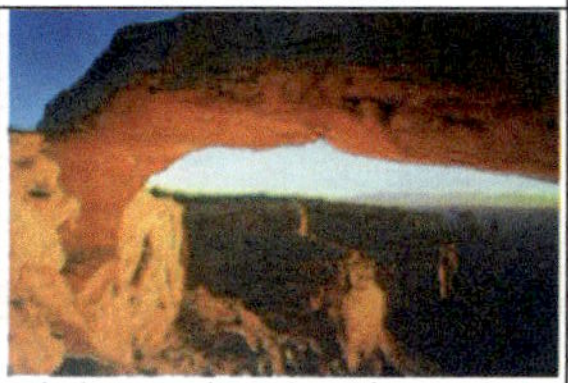
Fig.8 America Missouri stone circle By Shi Yingtian Nation Ethnic 1999.

Fig.9 America Sedona Slide Rock Park By Nicholas etc. Los angeles 2002.

Fig.10 America Dath Valley By Wang Zhihong. America West Coastline .1999

Fig.11 Australia Kings Canyon By Jocelyn Burt. Australia.1987

Fig.12 Australia Finke Gorge National Park By Huang Zhongzheng Australia .1995

Fig.13 Australia Victoria river valley By Jocelyn Burt. Australia.1987

Fig.14 Australia Dales Gorge By Jocelyn Burt.Australia.1987

Fig.15 UK Orkney Islands By K. Somerville UK .2002

Fig.16 UK Duncansby Head By Faang Hao UK .1979

Fig.17 Greece Kastraki By APA Publication *South Africa* .2001

Fig.18 Greece Thera Island By Gao Liang Greek .2000

Fig.19 Egypt Luxor By Henglaisifu, Egypt .2003

Fig.20 Mali Manding Plateau

10.3 Synthesis Conclusion of the Comparative Analysis

(1) As a serial heritage, the Danxia of SE China reveals a sequence of different, but related landscapes that explain the development from youth to old age of this landscape type. This cannot be done in any other place in the world.

(2) In order to undertake a comparative analysis, it should be necessary to separate the Danxia-type landscapes of the world into different geological and morphoclimatic types. However, good scientific knowledge on the ages and formation of the vast majority of these landscapes does not exist, and therefore, it is not possible to classify them, other than on simple morphological or climatic grounds. However, neither of these categories enables meaningful geomorphological comparisons between sites.

It is known that the Danxia sites forming this serial heritage are all located in the warm, humid zone of SE China and have been greatly influenced by fluvial processes. The sites of SE China appear to be unique, particularly in the density of the fracture systems which control fluvial incision, and the weathering processes working on the clayey clastic sedimentary rocks, to produce in many of the sites remarkable bell-shaped or cone-shaped mountain groups. Some other of the world sites also exhibit rounded slopes and groups of cone-shaped hills (e.g., Meteora, Purnululu, and Capivara), but these fall in different climate zones, and in any case it is suspected in one at least that quartzite dissolution and a long period of erosion helped determine the formation of the landscape we see today. Neither of these factors is considered to be of great importance in the formation of the Danxia in SE China.

Thus, while there are some other inscribed and non-inscribed red bed, landscapes around the world that are visually similar to China Danxia; these all have elements which are different from the China examples, and in any case are single, isolated sites, not representative of a geomorphological group, or series, as in China.

(3) No red bed, "Danxia" geomorphological landscape, has been specifically inscribed on the World Heritage List. This current nomination will fill this gap and could eventually be extended to also include China's arid zone Danxia landscapes.

(4) While there is some Danxia in China's existing World Heritage Sites, it is not well developed and does not rival the quality and importance of the Danxia of this serial heritage.

(5) The landscape character and scenic beauty of the Danxia of SE China are unique and of a higher quality than most other red bed geomorphological and biological landscapes on the planet.

(6) The Mesozoic stratigraphy of the China Danxia fills a gap in the planetary record of continental red bed formation.

(7) The relationships between geology, landscape, and culture in the heritage sites are different from those in other Danxia-type sites around the world in being more spiritually connected and less utilitarian.

10.4 Integrity of the Heritage Site

The sites of the serial heritage have great integrity, both individually and as a group. As explained in the preceding section, each individual site exhibits distinctive continental sedimentary geology, (lithology and structure), exogenetic geomorphological processes, and landforms. Other than small local differences in the factors controlling landscape formation, such as rock lithology, bedding, structure, climate, and the relative balance between the geomorphic process, it is mainly time or stage of degradation which separates them. Thus, each site is a scientifically integral system, while the same can also be said for the suite of complementary sites that make up the serial heritage.

These landscapes also have great conservation and management integrity. Because the Danxia sites of this serial heritage have distinctive geographic environments, they remain some of the few natural areas that have been barely touched by human activity. They, therefore, remain very special natural resources, which contemporary laws and management practices seek to maintain for the future. The heritage properties not only have outstanding, distinct earth science, ecological and natural esthetic qualities, and values, but also each region seeks to maintain the integrity of its heritage and the pristine and diverse condition of its natural environment.

10.4.1 The Clear Outline of Property and the Clarified Boundary

Each China Danxia heritage sites has a core zone that is distinctly divided by ridge line, valley line, river and other natural line, and also has a buffer zone. Usually, the boundary corresponds with the border of current protective denomination zone. Every heritage sites has undertaken a boundary survey. There are distinct borders in accordance with the heritage sites, the buffer zone, and strictly protected areas within the core zone.

In defining a property's core zone, the following principles have been used: the area must be typical and the most representative of the Danxia landscape; it must have low population, and no large urban development, industrial or mining enterprises; it must have a good forest ecosystem and ecological environment for endangered species; it should be an area not threatened by any future large projects.

Each heritage sites has a buffer zone. The principles of defining the boundary of the buffer zone are: the peripheral natural area must provide a buffering effect; there should be low population, no large urban development, industrial and mining enterprises (residential areas already established inside the buffer area will be strictly controlled, and industrial and mining enterprises will be steadily closed down or moved); the foreground area should reflect the significance of the heritage sites, not including potential air and water pollution source; the buffer area should have sufficient width.

All the heritage sites except the Xinjiang basin and the Jianglangshan region are developed to old age that distribution showing natural spot like the other youth and adult Danxia were displayed highly of continuity and integrity. While these landscapes are not very high, the walk is rather tough. This is because the positive landform has steep cliffs, while the valleys are also difficult to go through. The old-stage Danxia generally has wide and flat river valleys, and extensive mountain valley, with many developed into agricultural zone, but the landform completeness has not been damaged.

10.4.2 The Propery with Sufficient Area Benefits the Protection and Natural Development of Property

Except the old-stage Danxia landforms like Mt. Guifeng and Mt. Jianglangshan, the areas of the nominated sites are all over 6000 ha, which forms a large enough "ecological island". Most of the heritage sites have been hardly touched by human activity, providing a big natural scope for protecting endangered species. All the sites have more than one national protective designation, and all but one is national parks. The import of exotic plant has not been permitted, and restrictions dictate that each nominated site should develop based on the natural evolving law.

The basic attributes and characters of China Danxia are achieved in each site. First, a nominated area should present the most beautiful landform, water, and biological factors; second, the site and its landforms should illustrate the long the geological and geomorphological history, the palaeo-enviroment and palaeogeography of the area, which is the most typical area in Danxia development, with assurance that the factors that reflect the earth scientific value most completely preserved; third, the scale of the area nominated is of sufficient size to ensure protection of species diversity and the original biological environment, protection of endangered animal and plant distribution areas. As a group, therefore, the nominated sites provide completeness and a diversity of ecological environments, and excellent protection for the continued evolution of the biological and ecological system.

10.4.3 The Influence Exerted by Human on the Natures and Outstanding Universal Value of Danxia

Conditions for farming and other economic land uses in the Danxia sites are harsh, such possible developments being impeded by a terrain made-up of crisscross gulleys, intense rises and falls of the land surface, a thick weathering crust and poor soil nutrition in soil, and a fragile ecological balance. Therefore, development is usually limited. This kind of land use is restricted to traditional agricultural along the river valley plain, and does not change the natural attributes of Danxia. The ancient people constructed temple, mountain village, and graveyard based on the Danxia landform, involving the natural beauty of the environment, pursuing "the harmoniousness between human and heaven" state and formed natural but complicated beauty. Today, care is taken to ensure the Danxia landscape is free of harmful industry and mining, but allowing some dispersed small areas of traditional agriculture. This has little adverse effect on the natural environment.

In history, there were cutting firewood, hunting, and gentle slope reclamation activities in the nearby villages; however, since the imposition of a protective landscape zone, and with the intensify of the protection measure, the protection thought of farmers, the country economic growth, the above ways of developing just disappeared.

10.4.4 The System of Protection and Management and Measures of Protection

The common landscapes resource among heritage sites includes six classes: the landscape of red beds geological relics, Danxia landform, rare species, river scenery and waterfall, weather and meteorology, and historic culture. Such landscapes mentioned above are well protected. Since 1980s, the country has successively promulgated a series of laws and regulations on natural protection and clarified the nature of regional protection by establishing nature reserves, scenic spots, forest park, and geopark. Meanwhile, local governments have established authorities at all levels, teams for protection and legislated concrete regulations and measures on management. The villages have established relevant regulations and engagements, made it clear that the top leaders of governments at all levels are the first person liable and formed a comprehensive system of management for the heritage protection. Meanwhile, projects have been adopted regarding the preservation of water conservation, forest, natural forest, and ecological forest, which makes all the heritage factors of properties have been effectively protected.

The Jinjiang Valley of Danxiashan

11 Conservation and Forthputting

11.1 Ancient Exploitation

The exploitation of China Danxia has been a long history; there are many prehistoric cultural relics have been found in candidates. It has proved that human beings have developed river valley for farming and occupied natural caves as dwelling since ancient times. It has generally experienced three stages: the first is the stage of original fishing, hunting, and farming; the second is cultural and esthetic utilization in ancient times, and the last is modern conservation and tourism comprehensive utilization.

Human began to live in natural caves long ago, which is why that many peaks are named “fortified village.” Numerous ancient fortified villages of different historic periods have been preserved. Although most of the fortified villages are abandoned, they still serve as an indispensable feature of the cultural relics and an elapsed civilization. So, there goes a saying that there is always a village on each hill, each village has a gate of its own, and each gate is erected on a steep point. Because the Danxia caves are room-like, dry, and well ventilated, our forefathers used caves as coffins, which forms a special grave cultural landscape of cave living, cave coffins, hanging coffins, and so on in Danxia landform area. Till now, there are still some native residents living in cave rooms in Chishui; the custom of storing coffins in Danxia caves is still well preserved in Taining; there are more than 200 hanging coffins of Spring and Autumn and Warring Period preserved in cliffs of Xianshuiyan, Mt. Longhushan.

The most prominent shapes of Danxia landscape are block-like and fort-like forms, together with its distinct color of purplish red, making the Danxia landform a religious bethel. For instance, the history of Mt. Longhushan, the Taoism bethel of China, is as long as more than 2000 years. It is still regarded as the holy land of Taoism at present; the oldest temple of Taining was built in 1300 years ago, and there are more than 80 Danxia cave temples that have ever been found, 80% of which were built from Tang Dynasty to Yuan Dynasty. At present, they are more than 40 temples in use. Mt. Danxiashan is a famous place for religion in Lingnan region, where shaman began to practice Buddhism in caves in Sui and Tang Dynasty. The religious activities reached the peak in Ming and Qing Dynasties. More than 40 cave temples have been found, whose buildings were well complied with the mountains, forming a harmonious view.

The red cliffs of Danxia landform mainly consist of red sandstone, which is dense and even in lithology. This provides a good condition for carving, therefore, a good carvings, stone statues, and inscriptions have been found in Danxia landform region, such as rock carvings in Nanyan Temple of Guifeng Peak, Ganlu Temple of Taining, Jinshiyan Cave of Mt. Danxiashan, and inscriptions in Mt. Danxiashan. These carvings are precious for history and religion.

H. Peng, *China Danxia*, Springer Geography,
https://doi.org/10.1007/978-981-13-5959-0_11

11.2 Conservation History

Danxia candidates belong to remote region in ancient China, simply developing original fishing, hunting, and traditional farming, which has posed little influence on nature. As the increasing recognition of Danxia esthetic value and accessibility of religious culture, people generate a sort of awe and psychological dependence, and the principle of nature conservation is widely accepted. Meanwhile, this period is also traditional farming economic time which affects little on Danxia landform areas. In modern time, candidates still belong to undeveloped areas. Influenced by precipitous landform and sparse population, candidates remain the mode of production of traditional agricultural areas, and regional development is led by natural force. Accordingly, the resource environment of candidates is still well preserved.

Conclusively, the conservation history of candidates can be categorized into three stages.

(1) **The Stage of Conscious Conservation Affected by Naive Philosophy of Ancient China**

As what has mentioned above, affected by Chinese traditional culture, residents of candidates awe nature, worship nature, and protect nature consciously. This has become the creed of environment conservation for their livings in ancient time. Influenced by religion, geomantic omen, and cultural harbinger, the cultural tradition of nature conservation has been continued. It has objectively promoted the conscious conservation of local residents and sustainable utilization of natural resource. In this stage, the lasting time differs in different areas. Though it is not the leading power, the influence of culture has lasted till now and become the most crucial power of nature conservation.

(2) **The Stage of Village Rules Conservation**

As the growth of population and social development, the conflicts between human and nature are increased. Thus, there successively appears a sort of village rules to protect natural environment, such as mountain forest, woods, and water source. It has been regarded as the embodiment of original rules conservation consciousness. Residents make and obey it together. It is established by usage and considered as a sort of original form of law. Some village rules are made in 1950' to protect mountain stones, trees, and water sources. It has become an effective form of protection.

(3) **Conservation and Administration of Government**

After 1949, the resources conservation of candidates was highly valued by the nation. The government of candidates successively established forest farm and management organization to protect heritage site. Danxia landform areas in which candidates locate are not used as productive forest farms. Instead, there has built some bases for seeding cultivation, livestock farms, and cultivation bases; therefore, landscape resource has been well preserved.

Since 1956, the nation has established nature reserve and national scenery and resort area (1982). Meanwhile, relative laws have been made. Local governments in different levels formulated and promulgated corresponding management regulations for nature and scenery resource conservation. Thus, the conservation of scenery and resort resource has been brought into legal orbit. Each of the candidates in this application has national protected name, has established corresponding management organizations, and made planning for conservation and management. The conservation of heritage sites is legally guaranteed by the nation.

11.3 Present State of Conservation

(1) **Human exploitation in history never affects the outstanding universal value of nominated properties**.

In ancient times, China Danxia nominated properties have simply developed original fishing, hunting, and traditional agriculture, any of which poses little effect on the nature. As the accreditation of Danxia esthetic value and the access of religious culture, awe and psychological reliance emerge from people's cognition and the idea of nature protection becomes quite acceptable. Meanwhile, it is also belonging to traditional farming economic time when the effect on Danxia landform area was still weak. In modern times, nominated properties still belong to undeveloped area, together with the precipitous landform and sparse population, which makes nominated properties remain the production mode of agriculture area. Natural force dominates the regional development.

Although natural factors and human activities might inevitably affect the nominated properties to some extent, they hardly influence the integrity and outstanding universal value of nominated properties. In general, it is well preserved that the key elements of nominated properties, including not only Danxia landform, esthetic value of landscape, ecosystem, endangered species and habitats, but the integrity of species tendency, ecosystem and natural environment as well.

(2) **Nominated properties have been granted high-class protective denominations, which means that they will be under the protection of national laws.**

Since China's Reform and Opening-up, people have placed a growing emphasis on the scientific and landscape value of Danxia landform. Therefore, it is increasingly acceptable that it serves as the value of important natural heritage for the country and the whole people. Due to the significant scientific and landscape value, nominated properties have been successively granted protective denominations in every class by the government of province or the country, such as nature reserves, scenic spots, national heritages, forest parks, and geoparks. Accordingly, they are under the protection of relevant laws and regulations, such as Constitution of the People's Republic of China (PRC), Law of PRC on Forest, Environmental Protection Law of PRC, Water Law of PRC, Law of PRC on Protection of Wildlife, Regulations of PRC on Nature Reserves, Law of PRC on Protection of Wildlife, Enforcement Regulation on Forest Law of PRC, and Regulations of PRC on Scenic Spots. The state of conservation is favorable at present. The comparison between nominated properties and IUCN in the category of protection has been shown in table.

Comparison between China Danxia nominated properties and IUCN in the type of protection

Category of protected area	Sub-category or type	Examples of candidate protected areas	Potentially possible assignment to IUCN categories
Nature reserve	Ecosystem type	–	I, II, V, or VI
	Wildlife type	Chishui (Nature reserve of wildplant type)	IV
	Natural relic type	Mt. Danxiashan (Reserve of significant geological and geomorphic relics)	III
Scenic interest area (National Park)		Danxiashan, Langshan, Taining, Longhushan, Jianglangshan, Chishui	II (National park) V (Protected landscape)

Remarks The corresponding names of numbers in the table stem from IUCN's category of protected area (1994): I. Strict Nature Reserve; Ia. Strict Nature Reserve; Ib. Wildeness Area; II. National Park; III. Natural Monument; IV. Habitat/Species Management Area; V. Protected Landscape/Seascape; VI. Resource Management Protected Area

(3) **The extent to which the management system, staff, organizations and funds are guaranteed in nominated properties.**

An effective multilevel management system has been established in China Danxia nominated properties in which all aspects are included, such as the attention of government, the cooperation among departments, the support from society, the uniform coordination, and management of environmental protection. Besides, the local authorities on resource and relevant enterprises have accordingly established environmental protection organizations. Due to such acts, the pollution has been controlled and ecological environment has been well preserved. The atmosphere, water, soil, and noise are kept in a perfect state. It is especially favorable in nominated properties with sparse population that natural ecology is still kept in original state with virginity, marvelousness, and beauty.

All the China Danxia nominated properties have clarified the boundaries and established corresponding monitoring index, which will help timely monitor and solve problems in time. The staff, organizations, and funds have been sufficiently guaranteed.

(4) **The geological geomorphic elements and natural ecological elements of China Danxia have been well preserved**.

Firstly, in six Danxia landform areas involved in nominated properties, the outshirt hill and plain have been exploited for traditional farming, no quarry or mining industry occur in history. Especially with the economic development after China's Reform, farmers no longer use the trees or stones for money, which is a beneficial factor for the management of the parks.

Secondly, the intensity for tourism development of these areas is not strong. Furthermore, the protection of the tourism resources has always served as the precondition for any development and construction. No explosion for any purpose is allowed, and resources are well protected from the damage from tourists. Proper development guaranteed the preservation of geologic relics.

Thirdly, the population in typical Danxia landform areas is small, the villages and buildings are harmonious with the environment, with agricultural activities limited within the river plain, which, however, could not form threat toward the protection of geological heritage or environment. Therefore, the Danxia landform areas have always been well protected and remained in the natural status. For instance, the slightly affected and intact regions of Mt. Danxiashan sum up to 210.8 km^2, taking up 78% of the total regions.

Such nominated properties mentioned above locate in places where strata outcrop well, with clear tectonic structure and typical elements of the Danxia landform, which clearly display undergoing geomorphologic phenomena. These geological relics are evolving according to the natural rules; there would be no threat to the relics from any human factor or tourist activities except for natural collapses of some cliffs, stone arches, or caves.

(5) **Monitoring archives of main natural elements (properties) have been established in every China Danxia nominated property**.

Under the guidance of Ministry of Housing and Urban-Rural Development of the People's Republic of China, all nominated properties have made comprehensive monitoring broadcast on bio-diversity, ecological environment, state of vegetation, forest coverage, surface water quality, biology, air, natural disasters, settlement sites and villages, human landscapes and economy of society and community by means of satellite monitoring, air monitoring and ground itinerant monitoring, etc. Meanwhile, monitoring archives of properties have been established.

(6) **Nominated properties have been under the fork traditional protection**.

Natural factors and human activities have affected China Danxia nominated properties to some extent. Natural factors are manifested as landslide, debris flow, forest fire, forest pests, and so forth; while human activities display as farming, forest-cutting, and projects in nominated areas. Nevertheless, relevant authorities have taken effective action to control the situation. For example, there implements the project of returning farmland to bamboo on Danxia hills, especially on many slopes where farming is not suitable. Moreover, it has made good effect by planting some economical trees, adjusting local industrial structure of agriculture and developing diversified economy. Accordingly, external elements, no matter natural factors or human activities, have never basically affected the outstanding universal value of nominated properties.

11.4 Factors Affecting the Property

As nominated properties have been affected by human activities in varied forms, their Danxia landscapes, ecological environment, bio-diversity, and ethnic culture are suffering from pressures to some extent. Nominated properties, however, have been granted protective denominations (national nature reserve, scenic spot, geopark, etc.) and been under increasingly strict protection, which reduces the harmful effect exerted by human activities to some extent. Therefore, the negative effect on properties is weak at present.

(1) **Development pressures (e.g., encroachment, adaptation, agriculture, and mining)**

Each of nominated properties has experienced thousands of years of human activities and the population of both property and buffer zone are about 10,000. There always exists a contradiction between the demand of natural resource for human survival and development and the protection of resource. Some of the villages inside property and buffer zone are still poor, and the villagers are highly dependant on the nature environment, trees in the surrounding areas are often cut down by the villagers. The protection of Danxia landscape and ecology can also be affected by the construction of towns and infrastructure occurring near buffer zone, such as village construction and road construction. In addition, some water projects, which change the local hydrological environment, have influenced the inhabit and multiply of aquatic and land wild lives.

Modern civilization and exotic cultures have exerted significant influence on the preservation of traditional architectural style, the continuity of local folk customs and the inheritance of local art and national culture. Especially the application of modern architectural materials destroy the traditional architectural style in scenic spot, disturb the visual landscape of the nature, and burden the difficulty and the pressures in the preservation of traditional architectural culture.

(2) **Environmental pressures (e.g., Pollution, climate change, and desertification)**

Since all nominated properties are denominated as national scenic spot or national nature reserve, some problems, such as soil and water loss and forest-cutting, have been effectively controlled. Tourism development, especially the infrastructure construction of tourism reception, however, has brought about some pollution and visual effect on local water environment and landform. Meanwhile, it has also affected the animal inhabit and the integrity and stability of ecosystem. Furthermore, the protection for bio-diversity will be threatened.

There imposes great pressures on the protection of natural environment due to poor peasants' laggard mode of living and production and over-exploitation of natural resources as the rapid social-economic development in nearby areas. There are increasing possibilities of pollution in the rivers in every nominated area due to the urban and industrial development and dumping of residential areas in the upper reach of the rivers.

(3) **Natural disasters and risk preparedness (earthquakes, floods, fires, etc.)**

Main natural disasters affecting nominated properties include:

- Forest fire: forest fire occurs easily in drought season due to the feature of distinct dry-wet seasons.
- Mountain torrents, collapse, and landslide: historic records show that nominated properties are often threatened by collapse, landslide, and debris flow due to their particular geomorphic and ecological conditions. Small-scale collapse in cliffs is quite common.

- Freezing hazard: the rare freezing hazard happened in the early 2008 damaged a lot on the vegetation and infrastructures of some nominated properties.

Means of preparedness:

In order to control such natural disasters in scientific approaches and to positively prepare for any risk, the training for the sense and measures of disaster preparedness has been in progress in nominated properties. Besides, the caution board and other protective measures have been set in dangerous places of tourism area and patrolled and inspected regularly. In order to enhance the propaganda, preparedness, and remedy of forest fire, nominated properties have established forest fireproof preparedness, fireproof warning system of government-village-resident, fireproof headquarter, and professional team of forest fire protection. They help enhance the sense of fireproof among villagers by fireproof propaganda, such as slogans and entering household propaganda. Moreover, nominated properties have founded professional monitoring teams to scientifically compile the report of evaluation and forecast on nominated properties geological disasters and to strengthen the project construction, such as protective project on geological relics, the project on the river-dredging and levees reinforcement and the establishment of scenic forest fireproof channel and ecological fireproof. Finally, they monitor and control sudden diseases and pests, enhance the quarantine inspection of exotic timber and bamboo species, put an end to the biological invasion, lead the masses to plant mixed forest in a reasonable way, and enhance the tending management.

(4) **Visitor/tourism pressures**

In general, these nominated properties develop relatively late. The amount of tourists is far smaller than the capacities of scenic spot, and the environment would not be quite negatively affected probably because of the limits of location and transportation. The current tourism development mainly focuses on the sightseeing, leisure, and holiday-oriented projects in the river plains and outskirt of the hilly region is under-developed, which make the scenic spots very crowded, especially in holiday. The amount of tourists will be saturated, even overloaded in short time. Also, the pollution of trash caused by huge amount of tourists is increasing and insufficient capacities of the scenic spots become an outstanding problem.

In order to reduce the influence exerted by the peak of tourists on environment, nominated properties are planning to enhance the propaganda and management, clear up the trash and wastewater caused by tourism, and maintain a nice sanitation. Armed with the system of tourism information management and environmental monitoring information, authorities in nominated property could monitor the total amount of tourists in scenic spot. If necessary, they should evacuate tourists in a scientific way and limit the amount of tourists within the bounds of environmental carrying capacity.

(5) **Number of inhabitants within the property and the buffer zone**

The total area of core zone in China Danxia nominated properties is 821.51 km^2 with the population of 34,026 persons; the population density is 41 persons/km^2. The total area of buffer zone is 1362.06 km^2 with the population of 100,259 persons, and the population density is 74 persons/km^2. The area in sum is 2183.57 km^2 with the population of 134,285, and the population density is 62 persons/km^2. Among them, there is no population living in the core zone of Jianglangshan; the population density of Taining core zone is only 6 persons/km^2; Danxiashan is 9 persons/km^2. Longhushan in old age is relatively larger in population density. However, the population in these two properties are distributed in river valley.Residents in nominated properties (Year: 2007)

Residents in nominated properties (Year: 2007)

Province	Property	Residents in core zone (person)	Population density of core zone (person/km^2)	Residents in buffer zone (person)	Population density of buffer zone (person/km^2)	Total population (person)	Mean population density (person/km^2)
Guizhou	Chishui	4751	17	19,756	44	24,507	40
Fujian	Taining	691	6	10,872	87	11,563	49
Hunan	Langshan	3040	46	12,983	209	16,023	125
Guangdong	Danxiashan	1578	9	8724	70	10,302	35
Jiangxi	Longhushan	23,966	122	46,560	78	70,526	89
Zhejiang	Jianglangshan	0	0	1364	239	1364	115
Total		34,026	41	100,259	74	134,285	62

The total amount of residents in core zone and population density 34,026/821.51 = 41 (person/km^2)
The total amount of residents in buffer zone and population density 100,259/1362.06 = 74 (person/km^2)
The total amount of residents and population density 134,285/2183.57 = 62 (person/km^2)

11.5 The Use Form in History Embodies "Harmony between Human and Nature"

Danxia landform usually features crisscross gullies, violent fluctuation, thin weathering crust, barren soil, and fragile ecological balance. It is not suitable for industry, agriculture, transportation, and urban construction. Therefore, the exploitation of Danxia landform areas has been focused on their landscape, and the development intensity was low.

The ancients who exploited Danxia have pursued a state of "the harmony between human and nature", particularly a principle that structures should conform to the lie of mountains. Consider the Grotto Temple at Jinshiyan Rock, Mt. Danxiashan, the monks' dormitory, and Buddhist halls in which were all erected on steep cliffs in caves, each of such establishments backs on rocks and cliffs and faces deep gullies. Outside the rock, Mawei Waterfalls hang on cliffs with water droplets flying in the air. The mouth of each of these cave holes faces Jinjiang River, which winds through the feet of these cliffs, presenting a beautiful landscape. Architectures of other temples are all of their own marvelousness. For instance, the gates of Biechuan Temple and Xiaguan are erected on majestic and precipitous cliffs, backing on cliffs and facing deep gullies. Such architectural layout has made a perfect fusion of the great natures and artificial establishments, forming up a sort of marvelous esthetic beauty.

Ravine plains have been considered as farming areas since ancient times. Long-term development of farming culture and mountain-water culture of Mt. Danxiashan have made a harmonious fusion of human and nature, forming a sort of rural cultural landscape of Danxia area.

Part II

12 Chishui, Guizhou

12.1 Introduction to the Heritage

12.1.1 Summary

Name of Property: Chishui.
County, Province: Chishui City and Xishui County, Guizhou Province.
Geographical coordinates to the nearest second (Central):
Western part: 28°22′11″N, 105°47′39″E.
Eastern part: 28°25′19″N, 106°2′33″E.
Area of Property:
Heritage Site (ha): 27,364 (Western part: 10,142, Eastern part: 17,222).
Buffer Zone (ha): 44,814 (Western part: 25,341, Eastern part: 19,473).

12.1.2 Description of Boundaries of the Heritage Site

The boundary of the heritage area and proposed buffer zone of the Chishui Danxia Heritage Site is clearly labeled on the maps and actual spots. In the boundary of heritage areas, they mainly remain the integrities of natural geomorphology and ecosystem and prohibit all human activities with negative impacts on the heritage areas in exception of scientific researches. In buffer zone, the contents and patterns of human activities are also controlled to in some extent.

The delimitation of boundary for the Chishui Danxia Heritage Site mainly accords to the integrity of Danxia landform development and distribution, forest ecosystem, and habitat of rare and endangered species. Rivers and ridgelines are regarded as natural lines, and the delimitation of boundary tries to avoid enclosing the modern artificial construction, settlements, mining, and hydropower stations. Chishui Danxia Heritage Site mainly contains Shizhangdong and Bing'an scenic spot in Chishui National Park, Chishui *Alsophila spimdosa* Nature Reserve, Chishui Bamboo National Forest Park and part of the Xishui Subtropical Evergreen Broad-leaved Forest Nature Reserve. The delimitation of boundary for buffer zone not only accords with the integrality of Danxia landform development and distribution, but also accords with the factors of protection of the Heritage Site, such as avoiding the influence of human activities. As well as Heritage Site, rivers, roads, ridgelines and phase-change lines are regarded as natural lines. The western boundary keeps coherence with the boundary of Chishui National Park. The southern boundary keeps coherence with the common boundary of Guizhou and Sichuan, phase-change lines.

H. Peng, *China Danxia*, Springer Geography,
https://doi.org/10.1007/978-981-13-5959-0_12

12.1.3 Justification Statement of Outstanding Universal Value

(1) **Outstanding universal value of beauty**

In Chishui Danxia nomination, many rivers flowing among the mountains have developed typical danxia cliffs and canyons landform with many rundle spectacular cliffs and waterfall groups in these valleys. Danxia cliffs are so high, so wide and so enormous that they have strong vision punch power. Here, we can find the biggest waterfall in Danxia area, such as Shizhangdong waterfall, which are the biggest Danxia Waterfall landscape. Foguangyan cliff and Dabaiyan cliff make people's strong wallop. Due to wet climate and affluent rainfall, this place has developed into the luxuriant forests with more than 90% forest coverage which has been the primeval forest area for many rare and endangered species, living fossils, and dense plant, such as *Alsophila spinulosa.* Danxia nomination, with full range of landscapes together in an area of 72,178 ha, such as cliffs, Danxia peaks, Danxia caves, large collapsed stones, waterfalls and other landscapes, is a splendiful and beautiful picture combined mountains, and it is the typical one of the young Danxia.

(2) **Outstanding values of earth history**

Chishui Danxia landform is an outstanding representative of Danxia landform in young stage, which is in its early young stage due to geomorphological rejuvenation since the Quaternary. It results from the strong erosion of the Jurassic and Cretaceous red beds, which experienced a long-time denudation in Tertiary Era and then was rapidly uplifted by Quaternary neotectonic movement. Its evolution profoundly reflects the process of regional tectonic movements since Late Triassic Period in Southwestern China and reveals the important influence of differentiated uplifting of Guizhou Plateau and Sichuan Basin in neotectonic movement on geomorphological evolution, especially, the significant meaning of differentiated movement of Northern Guizhou Platform Uprise and Middle Sichuan Platform Subside in the development of Chishui Danxia landform.

The emergence and disappearance of Bashu Paleo-lake are an important event in the geological history of Southwestern China. The sedimentary formations in the Heritage Site and its surrounding areas at the Southern edge of the paleo-lake have truly recorded its life process and provided reliable and also valuable sedimentological evidence to reveal the geological change and paleo-geographic environment. Chishui Danxia is located in typical plateau–gorge Danxia landform area and has a sharp morphological difference from that found in Southeastern and Northwestern China.

Chishui Danxia World Heritage Site, on the one hand, the palaeontological fossils held in extensively outcropping Jurassic and Cretaceous stratums, truly records the lives which used to live here. On the other hand, large area of relic *Alsophila spimdosa* in the Heritage Site, in the way of "living fossils," has demonstrated the stubbornness and continuity of life evolution in this area and provides valuable living samples to study the paleo-geographic environment.

(3) **Outstanding universal values of biology and ecology**

The Chishui Danxia Heritage Site shows us an important and ongoing process of biology and ecology. The natural succession of vegetation has reached stage of climax community and formed a primary central subtropical evergreen broad-leaved ecosystem. The evergreen broad-leaved forest with Fagaceae, Lauraceae, Theaceae, Magnoliaceae mainly developed in the unique plateau–gorge Danxia landform is a place of plentiful plant species and a well-preserved forest and is a typical development type and evolution process of the central subtropical Danxia ecosystem. Local native vegetation and unique landforms' system produced types of habitat, including rivers, valleys, lakes, mountains, swamps, caves, cliff, and so on. For long stable temperature and water condition, as well as fewer artificial interference, the largest area and population quantity of tree fern in Asia were preserved well in the Heritage Site.

There are multiple vegetation types in the Heritage Site such as evergreen broad-leaved forest, aestidurilignosa, mixed broad-leaved conifer forest, conifer forest, bamboo, scrub that constitute the most well preserved and typical central subtropical forest system in Danxia area and occupy an important berth in the world Danxia vegetation with its unique feature. It has been an ideal and nature proving ground for studying "background nature" of Danxia forest and structure, function, and balance of forest ecosystem in Danxia landform. A warm moist and meta-acid ecosystem has been formed with long geological evolution for the subtropics moist monsoon climatic in the sites. As a result, the warm moist and meta-acid vegetation has constituted a typical and special ecosystem, fully showing a good ecological process in the area.

Chishui Danxia Heritage Site had high level of biodiversity with 2116 species of advanced plants, 1668 species of animal, of which 59 species of animals and plants are under national priority protection. The heritage area has provided refuge for many species of IUCN Red List and China Red List (totaling 483 species, 115 species of plants, 368 species of animals), endemic local species (27 species of Chishui local area, 25 species of fish in the headwaters of Yangtze River), and 71 species of CITES (35 species of plants, 36 species of animals). Moreover, the flora was with many ancient living fossils and old trees. It is clear that the Chishui Danxia World Natural Heritage Site has provided refuge and habitat for many rare, precious, and unique species.

12.2 Description of Chishui Danxia

12.2.1 Physical Geography

(1) **Geomorphology and relief**

The Heritage Site lies in the southern border part of Sichuan Basin, which is adjacent to Guizhou Plateau. Due to the rapid tectonic uplift, the Jurassic and Cretaceous red rocks outcropped extensively and was intensively eroded. The Danxia landform in the Heritage Site is being in its early young stage due to rejuvenation process since Quaternary. Generally speaking, the site is characterized by plateau–gorge and mountain–gorge Danxia landform with deep V-shaped gorges, lofty cliffs, brooks, and waterfalls well developed. The altitude of the highest point in the Heritage Site is 1730 m and the lowest only 240 m with several obvious summit surfaces of 1600–1700, 1400–1500, 1100–1200, 900–1000, and 300–500 m. The differences in lithology play active roles in the development of Danxia landform in the Heritage Site. The Cretaceous Jiading Series is mainly composed of thick and hard red rock comprising sandstone and siltstone with strong erosion resistance and lots of vertical joints. In outcrop areas of Jiading Series, breathtaking Danxia landforms are well developed, especially lofty red escarpment more than 500 m tall, deep V-shaped valley, and lots of amazing waterfalls. On the contrary, the Jurassic red beds are mainly composed of relatively soft mudstone and shale and thus are vulnerable to be eroded. As a result, the landforms in the outcrop areas of Jurassic red beds are mainly short hills less than 300 m tall with gentle slope.

(2) **Climate**

The Heritage Site is predominated by subtropical mountain monsoon climate with annual mean temperature about 18.1 °C. The monthly mean temperature in July can be up to 28 °C and the down to about 7.9 °C in January. The annual range of temperature change varies from 20.1 to 20.5 °C. As far as extreme temperature is concerned, the recorded lowest temperature in the Heritage Site ranges from −7.1 to −3.3 °C, and the recorded highest temperature in summer is generally from 35.7 to 40.5 °C. The climate change is relatively mild in valley areas, where the recorded lowest temperature in winter is generally above −4 °C. Due to the influence of altitude and landform, the annual mean temperature is relatively low in the core area of the

Heritage Site, generally ranging about 16–17 °C, and the recorded highest temperature is usually below 39 °C. By and large, owing to being of high altitude and covered by thick forest, harmful high temperature above 40 °C is rare in the Heritage Site. The annual precipitation that the Heritage Site receives generally ranges from 800 to 1700 mm. The annual rainfall mainly distributes in the period from April to October, in which the received precipitation generally accounts for more than 80% of annual precipitation. In four seasons, summer is especially rainy with 383–681 mm rainfall, about 40.12–44.7% of annual precipitation. Winter is relatively dry. The total precipitation received in the season is only about 58.3–166 mm, accounting for 6.5–10.5% of the annual precipitation. As far as wind is concerned in the Heritage Site, southeastern wind is prevailing in summer season, but northern wind in winter. The annual mean wind speed is about 1.6 m/s, but the extreme wind speed can be up to 27 m/s. The period from March to September is windy in a year with strong wind more than eight forces being frequent in July and August.

(3) **Hydrology and water resourcesc**

Chishui River, with a watershed of 20,440 km^2, belongs to Yangtze River system and is the largest stream passing through the Heritage Site. It is about 524 km long and drops about 1558 m from its headstream to outlet. Chishui River wanders through the middle part of the Heritage Site, and the passage of the river in Chishui is about 74 km long. Above Bing'an Town is the middle stream of Chishui River with a V-shaped valley and torrent flow, where the river bed is narrow and small-sized powerboat can go and come. From Bing'an to the outlet is its lower stream, where the river bed is wider and water flows slowly. The lower stream is about 20–200 m wide at standard water level, and it is available for ships with 100–300-ton load to pass easily.

(4) **Soil and creatures**

Juvenile soil growing on the purple sandstone and mudstone in the heritage area in Jurassic and Cretaceous periods. Much diluvium with chips were washed to the smooth terrain, which developed thick soil with neutral and acidity. Vegetation type was central subtropical evergreen broad-leaved forest, and the vegetation coverage was much more than 90%. The sites for World Natural Heritage listing have an extremely rich biodiversity that incorporates a total of 1964 species of tracheophyte. There are 20 protected plant species of national importance, and 27 endemic local species (i.e., endemic to Chishui Danxia) have been discovered. Complex landscapes' background, superior climate and hydrology characteristic, and various plant communities provide abundance environment for the animals. The nominate sites were rich in animal resources, including 404 species of vertebrate, 1264 species of insects, 39 protected animal species of national importance, 25 endemic fish species of the headwaters of the Yangtze River.

12.2.2 Geology and Geomorphology

12.2.2.1 Geological Background

(1) **Regional geology**

Tectonically, the Heritage Site belongs to the connected area shared by the two different units of the Yangtze Para-platform. To its southeast is the Bijie Tectonic Deformation Zone extending from northeast to southwest of Zunyi Fault Arc of North Guizhou Platform Uprise, and to its Northwest is the tectonic deformation area of Middle Sichuan Platform Subside. From the Late Triassic Era to the end of Cretaceous Era, the Middle Sichuan Platform Subside was occupied by Bashu Paleo-lake, as the results in the formation of red rocks holding coal seam thousands of meters thick, and separated from North

The lithological column in the Chishui Danxia heritage site

System	Series	Set	Symbol	Column	Thickness (m)	Lithological features
Quaternary			Q		0–11	Grey and purplish grey fine-grained sandy alluvial deposits, holding pebble layer at the bottom
Cretaceous	Jiading	Jiaguan	K_2j		505–922	Amaranth, brick red, grey purple thick fine-grained feldspathic quartz sandstone with layers of amaranth mudstone and conglomerate 2–3.5m thick at the bottom
Jurassic	Chongqing	Penglaizhen	J_3p		697–1195	Dark purple fine-grained sandstone, interbedding with dark purple and amaranth mudstone, grey green sandstone, holding copper about 5m thick at the bottom and amaranth sandstone and mudstone at the top
	Ziliujing	Suining	J_2sn		375–578	Dark purple sandy mudstone holding green grey sandstone, light grey fine-grained feldspathic quartz sandstone 10m thick at the bottom
		Shangshaximiao	J_2s		871–1230	Light grey, purple grey sandstone interbedding with dark purple mudstone, holding a brick red sandstone bed 12m thick at the upper part
		Xiashaximiao	J_2x		261–377	Amaranth mudstone holding light green grey and purple grey sandstone, dark grey fine-grained sandstone at the upper part
	Xiangxi	Ziliujing	$J_{1\text{-}2}zl$		290–476	Black shale and dark purple, amaranth mudstone holding green grey sandstone, fine-grained sandstone, grey marl holding black shale, dark purple mudstone holding green sandstone

Guizhou Platform Uprise along the southern edge of Cretaceous Jiading Series.

(2) **Stratums and lithology**

Stratums from Jurassic Era to Quaternary Era outcropping in the Heritage Site and its surrounding areas can be divided into six sets, all of which are sedimentary rocks. The Jurassic and Cretaceous stratums are mainly composed of red and amaranth sandstone, mudstone, and fine sandy shale deposited in rivers and lakes, and the Quaternary are mainly composed of alluvium and diluvium. The Jurassic stratum holds lots of fossils, such as ostracods, lamellibranch, conchostracans, fish, trionychid, and so forth. The fossils of dinosaur footprints were discovered and well protected in the Heritage Site's Cretaceous stratum. All these life records can help us to understand the Heritage Site's biological features in Mesozoic. The Jurassic stratum in the Heritage Site is mainly composed of meander alluvial sediments in dry and hot climate, part of which formed in lake and usually outcropped at the axis part of some short-axis anticlines. By and large, mudstone and shale are the main body of the Heritage Site's Jurassic stratum, which is relatively soft and vulnerable to be weathered and also is mainly responsible for the development of short and gentle red hills.

The Cretaceous Jiading Formation, due to experiencing dozens of intermittent sedimentary gyrations, is mainly composed of brick red sandstone holding amaranth mudstone. For the development of spectacular Danxia landforms in the Heritage Site, the extensive outcropping Cretaceous thick-bedded feldspathic quartz sandstone is the most important material foundation. The Cretaceous stratums are generally of gentle obliquity ranging from 5° to 15°. Since arkose and quartz sandstone are very hard and thus keep themselves away from being weathered and denudated rapidly. In the Heritage Site, erosive water sculpture out large area of ridge-shaped and table-shaped mountains which are usually surrounded by breathtakingly lofty red cliffs.

Fish fossils in Cretaceous (Jinshagou)

Conglomerate in Jiading series

Dinosaur footprint fossils in Cretaceous (Jinshagou)

Red rock in Jiading series

(3) **Geological structure**

Since Late Eocene, Sichuan Platform Subside experienced tectonic movements, which brought forth lots of folds and fractures, and led to a series of E–W and S–N tectonic systems in the Heritage Site, including Dabaitang syncline, Moziyan anticline, Tucheng nose-shaped anticline, Wanglongchang anticline, and Guandu anticline, so forth. Guandu anticline is the major anticline structure in the Heritage Site extending nearly from south to north into Xishui County, to the northeast of which is Yanziyan syncline. In the axis part of Guandu anticline outcrops sandstone of Penglaizhen set, and Cretaceous sandstone in its wing part with gentle inclination ranging from 4° to 6°. Yanziyan syncline is located in the northeast edge area of Chishui and extends by and large from NW to SE and stretches into Xishui County. Besides, Xiangbichang syncline is also one of the major synclines in the Heritage Site, which extends E–W and gets into Xishui basically along Bingan, Daqun, Jinsha, Heishenyan, Moziyan. The inclination of Xiangbichang syncline ranges generally from 3° to 10°, in the axis part of which outcrops thick purple fine-grained calcareous quartz sandstone of Jiaguan set.

Fold and fracture structures in the Heritage Site and its surrounding area are not so developed due to the influence of rigid Sichuan Basin. Be that as it may, vertical joint and various kinds of bedding structures develop well, as creates favorable conditions for the development of breathtaking lofty Danxia landforms in the Heritage Site. Besides, the cross bedding and muddy crack and also rippling structures are well developed, as typically reflects the sedimentary characteristics in rivers and lakes. The fracture structures mainly appear in the area to the west of

Brief description of Jiading Group in Xiaoqiao forest farm

Code and name			Lithological feature	Thickness (m)
Jiading Formation (894 m)	The second set (659 m)	17	Brick red medium-to-thick-grained feldspathic quartz sandstone (the ratio between feldspar and quartz is 1:7), holding argillaceous crumb and lentoid mudstone	>45
		16	Amaranth and brick red silt-grained and fine-grained feldspathic quartz sandstone holding rock debris and argillaceous siltstone	73
		15	Brick red thick-bedded and medium-thin-bedded fine-medium-grained chalybeate and calcareous (2:7) sandstone with ripple-shaped and ramous crack	85
		14	Brick red fine-grained feldspathic quartz (1:7) sandstone	256
		13	Brick red and bright red shale interbedding with fine-grained feldspathic quartz (1:4) sandstone	41
		12	Brick red and bright red block fine-grained calcareous feldspathic quartz sandstone	36
		11	Brick red and bright red silty and sandy mudstone interbedding with calcareous feldspathic quartz sandstone and siltite	41
		10	Brick red block fine–medium-grained feldspathic quartz sandstone	15
		9	Brick red silty and sandy mudstone	3
		8	Amaranth and brick red block fine-grained quartz sandstone holding calcareous debris	62
	The first set (235 m)	7	Dark amaranth medium-thick-bedded quartz sandstone holding calcareous debris and calcareous and silty mudstone, marl at the bottom	37
		6	Dark amaranth thick-bedded block calcareous and silty sandstone, argillaceous siltite, shale holding lenticle	103
		5	Amaranth fine-grained calcareous feldspathic quartz (1:8) sandstone holding silty mudstone	8
		4	Amaranth and gray-green thick-bedded calcareous block siltite holding calcareous and silty shale	56
		3	Dark amaranth calcareous siltite holding lentoid sandstone, with well-developed oblique beddings	12
		2	Dark amaranth calcareous mudstone and silty sandstone	13.7
		1	Dark amaranth thick-bedded and block medium–fine-grained sandstone, conglomerate 0.1–1.5 m thick at the bottom	5.3

Chishui River, most of them extend from E to W and from northeast to southwest. In Huangjin and Hongqi, a group of E–W and N–S faults come together and form lattice-shaped structure and have a good control of the development of valley and river.

Vertical joints (Wuzhufeng)

Cross bedding (Jinshagou)

Mud crack structure (Jinshagou)

Rippling structure (Wuzhufeng)

(4) **Geological evolution**

From Permian Era to Triassic Era, the Heritage Site and its surrounding areas had frequently experienced marine transgression and regression. However, this area had still been a relatively stable geosyncline almost without intense structures of fold and fracture. In the following tectonic movement, fold extensively happened at the Heritage Site and its surrounding area, which is actually a part of "Southeastern Sichuan folding Belt" also called "Chishui Folding Belt" in Guizhou.

After the marine regression at the end of Triassic Era, the above-mentioned Middle Sichuan Platform Subside, affected by the Indosinian Movement, became a huge inland lake basin. The Heritage Site also subsided consequently. From Jurassic Era to Cretaceous Era, Sichuan Basin had been covered by water and was so-called Sichuan Paleo-lake or Bashu Paleo-lake. In this long time, the climate was hot, as led to extensive oxidization of iron. As a result, the sediments at the bottom of the paleo-lake became red and further formed a large area of red beds. The Heritage Site lied at the southern edge of the lake basin, as still brought forth thick and extensive deposits of Jurassic and Cretaceous red beds, which are about 3000–4000 m thick. The Heritage Site and its surrounding area are Guizhou's largest sedimentary area in Middle and Late Mesozoic Era. Yanshan

Movement, happened at the end of Cretaceous Era, made Sichuan Paleo-lake uplift extensively and ended the Heritage Site's sedimentary history.

The Neotectonic Movement, beginning at Middle and Late Quaternary Era, caused the arc-shaped structures formed in Yanshanion Movement to be compressed intensely. The whole Heritage Site uplifted greatly with the uprising of Lushan Mountain Range and formed a red plateau which stood above Sichuan Basin. Due to the big difference in drop, rivers in the area strongly dissected the red beds and sculptured out lots of gulches. Later, with the long-time weathering, denudation, and collapse, the plateau–gorge Danxia landform was gradually formed and became an outstanding representative for young-staged Danxia landforms in China. Owing to relatively soft rock and limited uplift, to the northwest of the Heritage Site develops extensively red hills with gentle slope.

12.2.2.2 Danxia Landforms

(1) **Brief description of Danxia landform**

The typical Danxia landforms in the Heritage Site mainly distribute in the southern part of Chishui, and Chishui River divides its core area into two parts. The eastern part characterized by mountain–gorge landform deeply dissected by aggressive flowing water owns the largest and the most typical Danxia landform in the Heritage Site, covering an area of 17,222 ha, which is to the south of Hushi, to the east of Chishui River, and to the west of Xishui River. The west part lies to the west of Chishui River and to the east of Fengxi River and to the southwest of Chishui City, covering an area of 10,142 ha. The dissection in the west part is not as strong as that in the east part and is characterized by plateau–gorge. The deep gorges, lofty red escarpments, waterfalls, dense forests, and large area of bamboos get together and form unique Danxia landscape.

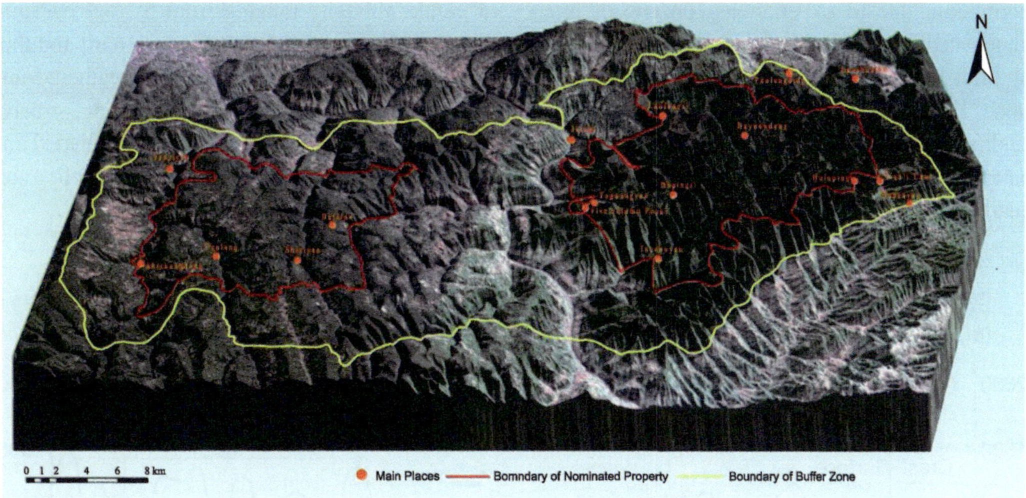

The 3D RS map of the Heritage Site

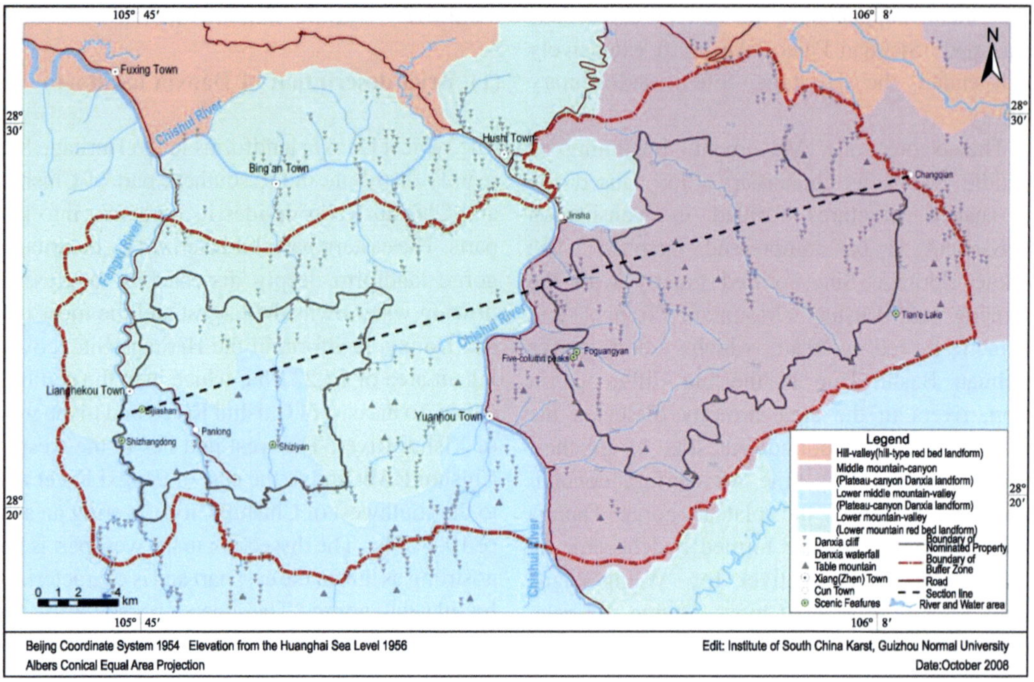

Classification in the Heritage Site and its surrounding areas

Geomorphological: The Cretaceous Jiading Formation, due to experiencing dozens of intermittent sedimentary gyrations, is mainly composed of brick red sandstone holding amaranth mudstone. For the development of spectacular Danxia landforms in the Heritage Site, the extensive outcropping Cretaceous thick-bedded feldspathic quartz sandstone is the most important material foundation. The Cretaceous stratums are generally of gentle obliquity ranging from 5° to 15°. Since arkose and quartz sandstone are very hard and thus keep themselves away from being weathered and denudated rapidly. In the Heritage Site, erosive water sculptures out large area of ridge-shaped and table-shaped mountains which are usually surrounded by breathtakingly lofty red cliffs.

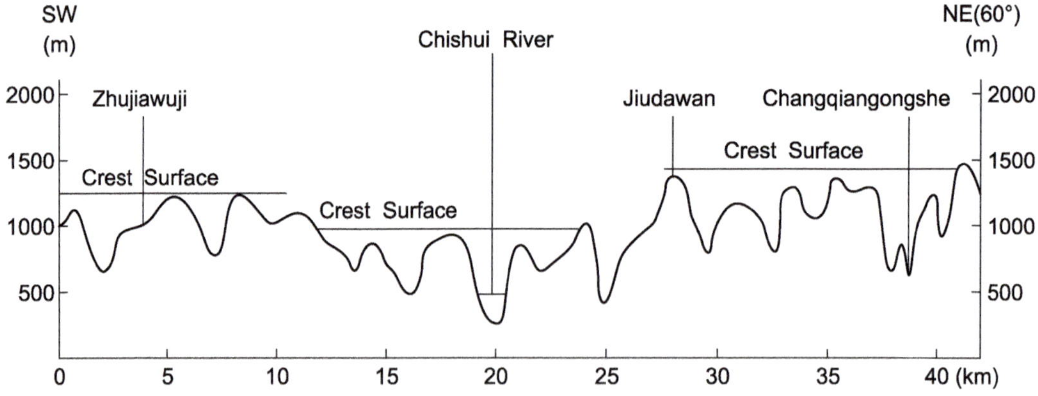

Cross section from Zhujiawuji to Changqiangongshe

(2) **The external forces for Danxia landform development**

- **Weathering**: Various kinds of physical, chemical, and biological weathering bring forth extensive and profound influences on the development of Danxia landform development. Firstly, owing to the differential weathering resulting from lithological difference between relatively soft Jurassic and hard Cretaceous, the two different geomorphologies of hill–gorge and low-middle mountain–gorge came into being. Secondly, some Danxia notches, Danxia natural bridges, and stone mushrooms develop on hard stratums holding soft rocks by differential weathering. On the west part of the Heritage Site, red beds have experienced long-timed physical, biological, and chemical weathering, as results in common development of spalling.

- **Erosion**: River system develops well in the Heritage Site due to lying in subtropical monsoon climate zone with rich precipitation and extensive outcropping of sandstone and shale with low penetrability. These rivers strongly dissect red beds and are the most important external forces for Danxia landform development in the Heritage Site, where river density is commonly beyond 0.77 km/km^2 with the extreme value can reach 1.37 km/km^2 and is in the top rank in Guizhou. The rivers in the Heritage Site are characterized by steep river bed with rapid flowing water cutting down the ground surface, which plays a very active role in the formation of deep gorges.

- **Gravity**: In the Heritage Site, the stratum usually has a gentle dip angle, steep-dipped or even vertical joints develop very well, and collapse process is prevailing, as is especially typical in Cretaceous stratum. Vertical joints are responsible for the development of some spectacular landforms, such as lofty red cliffs, deep valley, natural bridges, stone pillars. Collapsed rocks can be found commonly in river bed and fields.

Spalling in red beds (Malu)

Spalling in red beds (Jiujiaodong)

Biological weathering (Jiujiaodong)

Biological weathering (Wuzhufeng)

(3) **Classification nature for Danxia landform**

In the term of material base, the Danxia landform in the Heritage Site belongs to sandstone Danxia landform. Although the stratum outcropping in the Heritage Site includes sandstone, conglomerate, mudstone, muddy limestone, and even coal seam, sandstone distributes most extensively. Especially, the thick quartz sandstone of Jiaguan set of Cretaceous Jiading Series becomes the leading rocks of Danxia landform development in the Heritage Site. It is easy for steep and lofty cliffs to develop along vertical joints influenced by flowing water and frequent collapse.

From the angle of stratum dip, Chishui Danxia belongs to nearly horizontal Danxia. Except for the Jurassic stratum southeastern to the Heritage Site, the majority of Cretaceous stratum dips gently, usually no more than 10°. Summit surface nearly parallel with the bedding plane and Danxia table-shaped mountains develop in the Heritage Site.

As far as the climate condition and geomorphological forces are concerned, Chishui Danxia belongs respectively to humid Danxia and water erosion Danxia. Under the control of warm and humid subtropical mountainous monsoon climate, the Heritage Site is covered by dense vegetation. Flowing water is very active in shaping Danxia landform, resulting in the formation of group of potholes on river beds, and also amazing U-shaped red cliffs owing to headward erosion on straight cliffs.

In light of the development stage and geomorphological combination, Chishui Danxia belongs to the plateau–mountain, plateau–gorge landscape in the stage of early rejuvenated young stage. Similar as the edge areas of Guizhou Plateau, in the process of neotectonic movement, the peneplain formed in Tertiary has experienced strong uplifting and fluvial dissection. Therefore, the geomorphological development is typically characterized by rejuvenation. In the west part of the Heritage Site remains large area of plateau landform with relatively deep V-shaped valley extensively distributing, whose Danxia is in early rejuvenated early young stage and geomorphologically belongs to typical plateau–gorge landscape. The summit surface in the eastern Heritage Site well remains with an average altitude between 1400 and 1500 m. Table-shaped mountain with horizontal top can be found in some places. The gorges in the Heritage Site are usually steep shaped by Chishui River and its tributaries, as demonstrates typical characteristic of mountain plateau–gorge landscape in young stage.

(4) Major Danxia landform types in the Heritage Site

- **Positive Danxia Landforms**

Danxia cliff

Crest surface and huge Danxia cliffs in the east part of the Chishui site (Yuanhou)

Huge horseshoe-shaped Danxia cliff (Foguangyan)

Straight Danxia cliff (Shizhangdong)

Danxia Stone Wall

Danxia Stone Wall (Yuanhou)

Table Mountain

Table mountain (Foguangyan)

Danxia ridge

Danxia ridges (Yuanhou)

Danxia stone pillar

Danxia stone pillar

Three Buddhas (Bing'an)

Finger peaks (Wuzhufeng)

Collapsed rocks

Collapsed rocks

Collapsed rocks

Collapsed rocks

Collapsed rocks

Single stone

Danxia E.T (Yuanhou)

Fire balloon (Yuanhou)

Other unique Danxia landforms

Confess stone (Yuanhou)

Stone monkey (Wuzhufeng)

Stone mushroom (Lianghekou)

Danxia notch

Danxia notch (Yangjiayan)

Danxia notch (Xiaoyan)

Negative Danxia landforms

Danxia gorge

Deep Danxia gorge (Whuzhufeng)

Deep Danxia gorge

Danxia arc

Danxia arc (Shizhangdong)

Collapsed rock cave

Danxia honeycomb

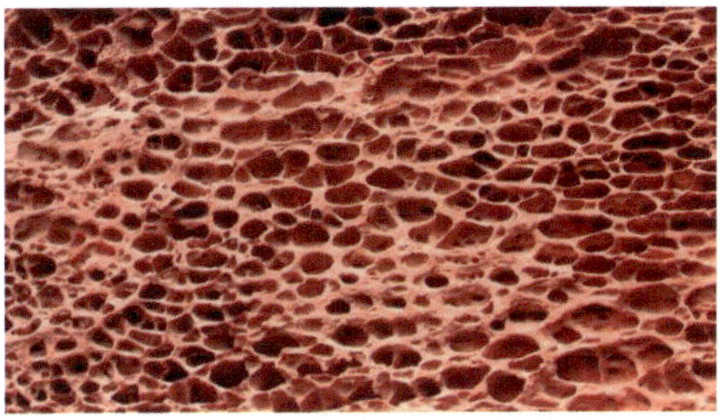

Danxia honeycomb (Xiaoyan)

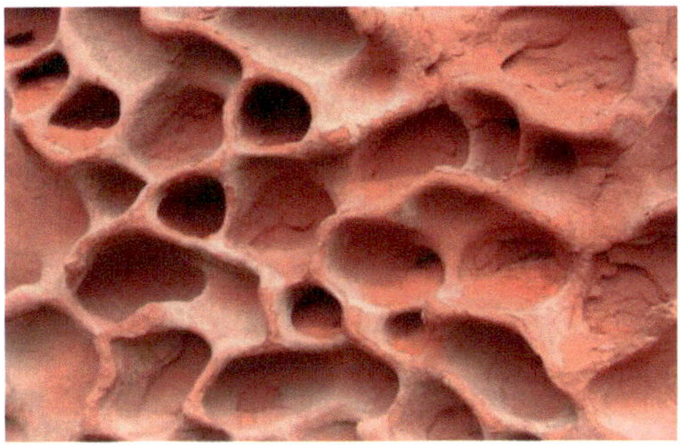

Danxia honeycomb (Xiaoyan)

Danxia pothole

String bead-shaped potholes

Danxia potholes (Shizhangdong)

Danxia potholes (Shiziyan)

Vertical notch

Vertical notch (Shizhangdong)

Danxia natural bridge

Danxia natural bridge (Bing'an)

Danxia natural bridge (Sidonggou)

12.2.3 Ecosystem and Biodiversity

12.2.3.1 Biogeographic Biota

According to the biogeographic biota system with Miklos D. F. Udvardy (1975), Chishui Danxia belongs to China subtropical forest province, Palearctic region. The site is located in subtropic moist monsoon climatic region, and the representative vegetation is central subtropic moist evergreen broad-leaved forest with characteristics of south subtropical transition. Tropical angiosperms make up 53.12% (284 genera) of all angiosperms in the Chishui Heritage Sites. The flora of spermatophyte is characteristics of tropics and subtropics.

The Chishui Danxia is situated in the core area of Rare Fish National Nature Reserve in the headwaters of Yangtze River, where a national reserve, a national forest park, and two national parks are situated. Typical evergreen broad-leaved forest spreads out above 700 m and forms climax community in the valley, while the forest including synusia with rainforest characters is dominant under 700 m. Evergreen and deciduous broad-leaved mixed forest, aestidurilignosa, and mixed broad-leaved conifer forest are distributed over the top of the mountain. In most part of the area, ravines cross each other and mountains are jagged; in the deep gorge, the vegetation spreads along the gill, undulating hill, and soil distribution. Different vegetation is distributed along the vertical dimension corresponding to different altitude and climate, but the continuity and integrity are still kept in the system.

There are 2116 species of higher plant in the heritage region, and the spermatophytes (1964 species) are dominant in front place among the central subtropical area both in the amount and type. Here are growing 25 endemic species and 21 endemic genera, such as *Gelidocalamus*, *Cyclocarya*, *Whytockia*, *Eomecon*, *Melliodendron, and Semiliquidambar*. At the same time, there are many monotype and little type genera spread in the area. For example, *Fokieniai*, *Sinomenium*, *Phrynna*, *Chaydaia*, *Toddalia*, *Thysanolaena, and Idesia* belong to monotype genus, and *Cunninghamia*, *Codonacamthus*, *Antenoron*, and so on belong to little type genus. The endemic local species named *Camellia luteoflora* with undifferentiation floral envelope are deemed to be a Primordial population in the Camellia phylogenetic development. From all the above, it could be concluded that the origin of the local flora was ancient and primitive. The flora analysis shows that this area is characterized by dominant woody plants, old origin, rich remaining plants, complicated distribution types, and many endemic species.

There are 1668 species of wild animal, including 287 species of terrestrial vertebrates. The dominant fauna found in the Heritage Sites are oriental and South China type, accounting for 39.02 and 23.34% of the terrestrial vertebrates, respectively, which exhibit a feature of the subtropical fauna.

Physiognomy of evergreen broad-leaved forest in Chishui Danxia

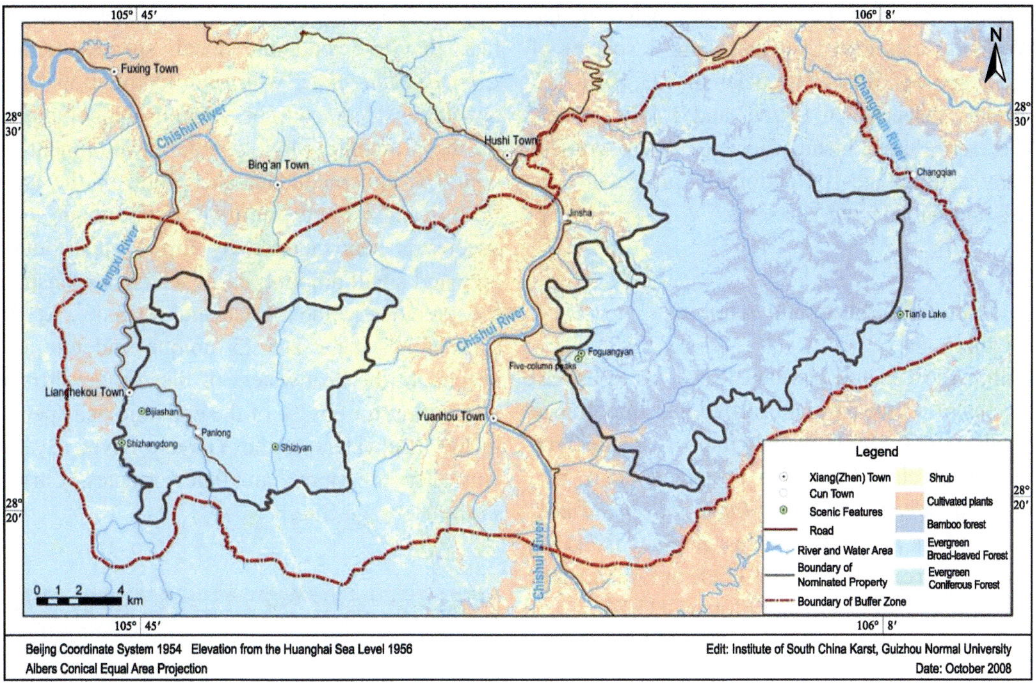

Vegetation of the Chishui Danxia Heritage Site

12.2.3.2 Species

(1) **Species of animals and plans**

The Chishui Danxia is located at subtropics moist monsoon climatic region with abundant heat and rain. Due to the elevation difference among many mountains in the area and the absolute altitude difference, there is a change of level of 1500 m across the site. The Heritage Site has become a natural refuge for the many animals and plants in geological environment evolution for the fine eco-environment and special geographical position. Resources of species are very abundant. There are approximately 1964 species of tracheophyte, including 249 species of fern, 13 species of gymnosperm, and 1702 species of angiosperm. There are 11 vegetation types, 91 phytoformations, and 117 associations in the core area.

The species of animal are about 1668, including 117 species of fish, 31 species of amphibia, 37 species of Reptilia, 147 species of bird, 72 species of mammals, and 1264 species of insect. The community types of animal include valley-reed-bamboo-valley flat fauna, evergreen broad-leaved forest fauna, miscellaneous bamboo fauna, shrub-grass fauna, farmland fauna, and cave fauna.

(2) **Red List**

Within the Heritage Sites, 115 species of flora are listed in IUCN Species Red List and China Species Red List, including 2 CR species, 14 EN species, 52 VU species, 18 NT/VU species, 27 LC species. The 88 seriously endangered plant species are characterized by obvious endemicity, such as *Taxus chinensis* var. *mairei, Fokienia hodginaii, Fokienia* hodginaii, *Dendrobium nobile*. Among these endangered species, 20 are listed in the World Species Red List, including 5 EN species, 3 VU species, 10 LR/NT species, 1 LR species, and 1 DD species.

There are also 368 species of fauna listed in *China Species Red List*, including 7 CR species, 12 EN species, 40 VU species, 20 NT/VU species, 11 NT species, and 278 LC species. Vertebrate accounts for 83.3% of the 90 threatened species. In these species, 25 are listed in the World Species Red List, including 1 species of CR, 5 species of EN, 9 species of VU, 8 LR/NT species, 1 species of LR/cd, and 1 DD species. Therefore, the Chishui Heritage Sites possess many assemblages of nationally protected species and endangered species.

(3) **Protected Plants and Animals**

Within the Chishui Heritage Sites, there are 20 species of protected plant of national importance (mother stock, 1999). Among them, there are three species of class I national protected plant, i.e., *Bretschneidera sinensis, Ginkgo biloba, Taxus mairei.* It is about 35 species of plant listed in Convention on International Trade in Endangered Species of Wild Fauna and Flora (CITES).

There are 39 protected species of animals of national importance (1989). Among them, 5 species belong to class I and 34 species belong to class II national protected animal. About 36 species of animal are listed in CITES of which ca. 9 and ca. 21 species are listed in Appendix I and in Appendix II, respectively, such as *Macaca mulatta, Macaca thibetana, Manis pentadactyla, Selenarctos thibetanus, Neofelis nebulosa, Panthera Pardus, Moschus berezovskii, Capricornis sumatraensis, Naemorhedus goral.* The largest population is *Macaca thibetana*, which is about 30 troops (700 individuals) in amount.

(4) **Endemic Local Plants**

To date, 27 endemic local species (i.e., endemic to Chishui Danxia) belonging to 17 genera and 9 families have been discovered, including 1 species of tree, 6 species of brush, 7 species of bamboo, 8 species of herb, and 3 species of macrofungi. The representative species are as follows: *Camellia luteoflora, Altingia multinervis, Huperzia chishuiensis, Indocalamus chishuiensis, Ampelocalamus scandeus, Impatiens chishuiensis, Camellia kweichouensis, Yushania chishuiensis, Fargesia rubignosa, Fargesia redicata, Dryopeteris hokouensis, Polystichum marginalisorum,* etc. Based on the fact that the endemic local species named *Camellia luteoflora* with undifferentiation floral envelope is deemed to be a Primordial population in the Camellia phylogenetic development, it could be concluded that the origin of the local flora is ancient and primitive. Each unique local species is characterized by limited distribution area and small amount, all of which demonstrate the precious value of Chishui Danxia. At the same time, there are 25 unique endemic fish species in the headwaters of the Yangtze River, accounting for 35.7% of the endemic fish species in the headwaters of the Yangtze River. By the way, there are more than 90 new species of insect in the Heritage Site.

Taxus chinensis var. *mairei.* (VU of China Species Red List. Class I National Protected)

Bretschneidera sinensis. (China Species Red List (VU); EN in World Species Red List. Class I National Protected)

Fokienia hodginaii. (LR/NT of China Species Red List. Class II National Protected)

Dendrobium wilsonii. (EN of China Species Red List)

(5) **Old trees**

The old trees in the Chishui Heritage Site are characterized by old, big, precious, and numerous. According to an initial count, there are 2208 plants of 27 species. Among them, about 241 plants are much older than 500 years, 517 plants are much older than 300–499 years and 1450 plants are much older than 100–299 years. Some national protected and China Red list species (e.g., 97 of *Manglietia insignis* and 41 plants of *Fokienia hodginaii*) are distributed in the area, demonstrating great scientific and conservation values. Among these old plants, Fagaceae and Lauraceae take a considerable proportion in amount.

Macaca mulatta. (China Species Red List (VU); LR/nt in World Species Red List. Class II national protected animal)

Fokienia hodginaii (footprint, photograph in 2007). (China species red list (EN), VU in World Species Red List. Class II national protected animal)

Endemic local species of Chishui Danxia: *Camellia luteoflora* and *Altingia multinervis*

Unique endemic balsam of Chishui Danxia: *Impatiens* sp. and *Impatiens spathuiata*

12.2.3.3 Ecological Processes

(1) **Community succession**

In large-scale distributed subtropical evergreen broad-leaved forest, plants of Lauraceae, Theaceae, Fagaceae, Magnolia, Hamamelidaceae, Symplocos are distributed in totally dominated place. Community succession may be classified into two sequences, i.e., primary succession and secondary succession.

Process of primary succession: Seasonal and sudden rainfall strongly erodes red rock constantly, and vertical erosion and source erosion of river contribute to the formation of steep cliff and deep valley. The participation of flow accelerates the process of weathering corrosion and collapsing process. All of the above-mentioned processes shape and reform current Danxia landforms constantly. Primary succession promotes a complete series of succession on time in different space. The communities developed by primary succession include early moss–herbaceous community, dwarf shrub community, and arbor community which are formed during mid- or late period.

Primary succession of plants on Danxia landforms in Chishui (Lichen invasion ⇒ Moss invasion ⇒ bush-grass stage ⇒ bush-arbor transition phase ⇒ Arbor stage ⇒ climax stage of evergreen broad-leaved forest)

Secondary succession process: Secondary succession series appear in valley and foothills. Evergreen broad-leaved forest is zonal climax vegetation which sustains through full succession. Succession series involve pioneer succession forest, transition succession forest, and basically stable succession forest. Secondary succession process includes several stages. In the situation that native evergreen broad-leaved forest has been felled or reclaimed by firing, the original vegetation and the species possible to evolve naturally are destroyed entirely, and then, the abandoned farmland will be occupied by different types of grassland. In the absence of interference by man, shrub will develop quickly and replace the grassland in three to five years. Along with the evolution of site condition and spread manners of trees, shrub forest may evolve successively from jungle stage (formed by invasion of quantities of trees, such as Lauraceous Da-Yeh Neolitsea forest) to subtropical coniferous forest stage (especially for Masson forest and fir forest), then to subtropical deciduous broad-leaved forest stage (e.g., Bestial forest, Quorums forest, Alums forest, and so on), and finally to evergreen broad-leaved forest. The time required for this evolution may be about 20–30 years. (Wang 1997, 1998, 2000, 2005).

(2) **Contact and evolution between present vegetation and ancient vegetation**

There is the largest area of Cyatheaceae in heritage region. Basing on the ancient geographical environment combining feature of pollen in this area, there are ancient vegetation of fern bushes with sparse vegetation, in the zone of 500–780 m above sea level, which mainly consist of fern, and Cyatheaceae exists there. The ancient vegetation of conifer-and-hardwood forest was dominated with woody plants. Groups of *Pinus, Podocarpus, Fagaceae Quercus* and beech located in the zone of 1330 m above sea level. Local paleoclimate changed in a tendency of warming and wetting, which can be proved by such a truth that pollen of Cyatheaceae tends to be increased with stratigraphic section. The side conclusion proved that the ancient geographical environment is similar to the present environment, and that Cyatheaceae has a long growing and developing history. Cyatheaceae community can be considered as a special layer of south subtropical rain forest, which reflects a feature of high temperature and humidity of habitat on the one hand, and reveals south subtropical property of the vegetation on the other hand. In heritage region, where developing stable communities for the special environment, well growing and developing condition of Cyatheaceae community and 'little human's strong interference'.

12.2.3.4 Ecological Characteristics

Chishui belongs to subtropical monsoon climate zone. In the region, mountains stagger with a huge difference on height, absolute value of elevation to 150 m. For the mountain with flat mountaintop, steep mountain-side, and gentle foothills, the temperature and water condition are poor on the mountaintop, which made subtropical evergreen hardwood forest and conifer-and-hardwood forest get developed. The communities mainly consist of *Fokienia hodginsii*, rhododendron, etc., and where exist a series difference of ecotypes, such as leaf area, leaves thickness, mechanical organization, and the height below branch and so on. At the foot of a mountain (gully), for well temperature and water condition, types of vegetation with the characteristic of quarter rainforest or rainforest got developed. The region has become the largest distribution area of *cyatheaceae* in Asia and the vegetation concomitance with *musa, alocasia, Angiopteris fokiensis*, and so on. Shrub on flood land, etc., got developed along the bank of a river; vegetations of evergreen broad-leaved forest with Lauraceae, Theaceae, Fagaceae, Magnolia, Symplocos, etc., developmented in the middle of the mountain. In addition, the well vegetation system maintained the wetland ecosystems, such as perched swampland and so on. The diversity of the environment supported local abundant biodiversity and triggered various characteristics on ecological adaptability, including Epipetreous, hygric, adnascent and parasitic appearances. On the special spongeous trunks of the tree fern were attached on many small-sized plant.

Generally, uniqueness and old originality were the key features of the ecosystems. The proper type of landscape ecology consists of subtropical ecosystems and Danxia landscape, formed the habitat for the species for the unique tropic plateau–canyon Danxia landscape, after long evolution and developing. For well temperature and water condition, as well as the isolation of high mountain in a long time, it becomes possible that triggering the formation of unique species, the unique species there reaches 27, for example, *Camellia luteoflora* and *Altingia multinervis*, typically. Which place also has become a refuge of many relict plants. There are not only large quantity of relict plants of *Jurassic cyatheaceae*, but some ancient relict plants like *Amentotaxus* argotaenia, *Cephalotaxus* from tertiary. The natural landscape and ecosystem still keep primeval nature for the big depopulated zone there,.

Hard evergreen forest landscape on mountaintop in Danxia landform area

Evergreen broad-leaved forest landscape in Danxia landform area

Rain forest valley landscape in Danxia landform area in Chishui

Perched swampland in the upper part of a mountain

Climbing of *Mucuna sempervirens* in forest

Epipetreous, hygric, and parasitic appearances of plants in Chishui Danxia

Epiphytes and semiparasitic plants on tree fern

12.2.4 Natural Landscapes and Beauty

Chishui danxia belongs to the subtropical humid zone's and plateau–valley type's danxia, developing a lot of various forms, such as danxia cliffs, table mountains, danxia columns, canyons, dutches, caves, and so on. Thanks to the superior nature environment and less human activities, it has developed and kept several integrated ecosystems with high percent forest coverage and a variety of species, such as *A. spinulosa* Reserve, Bamboo National Forest Park, Subtropical Evergreen–broad leaf Forest Nature Reserve; its main characteristics are high altitude and deep valleys and flying waterfalls hanging on cliffs, with red and bright water flowing, lively year-round evergreen bamboos and woods; all in all, the main landscapes are red mountains, blue waters, and green forests and hanging waterfalls. By complete types, unique and typical shapes and combination of static and dynamic elements, it shows extraordinary natural phenomena and beauty.

No wonder this area is named Chishui

Water flowing along red rock is red

Beautiful Danxia and its waterfall (the big Shizhangdong waterfall)

12.3 History and Development

12.3.1 Evolutionary Stages of Danxia Landform

Based on the regional geological evolution history and geomorphological features in the Heritage Site, the evolutionary process of Chishui Danxia landform can be divided into four stages, i.e., paleo-lake sedimentary stage, uplifting and denudation stage, Young Danxia stage I, and Young Danxia stage II.

(1) **Paleo-lake sedimentary stage**

The Indosinian Movement in the Middle and Late Trias resulted in the tectonic subsidence of Sichuan Basin and the formation of Bashu Paleo-lake. The lake did not disappear until the end of Cretaceous Era. As a result, extremely thick red beds deposited with a thickness of thousands of meters. Although there was no Danxia landform developed in this stage, Jurassic and Cretaceous thick red beds containing amaranth and brick red sandstone and mudstone and shale provide a favorable material foundation for later development of Danxia landform in the Heritage Site.

(2) **Uplifting and denudation stage**

Sichuan Basin finished its inland lake sedimentary stage due to Yanshan Movement beginning at the end of Cretaceous Era, and the thick red beds, deposited at the bottom of the Bashu Paleo-lake, was exposed to air. Chishui River emerged, and red bed landform began to develop. In Tertiary Era, the uplifted red beds and the adjacent Northern Guizhou Platform Uprise had experienced a long-time denudation and resulted in a peneplain's coming into being with smooth surface. The relief in the Heritage Site tended to incline from southeast to northwest.

Honeycomb holes (Yangjiayan): They look like honeycombs, but different from each other in size and shape, showing a kind of rhythm and innervation

(3) **Young Danxia stage I**

Influenced by the neotectonic movement in Quaternary Era, Guizhou Plateau uprose greatly and Sichuan Basin was extruded intensely, which resulted in the rejuvenation development of regional geomorphology in the Heritage Site. However, Sichuan Basin was lifted a little owing to its strong rigidity, where folds were not developed and stratum was by and large gently dipped in obliquity. The red beds in the Heritage Site, lying between Guizhou Plateau and Sichuan Basin, went up greatly owing to the rapid tectonic uplifting. As a result, the difference in relief between the two geomorphological units was further sharpened. Chishui River System continues to develop on the base of its previous stage and strongly eroded its river beds and began to bring forth shallow V-shaped valleys along large vertical joints of hard and thick-bedded sandstone in the Cretaceous Jiading Formation. The paleo-peneplain formed in Tertiary Era began to be dissected, and the whole geomorphological development came into its rejuvenation stage. Since it was not a long time, Chishui River and its tributaries dissected slightly the plateau surface and produced some shallow valleys, and large area of plateau surface remained extensively and table mountains were well developed in the Heritage Site. At the edges of remain, some lofty red cliffs came into being due to frequent collapse along vertical joints.

(4) **Young Danxia stage II**

With the constant dissection of Chishui River and its tributaries, the features of downward geomorphological evolution are more obvious. In the eastern Heritage Site, table mountain between two neighboring gorges was laterally eroded and gradually became narrow ridge-shaped and only limited small mesas remaining. Danxia landform in the Heritage Site entered its young development stage. In the west part of the Heritage Site, the extensive and relatively smooth plateau is remained. In its east part, the paleo-peneplain can be discernible basically according to the remained crest surface and V-shaped gorges increased greatly and became much deeper than its former stage with a maximal depth more than 500 m. Danxia cliffs, as one of the most typical Danxia landforms in the Heritage Site, have considerably increased and become much taller compared with the previous stage. In this stage, the plateau–gorge Danxia landscape is in its typical period. Meanwhile, due to the uncoordinated downward incision of mainstream and its tributaries and the differential erosion resulting from lithological difference, there are lots of waterfalls with different styles at the converging points of mainstream and their tributaries.

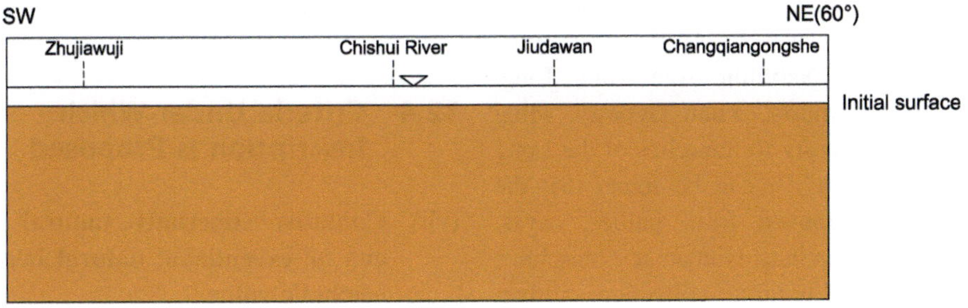

Paleo-lake sedimentary stage

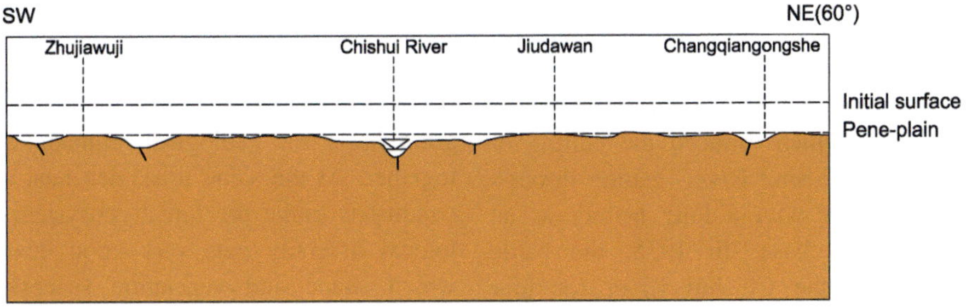

Uplift and denudation stage

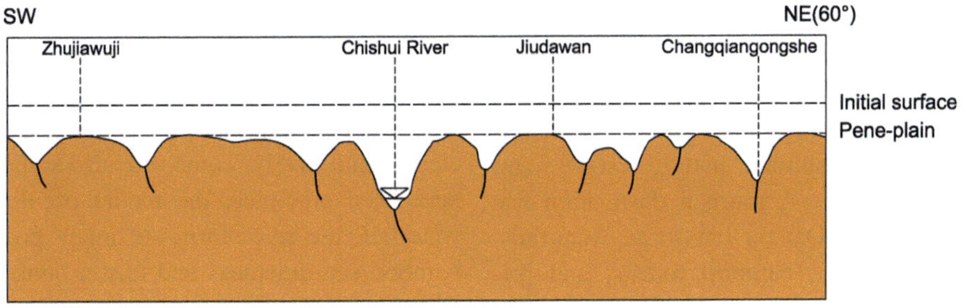

Young Danxia stage I

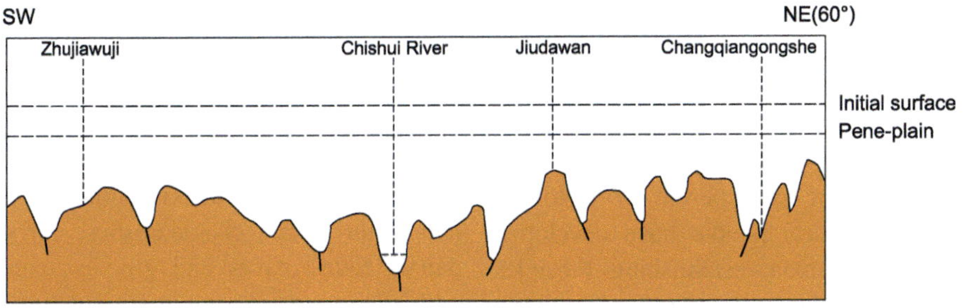

Young Danxia stage II

12.3.2 History and Development

There is a long history of human activities in the Chishui Danxia World Natural Heritage Site. Abundant relics of early human occupation were left, spanning Neolithic Age, Ceramic Age, Porcelain Age, Bronze Age, and Iron Age to the present. From the Neolithic Age, Han, Tang, Northern Song Dynasty, Yuan Dynasty, Ming Dynasty, Qing Dynasty to the relics of the Long March of Red Army, it can be found that the human activities moved from gullies, caves, riverbanks to Danxia flats. Human activities have included periods of hunting, collection, cultivation, settled agriculture, using-reamer and-fire agriculture, logging, tourism and scenic protection and tourism. Fuels for cooking and warming in modern times were firewood and charcoal from beeches and shaw. But after last mid-century, coal, natural gas, and electricity became popular which reduced the cutting of forest resources. Chishui River, getting through the city, is a river with a long history as an important military base. In 1935, the battle named "crossing the Chishui River for four times," directed by Chairman Mao Zedong, changed the situation of Chinese revolution history and recorded a fantastic chapter of military history both at home and abroad. In 1980s, laws, regulations, and systems were constituted and boundaries of the Danxia landscape protection area were defined. Chishui Danxia was ratified as a national park, national nature reserve, and national forest park. A tourism destination has been formed with Danxia landform, waterfalls, *Alsophila spinuloso*, bamboo forests, and primary forests as its main characters, accompanying with historic culture of long march, as well as some relics. Tourism became the pillar industry and the local government, communities, enterprises, and social groups developed some projects, such as planting trees, maintaining the eco-environment to protect the Danxia landscapes and the biodiversity.

During indeed a long period of historical involvement, the local people have developed close ties with Danxia, establishing lifestyles adapted to the Danxia environment and a culture relating intimately with the Danxia landscapes. Danxia has entered every aspect of the local peoples' life, including religion, celebrations, literature, painting, music and dance, and architecture, the form and connotation of which are highly harmonious with the Danxia environment.

12.4 Criteria Under Which Inscription Is Proposed

(vii) **Contains "superlative natural phenomena or exceptional natural beauty and aesthetic value."**

Chishui Danxia landscape shows typical and full individual forms such as red cliffs, caves, giant stones, the nature bridges, the narrow ridges, the danxia forts, danxia walls, danxia columns, and so on, because of the development with very good basis for geological, climatic conditions together. At the same time, rich heat and water conditions make the forest ecosystem and biological diversity gets very good development, which shape and vegetation succession have typical characteristic of humid plateau–valley Danxia landform, given the scenic thick aesthetic significance. Danxia mountain and blue water as organic group of emerald green forest congregate together, can be termed like this, landscape forest all has. Chinese landscape appreciation of the beauty experience such as get cloud and mist but elegant and lovely, come true also sufficiently in here. Red precipice, the model peculiarity, the waterfall, the river form are many posture, the bamboo sea quietness and turn a book over, all fresh work. Be grand and elegant and pretty, dynamic with static state, the best union and behavior of splendid and completely fresh, etc., natural color scenery form, and such aesthetic elements are so in accord with traditional Chinese landscape aesthetic attitude.

Chishui Danxia landscape, the characteristics of the shape diversity, uniqueness and harmony, the main form like danxia cliffs, danxia pillars and other detail characteristics, for example, danxia round caves and grooves, combination with rich vegetation types, water and color, meets the highest standards of Western Expert Paradigm concerning scenic quality assessment. In the nomination site, cultural basis is extraordinary and rich, including that the production factory of world famous China Wine is located here and Chishui River is called as the Fine Wine River; Bin'an Ancient Town on the Chishui River is a military fortress and the famous Red Army four times crossing Chishui beginning here, which exhibit shine each other with the beautiful danxia landscape. In a word, they have shown outstanding aesthetic value and unique cultural heritage.

(viii) **To be outstanding examples representing major stages of earth's history, including the record of life**...

Due to its unique geological evolution history and geomorphological features, Chishui Danxia landform is an outstanding example reflecting main stages of earth's evolution history, including significant geomorphological process and active geological forces, significant life records, unique geomorphologies, and integrated physical geographic landscape.

Chishui Danxia landform is an outstanding example of Danxia geomorphological evolution in subtropical huge red bed basin influenced by neotectonic movement and is also an outstanding example of earth history since the Mesozoic: Himalayas movement is the most significant tectonic movement since Cenozoic brings extensive and far-reaching influence on the current global geomorphological pattern and therefore is one of the important stages in earth history. Since the Mesozoic Era, the Heritage Site and its surrounding areas have experienced complex and significant geological and geomorphological change. The rapid uplifting of Tibetan Plateau in the neotectonic movement promoted the slantwise uplifting in its surrounding areas, and Guizhou Plateau and Sichuan Basin went up with different extent. Influenced by subtropical monsoon climate for a long time, a number of breathtaking landscapes came into being in the Heritage Site and its surrounding areas. Chishui Danxia is just the unique product of the above-mentioned earth history.

Sichuan Basin is the largest red bed basin with a area of 154,185 km^2 in China (Tong 2000; Zeng et al. 1978). Its red bed is about 2000–6000 m thick and deposited in Jurassic and Cretaceous periods. At the end of Cretaceous Period, the lake-facies sedimentary history in the Heritage Site was ended with the overall uplifting of Sichuan Basin and the disappearing of Bashu Paleo-lake (Liu et al. 2005). In Tertiary Period, a tectonically stable period, red beds in the heritage area began to be denudated and a relatively smooth peneplain came into being. In Quaternary Period, the Heritage Site was uplifted greatly and stood more than 1000 m above sea level due to the strong uprising of Tibetan Plateau and Guizhou Plateau. The sandstone of Cretaceous Jiading Formation is rigid, and thus, folds were not so developed. Gently dipped stratum is favorable for the development of vertical joints. Predominated by the wet and hot subtropical climate, these hard red rocks are strongly eroded, and the paleo-peneplain began to be decomposed rapidly. In the east part of the Heritage Site, table mountains are gradually replaced by ridge-shaped mountains and in its west part remains extensive plateau landforms. Red cliffs develop well, where there usually hang Danxia waterfalls. By and large, each of landscape components here makes the Heritage Site had a sharp contrast Danxia landscape with those in southeastern Chinese hilly areas. Therefore, it can be concluded that the Danxia landform in the Heritage Site is outstanding in its size, morphological features, and also its evolution process.

The plateau–gorge Danxia landform in the Heritage Site, which is being in its early young stage of rejuvenation, has rarely seen geomorphological features. According to the Danxia landform classification system put forward by Peng, Chishui Danxia landform is now in its early young stage. Due to thick and hard red beds, tectonic movement, and favorable climate conditions, Danxia landform of plateau–gorge (in western part) and mountain–plateau–gorge (in eastern part) develops perfectly in the Heritage

Site and contrasts with that in southeastern Chinese hilly areas and semiarid and arid areas of northwestern China. Although there are no Danform Fenglin and large lakes in the Heritage Site, lofty red Danxia cliffs and lots of waterfalls demonstrate another style of Danxia landform, which is well complementary with the Danxia landscape in rest China and makes the Danxia landform of China more typical and more integral.

The rich paleontological fossils and relic plants found in the heritage area are the significant life records of earth. The outcropping stratum is mainly Jurassic and Cretaceous and holds rich paleontological fossils including fishes, which are the true record of the lives ever living in the southern edge of Bashu Paleo-lake from Jurassic Period through Cretaceous Period. Especially, there remains a large area of Cyatheaceae in the Heritage Site, which used to coexist with dinosaur and thus enjoy the reputation of "living fossil." This kind of plant is world widely rarely seen considering its large area, strong primitivity, and good protection. The extensive existence of Cyatheaceae in the heritage area shows people an important aspect of life evolution on earth and enjoys a high scientific value.

The Heritage Site has rarely seen integrated physical geographical features. Chishui Danxia landform is a masterwork by different kinds of external and internal geomorphological forces and also is a valuable heritage that nature leaves to humankind. Compared with other Danxia landform area, the Heritage Site is no doubt a gold basin overlapped with Chishui National Park, Bamboo National Forest Park, Jinshagou State Nature Reserve for, and Xishui State Nature Reserve for Subtropical Evergreen Broad-Leaved Forest, as makes the Heritage Site own an uncomparable physical geographic connotation. In the Heritage Site, there is not only typical plateau–gorge Danxia landform, but also strongly primitive and well-protected subtropical evergreen broad-leaved forest, large area of Cretaceous relic plant, *A. spinulosa*, and bamboo forest. In those, intact canyons lie lots of waterfalls, some of which are spectacular, such as Shizhangdong waterfall and Foguangyan waterfall mentioned above. In one word, the perfect combination of Danxia landforms, waterfalls, cyatheaceaes, and bamboo forest in the Heritage Site demonstrate the unusual scientific and esthetic value.

(ix) **To be outstanding examples representing significant ongoing ecological and biological processes in the evolution and development of terrestrial, freshwater, coastal …**

Chishui Danxia area is a typical field which went through a series of tectonic movement—deposition of ancient lake, crustal uplift and drying up, folding movement. From conifer-and-hardwood forest recorded by ancient pollen to present evergreen broad-leaved forest in which ancient vegetation and modern vegetation stemming from many species coexist, all of then suggest that the heritage area where own typicality of ecological succession and integrity of ecosystem, which are rare at the same latitude area. Excepting the evolution of from ancient vegetation to modern vegetation and primary succession of bare rock, in the terms of relative close landform and humid and hot climate, integrity, typical, broad evergreen broad-leaved forest ecosystem got developed in heritage area, in which the vegetation with strong originality, biota producing special differentiation of symmetry and asymmetry stemming from the difference of topography, climate, hydrology, and soil. For the mountain of the vertical difference in elevation or drop to 1500 m, approximately below 700 m is the zone of evergreen broad-leaved forest with quarter rainforest; between 700 and 1000 m is the zone of evergreen broad-leaved forest; above 1000 m is conifer-and-hardwood mixed forest. And the vegetation also produced some horizontal differentiation along valley, mountain, such differentiation which resulted from local habitat differentiation can be seen frequently.

(x) **To contain the most important and significant natural habitats for in situ conservation of biological diversity** …

The heritage area owns below characteristic: with large area, high vegetation coverage—reaching 90%, with two big ecosystems of land and water, with strong heterogeneity for the habitat, with small human interference in most area, with well environment, the ecosystem still keeping their originality, adaptability and integrality, with higher biodiversity, comparing with adjacent region and other areas in China with Danxia landforms of ecosystem.

It was known that in Heritage Danxia landform area in Chishui, there has been 1964 kinds of vascular plants and 20 kinds of which have been listed in IUCN Red List, as well as 115 kinds of which has been listed in Species Red List of China. There are 21 kinds of bamboo here, 12 kinds of which are wild bamboo. Ancient vegetation in which cyatheaceae is a representation is abundant, and with such below features, with more single categories and rare categories, with more variations, in addition, with more rare, specific categories of plant partly of which are abound in number, especially for Camellia luteoflora which to be known as "panda in flowers," it has been regarded as a category one in a thousand for genetics and breeding research and ornamenting a garden, for its unique status in evolution and its ornamentality. The number of ancient trees with 100 years is 2208, which reflects the originality of the vegetation. There are 1668 kinds of animals, 370 kinds of which are endangered animals, and 25 kinds of which has been listed in IUCN Red List, as well as 368 kinds of which has been listed in Species Red List of China; what's more, the species of big animals in protection are abundant, with more than 90 kinds of new insects and 25 kinds of specific fish in the upper reaches of the Yangtze River.

All above, Danxia area in Chishui is an important source and habitat of rare species, which exist in subtropical regions in China, and is an ideal base for protecting wildlife in situ and carrying out comprehensive research.

12.5 Protection and Management

The Heritage Site of Chishui Danxia has enjoyed protection for a long time. Protection methods have changed from civilian self-protection to legal protection, and the level of protection has been upgraded; the means of protection have been enhanced and the responsible agencies had been gradually improved.

The Chishui Danxia Heritage Site is endowed with different protective designations including national park, national nature reserve, and national forest park. Therefore, it is protected by national laws such as *the Constitution of the People's Republic of China, Law of the People's Republic of China on Forest, Environmental Protection Law of the People's Republic of China, Water Resource Protection Law of the People's Republic of China, Law of People's Republic of China on Protection of Wildlife, Regulations of the People's Republic of China on Nature Reserve, Regulations for Wild Plants Protection of the People's Republic of China, Rules for Implication of the Law of the People's Republic of China on Forest, Regulations of the People's Republic of China Concerning the Places of National Park.* At present, Chishui Danxia Heritage Site is well protected.

Efficient administering systems of different levels have been built in Chishui Danxia Heritage Site and regarded by government, cooperated by departments, sustained by society. They protect and manage the environment in a unified and harmonious way. Organizations for environmental protection are established in concerned departments and enterprises of resources' management departments. The pollutions were controlled, the environment was protected, and atmosphere, water, soil, noise, and natural environment are in very high quality. Especially in the Heritage Site, a majority of area has few signs of human habitation, and the environment keeps aboriginality still.

The Chishui Danxia Heritage Site has been compartmentalized definitely, and the monitoring indicators have been built accordingly with timing monitoring. If problems have been discovered, it will be solved in time. There are enough staff and finance in the areas.

Outstanding universal values of Chishui Danxia World Natural Heritage Site, including Danxia landforms, aesthetic importance, ecosystems, endangered species, and habitats, are well kept. The integrality of trend of species, ecosystem, and natural environments are undamaged.

Chishui Heritage Site is influenced to varying degrees by natural factors and human activities. Natural factors include landslides and mud–rock flow, forest fires, forest diseases, and insect pests, etc. Human activities' impacts include cultivation, deforestation, and building projects in the Heritage Site. But relative administrations have taken effective measures to counteract impacts, including establishing fire prevention and warning system and a forest fire fighting team covering government–community–corporation–resident–scenic areas. For example, the People's Government of Chishui City adopted measures of positive preventing and extinguishing forest fires in time. In August 2005, the people established and promulgated *Contingency Plan on Forest Fire in Chishui City*, and Command Post was set up. Signs promoting fire prevention were put up, and villagers were educated to enhance fire prevention consciousness. On grassy slopes and forests with high fire danger, fire breaks were built and prohibit the lighting of fires in the open; in periods of high fire danger, the resulting effects are beneficial. Thus, the outstanding universal values of Chishui Danxia Heritage Site are not influenced radically.

13 Taining, Fujian

13.1 Introduction to the Heritage

13.1.1 Summary

Name of Property: Taining.
Country (City), Province: Taining County, Sanming City, Fujian Province.
Geographical coordinate (center point):
North part: 27°00′37″N, 117°13′07″E.
South part: 26°51′56″N, 117°02′22″E.
Area of the Property:
Heritage Site (ha): 11,087.
Buffer Zone (ha): 12,401.

13.1.2 Border Description of the Heritage Site

Taining Danxia Heritage Site includes typical Danxia landform area and natural geographic elements in Taining County.

Taining Heritage Site not only appears on the map, but also has its borderline indicating the heritage site and buffer zone on the spot. Taining heritage area is divided into South subarea and North subarea, including Taining national scenic spots, Shangqing River Scenic Spot, Golden Lake Scenic Spot, and Taining Cat Mountain (Mao'er Shan) National Forest Park. The northwestern borderline of north subarea is the faulted ravine, where the red bed contacts the base strata, the northern borderline is the borderline of Taining and Shaowu Counties, the eastern borderline is the ravine, while the southern borderline is the steep cliffs constituted by red sandy conglomerate; the northwestern borderline of the south subarea is the faulted ravine, where the red bed contacts the base strata, the northeastern borderline is the steep cliffs constituted by red sandy conglomerate, the southeastern borderline is the water surface of Golden Lake, while the western borderline is the ravine and mountain ridge. The Heritage site well preserves the integrity of Danxia distributed area, forest ecosystem and the habitats of rare and endangered animals. The buffer zone borderlines were taken using natural borderlines, mountain ridges, valley, river, road, etc. The borderlines were taken after considering the elements benefiting the protection of Heritage site, for example, trying to avoid the impact of human activities, etc.

13.1.3 Description on Distinct Values

Taining Heritage site is the outstanding representative of juvenile Danxia landforms in the southeastern China. It includes complete Danxia landforms formed after the Cretaceous red beds experienced multiple uplifts, were denudated and leveled and then were cut and eroded by flowing water since the Tertiary Period and it is still developing.

Taining Danxia landform area is the typical region of Mesozoic active continental marginal

H. Peng, *China Danxia*, Springer Geography,
https://doi.org/10.1007/978-981-13-5959-0_13

zone to West Pacific Ocean, recording and revealing the formation, development, and evolution of Mesozoic active continental marginal zone to West Pacific Ocean in the eastern China and the climate change processes of southeastern China since the Cretaceous. The formation of split-sunken red basins and the formation and evolution of Danxia landform are the typical representatives.

The quite complete distinct planation surfaces, huge gorgeous deep-cutting canyon clusters, and the exceedingly well-developed Danxia caves are the major features of Taining Danxia landform.

The Heritage site has exceedingly well-developed fissure valleys, lane valleys, canyons, and red cliffs, together with the unique canyon network formed by more than 400 deep-cutting canyons as well as red mountains. The density of canyon, the curvature of incised meander, and the primitiveness of canyon ecosystem are special in Taining, rarely seen in other China Danxia landform areas. The canyons are characteristic of the deepness of canyon, height of red cliffs, various caves, and natural ecology.

The large number of rock groove and cave, the large scale of cave, the shape, the combination, and admiration of cave are unique in Taining, no one like them can be found in other area as well as Taining Danxia cave culture has a long history.

Taining Danxia has a rarely seen natural beauty, mainly represented by the deep-cutting canyons, exotic caves, beautiful mountains and water landscape and the primitiveness of ecology.

Taining Danxia area has perfect biological and ecological diversity, preserving the unique dynamics alternation process of biological communities; in addition, it is the habitat for many rare and endangered fauna and flora species.

Taining Danxia area has 1276 species of vascular bundle plant, 380 species of vertebrate, and 1509 species of invertebrate of insects (including Acari of Arachnida).

The Danxia Heritage site has many diverse ecosystems. It has 8 vegetation types, 24 formations, 40 associations. The vegetation cover is up to 78.3%. The site is the representative of middle subtropical forest area.

The Heritage site has 3 plants under state first-grade protection, 11 plants under state second-grade protection, 10 plants listed in IUCN Red List, 65 listed in CITES Appendices, 77 plants listed in China Species Red List. It has 2 animals under state first-grade protection, 35 animals under state second-grade protection, 7 animals listed in IUCN Red List, 47 animals listed in CITES *Appendices* indicated no international trading allowed of these animals, 43 animals listed in *China Species Red List*. The Heritage site has become shelter for many rare raptors due to the steepness of mountain and a large number of caves.

Some rare plant, regionally special plants and arethusa are under protection in the Heritage site.

13.2 Description of Taining Danxia

13.2.1 Physical Geography

(1) **Geology and Landform**

Taining basin including two small red basins, Zhukou and Meikou trending in NE direction, is the Cretaceous faulted basin developed in Wuyi Mountain rise of the Cathaysia Land. The rocks composing the landform are conglomerates and sandy conglomerate of the Middle–Late Cretaceous Chong'an Formation. The faults, striking NE, NNE, NW, and S-N, are well developed in the area. Taining lies in the southeast side of the middle Wuyi Mountain Range, and the general topography inclines southeastwards and is higher in the west and north, gentle in the southeast and lower in the center. The Jizi top is the highest one in the area, with an elevation of 674 m. The maximum elevation difference in the area can reach to 400 m.

(2) **Climate**

Taining belongs to the middle subtropical monsoon humid climate. It is mild and humid with distinct four seasons without extremely hot in summer or cold in winter. The average annual temperature is 17.1 °C. January temperature is averaged at 5.9 °C, and the extremely lowest temperature can be −10.6 °C; July temperature can reach 33.7 °C, and the extremely highest temperature can be 38.9 °C. Taining enjoys abundant rainfall with high atmospheric humidity. Average annual rainfall is 1788 mm, and perennial mean relative humidity is 84%. Rain season is from March to June; while shower and thunderstorm often occur in July and August, dry weather can be expected from September to February next year.

(3) **Hydrology**

The water system running through Taining is the Jinxi Stream, belonging to the upper part of the Futun Stream. The watershed's two other branched streams, Shuixi and Shanxi, are converged together with the Jinxi and run into the Golden Lake. The surface area of the Golden Lake is 3600 ha, and the storage capacity is 870 million m^3. It is the largest inland wetland in Fujian Province and the most important water landscape of Taining. The quality of surface water is quite good, which meets the criteria of the Type II water quality of the *National Water Environmental Quality Standard for Surface Water* (GB3838-2002), and the quality of all drinking water meets the national standard.

(4) **Soil and vegetation**

The soil types of Taining mainly include red soil, purple soil. The red soil is widespread in the lowland and hilly areas at an altitude below 800 m. The soil has medium fertility and a pH value ranging from 4.1 to 5.0. The purple soil is mainly distributed in the red basins and has high potassium content. The Heritage site generally has a forest coverage rate of 78% but reaches to over 90% in its core area. The Heritage site has 8 vegetation types, 24 formations, and 40 associations, featuring Danxia herbosa, Danxia sclerophyllous evergreen broad-leaved forest, ravine evergreen broad-leaved forest. Drought enduring plants exceedingly well developed on the cliffs, and *torreya jackii* is only found in the Danxia landform area in southeastern China. While the waterlogging tolerant flora well developed in the ravines, the Heritage site has a lot of rare or endangered species of wild fauna and flora, being one of the areas rich in wild fauna and flora resources in a unit area within a small range in China.

13.2.2 Geological Structure

(1) **Regional geological background**

During the Mesozoic and Cenozoic, the geotectonic location of Taining was in the southeastern edge of the Eurasian Plate, southeast of the active continental marginal zone of West Pacific and southwestern Wuyi rise of the Cathaysia Land. The basement of ancient land was formed in the middle–late Proterozoic; the Paleozoic strata were missing, and the Mesozoic strata directly covered the Proterozoic strata; the Mesozoic continental facies basins overlaid, including the widely distributed upper Jurassic–Cretaceous continental facies volcanic rock, and the upper Cretaceous red clastic rocks, which are for the formation of the Danxia landform.

Taining and its surrounding regions have witnessed five major tectonic development stages, which are Cathaysia Land, Caledonian folding and orogeny, Tethys, transition and active continental margin. The Cathaysia Land formed metamorphic crystal basement of the area in the Late Proterozoic. Later, the land ever split and sank. The Caledonian Movement in Late Eopaleozoic led to the fold and uplift of the land. The Indo-China movement in late Early Triassic influenced this area. However, Taining now has no sediments of this period. The deposition of the Late Triassic–Middle Jurassic formed in the stretch-generated depressions or faulted basins within the plate. During the latest stage of the

Middle Jurassic, the Pacific Plate began to collide and crush the Eurasian Plate. The development stage of the regional geological structure shifted to an active continental margin stage. The duration from the Late Jurassic to early Early Cretaceous was the strongest crushing stage between the two plates. The crushing resulted in the extensive volcanic eruption accompanied by massive magma intrusions. The volcanic and intrusive rocks formed in this stage belong to calc-alkaline series. The structure background changed to a loose extension stage from the Early Cretaceous and thus formed the stretched- and faulted-induced depressions. The volcanic activities decreased and were mainly along fault zones. Therefore, the volcanic rocks are dominated by basic–acidic assemblage with a bimodal characteristic. In later Cretaceous, the basins had developed exceedingly thick red continental clastic rocks of Shaxian Formation and Chong'an Formation. The crust uplifted and began to be eroded in Late Cretaceous or Early Paleogene.

(2) **Stratum and Lithology**

The red beds in Taining basins include Shaxian Formation below and Chongan Formation above. The lower part of Shaxian Formation is fluvial deposition; the middle part is shallow lacustrine sedimentation; and the upper part is again fluvial deposition. The Chongan Formation is mainly alluvial–diluvium fan deposition. The debris flow and mud–sand flow facies are often seen. Piedmont deposit and fluvial and lacustrine facies are locally developed.

The Chongan Formation (K_2c) in Taining is dominated by coarse clastic deposits set down in terrestrial basins. The formation is composed of purplish-red massive or very thickly bedded conglomerates and sandy conglomerates with sandstone and siltstone interbeds. The gravel composition of conglomerate is complex, mainly metamorphic rock, granite, volcanic rock, quartz, etc., with ferruginous or argillo-arenaceous cements as base, and the gravels are in much different size, and they are generally in subangular or subrounded shape; the rocks, which are tough and of strong weathering and denudation resistance capacities, are the main rocks contributed to the formation of Danxia landforms as the soft sandstone, and siltstone interbeds were weathered leading to gravity collapses.

(3) **Geological structure characteristics of Heritage site**

Taining basin now has two subbasins, Zhukou and Meikou. They were derived originally from one dustpan-shaped faulted basin.

Faults of the basin edge: The NE-striking Zhaixia-Aoshang fault and SN-striking Maoershang-Chenkeng fault are two important basin-edge faults for the formation and development of Taining basins. The faults acted several times and have a series of fault facets, fault cliffs, fault valleys, and silicified zones.

With the extension in the early–middle Late Cretaceous, the Shaowu-Heyuan fault zone in Aoshang-Zhaixia area was characterized by the uplift of the fault's northwest side and the depression of its southeast side. This formed the Taining dustpan-shaped fault basin along east-side of the fault. Due to a long active duration of the fault, the altitude difference between two sides of the fault is great. The faulted amount of the basin is more than 2000 m. The rock fragments and sands from west mountains have been transported along the opening of the mountains to the basins by the flood flows, forming alluvial–diluvium fan deposition with huge thickness more than 1900 m.

Secondary faults in the basin: The secondary faults and joints are well developed in the area due to multiple activities of the basin-edge faults. The field survey and interpretation of remote sensing image have revealed that there are two groups of the secondary faults, one striking NE and the other striking NNE, and five groups of joints, striking SN, NW, NWW, NE, and NNE, respectively. The joints are crossed each other, leaving complicated networks in the remote sensing image.

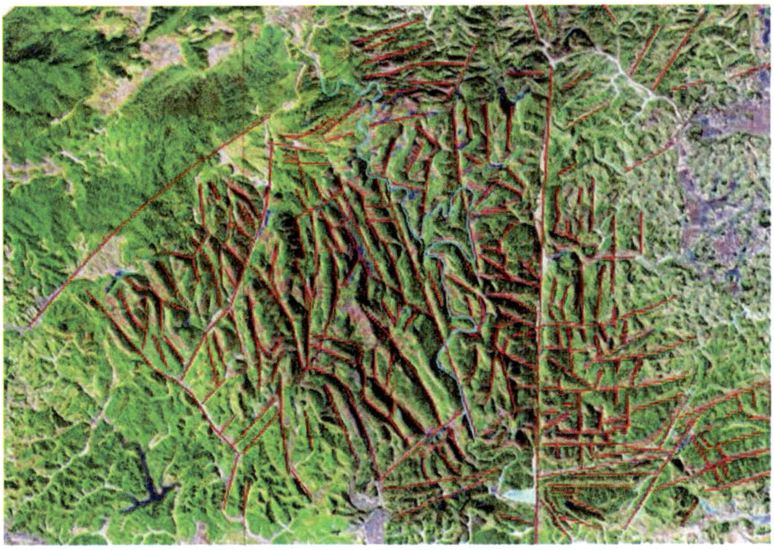

Remote sensing image of Shangqing River area

(4) **Relationship between Geological Structure and Landform Development**

Control of Danxia Landform Development by Lithological Difference: The Chong'an Formation (K_2c) in Taining is dominated by coarse clastic alluvial–diluvium fan deposits set down in terrestrial basins. The formation is composed of purplish-red massive or very thickly bedded conglomerates and sandy conglomerates with sandstone and siltstone interbeds. The rocks are poorly bedded, which is a main feature of Taining's red beds. Therefore, the rocks of the Chongan Formation can easily form red cliffs, deep-cutting canyons and variously shaped caves. The Shaxian Formation is composed of purplish-red siltstone and mudstone with sandstone and sandy conglomerate interbeds and the rocks are quite soft and generally have a gentle hilly appearance.

Alluvial–diluvium sandy conglomerate and conglomerate are the material foundation for the cave formation. These rocks are usually massive or very thick and poorly bedded. The huge thickness and heterogeneity of the rocks increase the randomness of the cave shapes. The thickness of rock beds is a major factor controlling the formation of the Danxia landform. The thickly bedded rock can form forehead-like cave, arched cave, or cave groups while thin rock bed can only produce flat cave and honeycomb-like cave.

Leading role of fault system: The fault system has played a leading role in the formation of Taining landform. The Taining basin is a fault-generated basin developed in the Wuyi Rise of north Fujian. The formation and evolution of the basins have been controlled by the regional NE-striking Chong'an-Shicheng fault and SN-striking Taining-Longyan fault. Multiple tectonic activities of the major fault led to the formation of multi-directional secondary faults. The conglomerate and sandy conglomerate of the Chongan Formation have been cut into many blocks by these faults and joints. This made the rocks more easily to be weathered and eroded and has laid the foundation for the development and distribution of the mountains in the red basins. The general distribution directions of the mountains in Taining red basins are obviously controlled by the NE-striking fault and SN-striking fault. While the secondary fault systems control the strike of a mountain, the density and orientation of canyon and the extension of incised meander running across the deep-cutting canyons.

Impact of neotectonic movements on landform development: The crust of the Heritage site has intermittently lifted several times in the Quaternary. The lift extent is about 400–500 m since the Late Neogene according to the heights of different planation surfaces and river terraces in the area. Due to the difference of lift extent, the high and fast lifted areas have mature Danxia landform developed. On the other hand, the low and lately uplifted areas developed peak plain–canyon type of youth Danxian landform in Shangqing River area.

13.2.3 Types and Features of Landform

13.2.3.1 Basic Features of Landform

Due to the differences in time, extent and velocity of the crust uplift, the landform developments in Taining have quite region differences. The Danxia landform in the south of the Golden Lake scenic area is in the mature developing stage of the landform and is characterized by Danxia peak cluster landscape. While the landform in north of the scenic area and in the Shiwang scenic area is in its young developing stage and has an old planation surface at an altitude of about 450 m. It has more than 400 deep-cutting canyon groups, forming distinct networked valleys and red mountain blocks.

(1) **Distinct Danxia canyons**

The Danxia canyon groups are composed of more than 70 linear valleys, over 130 lane valleys, and more than 220 canyons with different directions. Some are crossed each other, some extend in parallel, forming complicated networked valleys. The Shangqing River wanders in a deep-cutting canyon. It has a great curvature and is extremely beautiful, forming one of the distinct features of the Danxia landform.

(2) **Various Danxia caves**

The Danxia caves are well developed in the Heritage site. According to incomplete statistics, in Taining area there are more than 60 huge single caves and more than 100 niche-like cave groups. Numerous differently sized and shaped caves form the unique Danxia microlandscape, which makes Taining a Danxia Cave Museum. The Danxia cave has gestated the Taining's long-standing Danxia cave culture.

(3) **Grand and spectacular Danxia mountains**

The Danxia Mountains in the Heritage site are distributed mainly in the areas surrounding the Golden Lake, represented by various forms like peak forest, peak cluster, rock pillar, rock wall. They are either spectacular or delicate and making beautiful scenes such as upside-down red walls (reflected image of the red walls on water) and towering steep cliff.

(4) **Beautiful scenery of Danxia on water**

The well-integrated Danxia landform with various water scenes, e.g., Shangqing River, Golden Lake, and Jiulong Deep Pool, constructs the beautiful scenery named Danxia on water.

Landform perspective drawing in Shangqing River area

Landform perspective drawing in Zhaixia area

13.2.3.2 Landform Types

See the classification of Taining Danxia landform in Table 13.1.

Table 13.1 Classification of Taining Danxia landform based on individual landform unit

Class			Main origin	Example
Positive landform	Danxia cliff wall	Flat type	Rock beds collapse along a vertical fault in a relatively short time	Tongtian Stele
		Vertical trough type	Erosion by vertical water flow	Mirage
		Trough along bedding distribution type	Differential weathering of rock beds	Mountain orchard field
		Wavy type	Lateral erosion and down-cutting of water flow	Mile Stone
		Cave type	Weathering and collapse	Baitai Stone
		Mottle type	Lichen-induced differential weathering	Great Red Wall
		Combined type	Combined with two or more of the above types	Longtan Wall
	Danxia Rock fort		Eroded residue of near square rock-cut by faults	Perking Male Lion
	Danxia Rock wall		Eroded residue of near strip-like rock-cut by faults	Shiwang Qishan, Sail Welcoming Guests
	Danxia Rock pillar		Eroded residue of crushed rock-cut by faults	Masculine Pillar in Golden Lake, Lover Peaks, etc.
	Cone-like mountain		Top of thick massive rocks quickly eroded	Pyramid
	Bell-like mountain		Slop of thick massive rocks quickly eroded	Double-Breast Peak
	Danxia isolated rock		Residue by weathering or erosion	Stone Eggs in Golden Lake area
	Accumulation of Collapsed Materials, Huge Collapsed Rock Block		Accumulation of collapsed massive rocks	Wind-movable Rock in Jiulong Deep Pool area and Lucky Peach Rock in Shangqing River area
Negative landform	Canyon	Linear valley	Flow erosion or fault stretching along fracture	A-thread-of-sky in Golden Lake area
		Land valley	Flow erosion, collapse	Golden Dragon Valley in Zhaixia, Lane Valley Groups in Zhuangyuan-Rock area
		Canyon	Flow erosion, collapse	Shiwang Grand Canyon
		Deep-cutting Canyon with Stream	Flow erosion, collapse	Shangqing River

(continued)

Table 13.1 (continued)

Class			Main origin	Example
Huge single cave	Trough-like cave		Rock trough be further eroded and collapsed	Zhuangyuan Rock, Feng Rock
	Arched cave		Collapse of thick massive rocks cut by vertical joints	Sweet Due Rock
	Dome-shaped cave		Collapse of thick massive rocks	Tianqiong Rock
	Irregular cave		Weathering, collapse, sheet-flow erosion	
	Penetrating cave	Horizontal penetrating cave	Weathering, collapse, flow erosion	Lingzhi Rock
		Vertical penetrating cave		Jiulong Deep Pool
	Nature-borne bridge			Dingle Maze
Cave groups	Infested cave group	Horizontal bead string-like cave group	Water lateral erosion and sheet-flow erosion	Shangqing River
		Vertical bead string-like cave group	Sheet-flow erosion	Jingui Temple
		Calabash- or bamboo shoot-like cave group	Vertical water flow erosion	Lijia Rock, Zhuangyuan Rock
		Overlaid cave group	Sheet-flow erosion, collapse	Jinlong Valley
Honeycomb-like cave			weathering, water later erosion and dripping	Lijia Rock
Groove	Bedding-along groove		Weathering, sheet-flow erosion	Mountain orchard field
	Vertical groove		Sheet-flow erosion	Green and Water Spraying Valley, Sky-facing Well
Rock trough	Horizontal trough		Weathering, collapse	Lijia Rock
	Inclined trough			Jiulong Deep Pool
Karst landform	Stalactite		Chemical deposit	Ganlu Temple, Golden Bell Cave
	Stalagmite		Chemical deposit	Golden Bell Cave
	Rocky Curtain		Chemical deposit	Golden Bell Cave
	Rocky Flower		Chemical deposit	Golden Bell Cave

13.2.3.3 Typical Landscape

(1) **Danxia cliff walls**

Mottled cliff wall: Great Red Wall

Flat cliff wall: Tongtian Stele

Combined cliff Wall: Dragon Pool Wall

Wavy cliff wall: Evening Glow-Covering Wall

Wavy cliff wall: Mile Rock

Flat cliff: Qitian Cliff

(2) **Danxia rock walls and Danxia hoodoos**

Danxia rock wall: Sail Welcoming Guests

Hoodoo: Lover Peaks

(3) **Danxia hoodoo-peak rock wall cluster**

Danxia rock wall and hoodoo groups: Three Sword Peaks

Rock wall and peak clusters: Qiang, Qi, and Pai Mountains

(4) **Danxia canyons**

Danxia lane valley: Jiulong A-thread-of-sky

Danxia lane valley: Tranquil Valley Water Lane

Linear valley: Right-Angle, A-thread-of-sky

Inclined linear valley

Lane valley: Green and Water Spraying Valley

Lane valley: Diving Dragon

A-thread-of-sky on water in the lane valley: Celestial Rain Valley, Longshan Valley

(5) **Incised meander**

Shangqing River runs across the densely distributed peaks from the north to the south. The narrowest part of canyon is 25 m, while the canyon is 80–200 m deep. There are steep cliffs on both sides, decorated with differently shaped caves.

Main stream of Shangqing River: crossing the peaks

Valley landscape in Shangqing River

Valley–stream landscape in Taining Shangqing River

(6) **Typical cave landscape**

Forehead-shaped trough-like cave: Zhuangyuan

Enclosed valleys of Feng Rock (Li Gang, Prime Minister of the Dynasty, once studied here.)

Arched cave Sweet Dew Rock

Arched cave-linear valley combination: Heavenly Gate

Huge dome-shaped cave (Boat Rock is 100 m wide, 42 m deep, about 32 m high, and it is one of the biggest Danxia caves ever found in China.)

Niche-like cave groups: Tianqiong Rock

Niche-like cave groups: Liquan Rock

Penetrating cave: Linzhi Cave

Nature-borne bridge: Xianren Bridge

Niche-like cave groups: Baitai Stone

Niche-like cave groups: Bananas on Crocodile

Niche-like cave groups: Rock with multi-mouths expressing different feelings

Niche-like cave groups: Xianren Rock

Vertical groove: Sky-facing Well

Bedding-along groove: Lijia Rock

13.2.4 Species Diversity

(1) **Biogeographic Province**

According to the Udvardy (1975) biogeographic system, Taining is categorized habitat rich in biodiversity in the southern part of biogeographic province of China subtropical forest of Palearcitc region, locating in the subtropical forest zone in southeastern China in Indian–Malaya region in the Global 200 identified by the World Wildlife Fund (WWF).

(2) **Species**

Taining Heritage site has plenty of species, vascular bundle plants 192 families, 608 genuses, 1276 species, including fern 37 families, 70 genuses, 162 species, gymnosperm 6 families, 8 genuses, 8 species, angiosperm 149 families, 530 genuses, 1106 species.

The Heritage site has spinal animals 34 orders, 105 families, 380 species, including fish 5 orders, 13 families, 48 species, amphibian 2 orders, 7 families, 26 species, retile 3 orders, 12 families, 63 species, bird 17 orders, 53 families, 199 species, mammal 7 orders, 20 families, 44 species,

And it has invertebrate of insects (including Acari of Arachnida) 25 orders, 231 families, and 1509 species.

(3) **Biodiversity**

The Heritage site not only has various species, but also has global and regional rare and endangered species.

There are 3 plants under state first-grade protection, including gingkgo and *Taxus mairei* and *Bretschneidera sinensis*, 11 plants under state first-grade protection, including *Phoebe bournei, Ormosia henryi, Ormosia hosiei, Eurycorymbus cavaleriei, camptotheca acuminate*; 10 plants listed in the IUCN Red List, including officinal dendrobium stem, gingkgo, silver bell–tree; 65 plants listed in CITES Appendices indicated no international trading allowed of these plants, including lamb of tartary, *Anoectochilus roxburghii*, 77 plants listed in China Species Red List, including officinal dendrobium stem, *gingkgo*, etc.

The Heritage site has its special biodiversity of Danxia landforms that we can see drought enduring plants exceedingly well developed on the cliffs, including gesneriaceae plant, drought enduring oak species, drought enduring arethusa, herba selaginellae, drought enduring dayflower, torreya jackii, while the waterlogging tolerant flora is well developed in the ravines, including *camptotheca acuminate*, silver-leaved willow. Because of good protection, the Heritage site provides shelter for the endangered officinal dendrobium stem and phoebe bournei. The torreya jackii only exists in Danxia landform area in southeastern China.

There are 2 animals under state first-grade protection, including python and white-necked reeves pheasant, 35 animals under state second-grade protection, including tiger frog, turtle, mandarin duck and sparrow hawk, goshawk, 7 animals listed in IUCN Red List, including python, white-necked reeves pheasant, pangolin, jackal, oriental civet cat, Asian golden cat, serow, 47 animals listed in CITES Appendices indicated no international trading allowed of these animals, including falcon, white-necked reeves pheasant, golden cat, serow, 43 animals listed in China Species Red List, including falcon.

There are lots of crisscrossing ravines, steep mountains, and caves, the site has become the paradise for rare animals that the animals listed in CITES Appendix II and animals under state second-grade protection often can be seen, including black-eared kite, crested serpent eagle, black eagle, common buzzard, Hodgson's hawk-eagle, Bonelli's hawk-eagle. There are more than 20 black-eared kites living in the Zhaixia Village and more than 50 silver pheasants under state second-grade protection in Shangqing area.

(4) **Ecosystem**

In aspect of composition and structure, Taining, in general, has the features of evergreen broad-leaved forest in humid area of eastern middle subtropical zone, featuring special vegetation of middle subtropical zone. The complex habitat of Taining Danxia landform has various ecosystems developed. Especially in the small scope, the ecosystems are very complicated that on the same mountain, there are drought enduring sclerophyllous evergreen broad-leaved forest growing on the top, and there are torreya jackii, fern, etc. forming the ecosystem on the cliff, and there is special evergreen broad-leaved forest ecosystem developed at the mountain foot, while there is ombrogenous evergreen broad-leaved forest ecosystem developed in the ravines. The diversity of Danxia landform, structure and vegetation community provides complex varied environmental conditions, preserving perfect ecologic environment for the wildlife. According to different environments, communities can be divided into forest, cave, farmland, and aquatic animal communities.

13.2.5 Natural Landscape and Beauty of Nature

Taining Danxia landform has deep-cutting canyons, exotic caves, clean water, and primitive ecology as its special landscape.

(1) **Canyon Landscape**

Taining Danxia landform is characterized by the tall cliff, excellent ecosystem and the innumerable caves and thus has a high value of view and admiration. Some canyons cross each other, some extend in parallel but some are extremely curved which construct a marvelous spectacle of wondering river in deep-cutting canyon. Most of Danxia Canyons are tortuous and quiet with luxuriantly green trees and bamboos, dense vines, clear stream, and singing birds. Traveling through it, you would imagine a lot. Rafting along the wandering stream in the canyon, you would feel like in a beautiful landscape painting and have a dynamic aesthetic feeling.

(2) **Cave Landscape**

The caves are Taining Danxia landform's marvelous spectacle. The large caves can hold thousands of people, while the small one looks just like a honeycomb. Caves are either in single or in groups overlaid layer by layer. The shapes of the caves are various and many of them look just like human being, bird, animal, or something else. The decoration of red cliffs with caves greatly enhances the attraction of the red cliffs. Some of the Taining's Danxia caves are so large that make tremendous shock to the people; some are pictographic and can easily trigger people's imaginations, and some are so fancy that make people surprised at nature's exquisite craft. Moreover, Taining Danxia caves carry the substantial accumulation of Taining's human culture. It is really a magical and wonderful experience when visiting the Taining Danxia caves.

(3) **Mountain and Water Landscape**

The mountain peaks of Taining Danxia landform take a myriad of forms, and most of them are characterized mainly by their beauty. The elegant peaks, water in abundant, the red mountain cliffs, green trees, and clean waters are always combined together to form various lovely scenes, which enjoy people not only in eyes and but also in minds. The scenes showing to the people are just like a number of lyric Chinese landscape paintings, e.g., deep and bluish-green water of the Golden Lake, connected islands and lakes, spaced water ways, and beautiful peak clusters. People cannot help praising and just forget to return when facing these beautiful scenes.

(4) **Ecological Landscape**

Taining people have a long tradition of protecting forests, which can be traced back to ancient time. Careful protection year after year has created an excellent ecological environment for the Taining Danxia landform area. The core area of the Heritage site has rare traces of people and the integrity of its ecosystem is well conserved. The forests are flourishing, and trees and rocks are decorated by vines. The rare tree species and wildfowl are often seen. The air is so fresh, just like a natural oxygen chamber. It is far away from the crowd and is really an ideal resort for relaxation.

13.3 History and Development

13.3.1 Evolution of Taining Danxia Landform

During the Indo-China movement in late Early Triassic, the basins began to take its form but not sank largely, mainly developed fluvial deposit; in late Jurassic, the Yanshanian Movement broke out, and deep magma sprayed out that such process lasted until the early Cretaceous, leading to the volcanic deposition and fluvial and lacustrine facies deposition. The basins were not largely sunk by the activities in the above-mentioned stages, and the total thickness of the sedimentations is 272 m. Of them, 126 m is volcanic sequence. In the Late Cretaceous, the sinking intensified enlarging the scope of deposition, the red bed of Shaxian Formation dominated by fluvial and lacustrine facies deposited first, and then the surrounding mountains uplift strongly, the southeastern side of Zhaixia-Aoshang Fault and Maoershang-Chenkeng Fault sink quickly, forming large thick alluvial fan-facies of Chongan Formation.

At the end of Cretaceous and the Early Paleogene Period, the Himalayan tectonic movement began, and the crust uplift leading to the revival of the old faults, and Taining basins uplifted differentially and was cut by the fault, developing multiple faults and joints of NE, NNE strike. The red depositional sequences have been uplifted and exposed in the areas above the regional erosion basis and were eroded. According to the thermoluminescence dating data, the uplift and cutting of the red beds in the Zhukou Basins took place about 4.3 million years ago, earlier than the rise of the earth's crust, and the landform is quite young. All of these are the features of typical young developing stage of the Danxia landform characterized by the negative landform such as the Shangqing River in Shiwang District wandering in deep-cutting canyon.

Today, the water down-cutting and erosion are still ongoing, the caves on the cliff walls are undergoing weathering and the collapse are also happening. The Taining Danxia landform area is still active in its evolutionary stage.

Planation surfaces in Shangqing River

Planation surfaces in Zhaixia area

13.3.2 History and Development

According to the archaeological excavation, human activities have been found in Taining in neolithic age 5000 years ago.

Taining was called Qimindi in Zhou Dynasty, and then it was under Yue Country in the Warring States Period and then was under Minyue Country during the early Han Dynasty, and then later, it was under Kuaiji Shire. At the end of Han Dynasty, it was under Jian'an County and then was under Sui'an County in the Three Kingdoms Period. In the second year of Qianyuan Period in Tang Dynasty (759), Guihua Town was found in today's Taining area, and in the first year of Zhongxing Period of Nantang Dynasty (958), Guihua County was found that it was the first county found in today's Taining. In the first year of Yuanyou Period of Beisong Dynasty (1086), the emperor, Song Zhezong, granted Taining, the similar pronunciation as Tai Min in Guo Tai Min An (peaceful country and carefree people) that Taining County was called from that time on. Later on, its territory has no big change.

Since Tang Dynasty, the closed one crop farming was popular that Taining people planted crops and weaved for self-supply and had their own food storage. People lived happily. Taoism has nearly 2000 years' history in Taining; in the first year of Yonglong Period in Tang Dynasty (680), Buddhism was introduced to Taining.

At the end of Five Dynasties, the fields were plowed and the population increased in Taining. During the Yuanfeng Period of Beisong Dynasty, Taining became one of the big counties in the northwest of Min that the forestry, fishing and crop-planting flourished, and 30,000 households settled down.

In Yuan, Ming and Qing Dynasties, the population reduced due to the wars, but it still remained a closed one crop-farming county featuring cropping, weaving for self-supply and food storage.

Because of the development of the agriculture, in Ming Dynasty, Taining completed abstraction works and dams 38 in total. Since 1950, Taining has been paying much attention to the water conservancy construction that more than 1000 new or modified hydroprojects were done.

Taining's mining and smelting began in Tang Dynasty and the five Dynasties. Today, 28 mineral products have been found. Hebaoshan gold mining has been done that it brought quite handsome profit. Taining's handicraft industry has a long history.

Only a few hydroprojects mentioned above were done in the Heritage site, and mining was done outside the site.

In 1994, Taining Golden Lake was awarded National Key Scenic Spot, and in 1996, the Management Committee of Golden Lake National Key Scenic Spot was established in Taining that laws and standards were legislated for Taining natural heritage protection and management. In 2005, Taining World Geopark was approved to establish, providing unprecedented opportunity for the county's economic development.

13.3.3 Cultural Landscape

Taining has rich cultural resources. Traces of human activity, sites, ancient buildings, cliff carvings, stone carvings, ancient burials, relics left by talents, memorial sites were found here and there, in addition to plenty of literature heritage, poems, and records of custom.

Since Northern Song Dynasty, Taining has 2 Zhuangyuan (number one scholar), 54 Jinshi (a successful candidate in the highest imperial examinations in feudal China), and 101 Juren (a successful candidate in the highest imperial examinations at the provincial level in the Ming and Qing Dynasties). The ancient famous scholars, such as Li Gang, Zhu Xi, Yang Shi ever studied, giving lectures, and staying here. The splendid ancient culture endows us with precious cultural relics and rich folk arts. The main cultural activities are opera, folk song, folk music using gong and drum and folk dancing.

Temple built in Cave: Buddhism has a long history in Taining that the earliest temple recorded is Zhukou Ganhua Temple built in Yonglong Period (680–687 AD) under the emperor, Gaozhong in Tang Dynasty, more than 1300 years old. Taining temples are not famous for their spectacularity but their wide distribution and large number that are rarely seen in China. Most of Taining temples were built according to the natural terrain in the caves in the red cliffs that they are the special masterpieces showing people taking advantage of Danxia landforms. The most famous temples are Qizhen Rock Temple built at the end of Western Han Dynasty, Sweet Dew Rock Temple built in Song Dynasty, Hebao Nunnery built at the end of Yuan Dynasty, Liquan Buddha Temple, Lijia Rock Temple, Baogai Rock Temple, Feng Rock Temple, Danxia Rock Temple, all of them built in Ming Dynasty.

Sweet Dew Rock Temple

Part of Taining Ancient City

Taining Minster Mansion

Ancient buildings: Minister Mansion (Shang Shu Di) was built between end of Wanli Period and Tianqi Period in Ming Dynasty, more than 370 years old. It is the mansion of Li Chunhua who is the minister in charge of troops and grand tutor to the crown prince in Ming Dynasty. The mansion is exceeding big and carefully fit up, featuring wood, gray brick, granite being the building materials that it reveals the Jiangnan architectural style in Ming Dynasty. The mansion is the best-preserved Jiangnan ancient architectural building group built in Ming Dynasty that it was granted National Key Protected Relic in 1988. Another architectural masterpiece is Baoshide Mansion built in the early or middle Ming Dynasty. Raw building materials are used and the style is simple; it is as good as Minster Mansion regarding its history, art and scientific values, and it was granted National Key Protected Relic in June 2001.

13.3.4 Local Arts

Meilin Opera is the unique local opera, which is more than 300-year history. the performance, vocal music, name of the tune, list of plays and skill have unique style, showing rich features of the northwestern Min. the opera is simple to understand with lingering charm.

The primitive dancing dating from Yin or Shang Dynasties 3000 years ago is not forgotten, which is taken as the living fossil for anthropological, historical, and artistic study. It is the witch dance for disease healing, exorcising, and sacrificing.

13.4 Physical Features and Values in China Danxia Serial Heritages

13.4.1 Physical Features

Taining is located in the west part of volcanic rock zone in the southeastern China in the Mesozoic active continental marginal zone to the west part of West Pacific, revealing the formation, development, and evolution of the Mesozoic active continental marginal zone to the West Pacific. A series of overlaid Mesozoic basins (volcano) developed along the Cathaysia Land are rather complete records of the frequent magma and intensive tectonic activities. The exceedingly thick red clastic alluvial–diluvium fan deposits of Chong'an Formation in the linear faulted red basins are also the products from the continuous intense uplifts of the source area in the basins. The much-complicated fault crevices system in the red basins is the result of multiple tectonic activities.

Taining is the representative of Danxia landform featuring juvenile peak plain (flat plain)-canyon combination belonging to the Chinese middle tropical humid area.

Taining has a diversity of the Danxia landform and the peculiarity of landscape. Different from the other Danxia landscape featuring mountains, the remarkable excellences of Taining Danxia landscape are the large-scaled complex canyon clusters, deep-cutting rivers running across the canyons, colorful Danxia caves in plenty as well as the ecological environment having close relation with the Danxia landform that all of them make Taining one of the most outstanding representatives among China Danxia landforms and the global red beds.

As for biodiversity, in Danxia landform area, the drought enduring plants exceedingly well developed on the cliffs, including Gesneriaceae plant, drought enduring oak species, drought enduring arethusa, herba selaginellae, drought enduring dayflower, *Torreya jackii*, while the waterlogging-tolerant flora well developed in the ravines, including *Camptotheca acuminate*, silver-leaved willow etc. Because of good protection, the endangered *Torreya jackii*, officinal dendrobium stem and *Phoebe bournei* have found shelter here. The site has become paradise for rare animals due to a large number of criss-crossing ravines, steep mountains, and caves.

As for distribution of forest vegetation, Taining is the area distributed with typical evergreen broad-leaved forest in humid area of the eastern middle subtropical zone. The ecosystem is diverse, remaining the typical primitive ecological processes of forest vegetation development in the Chinese middle subtropical area.

13.4.2 Status and Values Among Serial Heritage

Taining among the serial heritages has the same or similar geological setting and evolvement to the other Heritage sites, showing landform features and features of physical geography in common. It meets the requirements (III.C, Item 137) for heritage serial nomination that it can not be replaced. The following gives the values:

- Taining is the representative of Danxia landform featuring juvenile peak plain (plat plain)-canyon combination belonging to the Chinese middle tropical humid area, and it is the non-replaceable part in the series nomination.
- Taining Danxia landform is the typical region of Mesozoic active continental marginal zone to the West Pacific, recording and revealing the formation, development and evolution of the Mesozoic active continental marginal zone to the West Pacific as well as the climate changes of southeastern China since the

Cretaceous. The formation of its dustpan-like faulted red basin and the formation and evolution of Danxia landform are typical among China Danxia Heritage sites that it cannot be replaced.

- Taining has linear valleys, lane valleys, canyons, and red cliffs exceedingly well developed, together with 400 deep-cutting canyon groups, forming distinct networked valleys and red mountain blocks. The density and narrowness of canyon, the curvature of its embedded meander and the primitiveness of canyon ecology are rarely seen in the Danxia landform areas. The large number of Taining Danxia rock grooves and caves, the large scale of cave clusters, the peculiarity of shape and combination of cave and the admiration of cave are rarely seen. Therefore, Taining Danxia landscape is the important and non-replaceable heritage regarding the diversity of China Danxia landform.
- Taining generally is in the juvenile stage of landform development, but with multiple stages. Since the Neogene Period, the basins uplifted differentially for a few times, leading to the formation of landform in different evolution stages. The process of geomorphic development is distinct. What's more, the water (rain, river water) erosion in the Heritage site is distinct and typical.
- The Heritage site lies in the southeast side of the Wuyi Mountain Range, having humid and rainy climate. The site shows rich biodiversity and ecological diversity and preserves dynamics of a unique alternation process of biological communities. In addition, it is the habitat for rare and endangered species. It is the typical representative regarding the middle subtropical biocommunity among the serial Heritage sites.
- The culture of Taining Danxia cave has a long history; besides, the Heritage site has become habitat for many rare animals due to the steepness of mountain, primitiveness of ecosystem. Taining is a good example showing human living in harmony with wildlife and Danxia landscape.
- Taining is one of the most charming places showing China Danxia landscape, having diversity, uniqueness, rareness, and naturality of Danxia landscape, and is the representative of such landform in the world.

13.5 Criteria Under Which Inscription Is Proposed

13.5.1 Taining Danxia Exhibits Exceedingly Fantastic Natural Phenomenon and Rare Natural Landscape

Taining Danxia landform exhibits Danxia canyon clusters containing more than 70 linear valleys, more than 130 lane valleys, more than 220 canyons. The number, density (maximum 23/km^2), narrowness, curvature of the deep meander are rarely seen in the world. Caves are the spectacle of Taining Danxia landform that their number, combination are rarely seen among Danxia landforms.

The complicated geological structure background and strong endogenic and exogenic geological processes of the Taining Danxia landform area have led the Danxia landform to have been combined together and become more complicated, which have derived a great deal of wonderful and incredible natural landscapes beyond people's imagination. The beauty of Taining Danxia landscape comes from the deep-cutting canyons, exotic caves, enchanting water, and mountains as well as the ecological primitiveness, endowing with much high aesthetic value, and they satisfy people's need who are looking forward to returning to nature, enjoying nature and enjoy themselves not only in eyes and but also in minds. It is not only the gift for China from the earth but also the aesthetic heritage for the whole nations.

13.5.2 Taining Danxia Area Is the Proof of Different Types of the Earth's Evolution and Geological Processes

The Heritage site has rather complete Mesozoic geological records, revealing the formation and evolution of the Mesozoic active continental marginal zone to the West Pacific. The development and depositional model of Taining basins are the typical sedimentation faulted basins caused by rifting and sinking of the active zone of the continental margin.

The Cretaceous red bed reveals earth history regarding the Cretaceous dry and exceedingly hot climate. The formation of Taining Danxia landform has close relation to such humid rainy climate. The development process of Taining Danxia landform shows that since the Cretaceous, especially since Palengene, the uplift of Tibetan Plateau leads to the climate change in the southeastern China.

13.5.3 Taining Danxia Preserves Dynamics of a Unique Alternation Process of Biological Communities

The Heritage site has dynamics of a unique alternation process of biological communities of the Danxia landform area. The continual collapses of the cliff face in geological history reveal the different evolutionary stages of the organisms in different environment.

13.5.4 Taining Danxia Area Is the Habitat of the Rare and Endangered Species

Being a ravined area with few people traces, this area has become the shelter of *Phoebe bournei,* which used to be wide distributed in China and now is in danger of extinction and under state second-grade protection. *Torreya jackii* is only found in the Danxia landform area in southeastern China. In addition, Taining has other plants, which are under state first-grade protection such as *Ginkgo biloba* and *Taxus maire*i, and under state second-grade protection like *torreya jackii, Camptotheca acuminate, Cibotium barometz,* Camphor tree*, Ormosia henryi, Ormosia hosiei, Eurycorymbus cavaleriei, Glycine soja*. Its most characteristic animal is the raptors of falconiformes. The area is also the preferred habitat to various birds of falconiformes and serow owing to the steep cliffs, innumerable caves and the completed food chains of the area.

Taining's unique evolution history of the earth and geological process, outstanding landform diversity and biological diversity and the rarely seen natural beauty meet the criteria vii, viii, ix, and x for being included in World Heritage list as natural heritage.

13.6 Protection for the Heritage Site

The Taining Danxia Heritage site and the buffer zone have clear borderlines, which are the natural borderlines such as mountain ridge, valley, river, and road. The borderlines were taken after considering to protect the Heritage site, e.g., try to avoid the impact by human activities. Owning that the landform is precipitous and the traffic is not developed, the Heritage site preserves the integrity of Danxia landform area, forest ecosystem, and habitat for rare and endangered species.

The local people have been protecting the forest from fire that most of the villages and clans have stipulations for the forest protection and fire prevention. Rules were made, which is very strict against logging and burning occurred in the fengshui forest, temple forest, and protection forest that any person breaking the rule are punished depending on the seriousness of the case. A stone against logging made in 1659 was found there in front of fengshui forest in Zhaixia Village. In addition, local people pay much attention to the protection for water sources that catching the wild fishes are prohibited, for instance, a stone with penal terms are written against fishing was put in front of Shangqing Village.

Since the establishment of the People's Republic of China, the government in the Heritage site has been paying much attention to the protection for the sources in the Heritage site. In 1980, Taining government conducted projects benefiting the villages and immigration for the sake of ecological protection that most of the villages in the southern part of Heritage site moved out. In 1990, the villages in the core area of Heritage site moved out. Up to date, the Golden Lake and the northern Heritage site have no dwellers.

Management Committee of Fujian Taining National Scenic Spot, which is the supervising office, was established at the end of 1986. In 1987, the Golden Lake was granted scenic spot at provincial level, and in 1994, Taining Golden Lake was granted national scenic spot, later in 1996, Management Committee of Taining Golden Lake Scenic Spot was approved to establish by provincial government of Fujian Province. In February 2005, it was granted Global Geoparks Network member by UNESCO. In November 2007, the Management Committee established the Management Bureau of Resource Protection. In 2008, Taining Golden Lake National Scenic Spot was renamed Taining National Scenic Spot. According to the principles of protection requiring scientific planning, unified managing, strict protection and continuous utilizing, the Management Office constituted a series of management policy and it carefully fulfilled its function of protection and management. A series of rule was made by the Management Committee who is playing a leading role, assisted by different sectors, Team of City Supervision, forestry police station, forest rangers that they formed a network for supervision and protection, such as *Measures for Resource Protection for Golden Lake Scenic Spot in Taining County*, *Stipulation of Management and Punishment of Golden Lake Scenic Spot in Taining County*, *Notice on Strengthening the Protection for the Natural Woods*, *Temporary Provision on Strengthening Development and Management on the Tourist Resources*, *Notice against Picking and Selling Wile Landscaped Plants and Collecting Resin in the Scenic Spot*, *Protecting Measures for China Danxia landform Natural Heritage in Fujian Province (draft)*. In addition, more communications were given for resource protection, and laws were strictly executed as well as effective protection policies were implemented that the heritage resources are under good protection and a beneficial cycle has been created for protection and management for the Heritage site.

Forest fire prevention, forestry pests, tourist population and structure, tourist service facilities and quality, population and growth, etc., are under supervision by the Management Office in the Heritage site, who has entrusted the relative functional departments or scientific institutions for regular and irregular supervision. In the Heritage site and the buffer zone, 12 monitory points have been fixed, including 2 for traffic management, 3 for ticket and tourist management, 5 for forest fire prevention, 1 for relic protection, 1 for landscape and tourist route supervision. Tourist service quality and the overall management level are improved in five aspects, resource protection, tourist services, management control, security and prevention, sustainable development.

Up to date, the water in Golden Lake, Shangqing River, and so on is quite good, and no distinct source of pollution is found around the water areas. Construction projects that are possible for water pollution are prohibited in the

headstream areas of Golden Lake, Shangqing River, etc., and the sewage from the local dwellers living along the rivers is processed until meeting standards before discharged to the water bodies nearby or for irrigation. Dustbins and garbage compression transfer stations were fixed in the nominate site that a perfect system, garbage collecting–transporting–transferring was built.

14 Langshan, Hunan

14.1 Introduction to the Heritage

14.1.1 Summary

Name of Property: Langshan.
County (City), Province: Xinning County, Shaoyang City, Hunan Province.
Geographical coordinates(Central):
26°20′24″N, 110°46′45″E
Area of Heritage site:
Heritage site (ha): 6600
Buffer Zone (ha): 6200

14.1.2 Description of Boundaries of the Heritage

The boundary of the heritage area and proposed buffer zone of the Langshan Heritage site is clearly labeled on the maps and actual spots. In the boundary of heritage areas, they mainly remain the integrities of natural geomorphology and ecosystem and prohibit all human activities with negative impacts on the heritage areas in exception of scientific researches. In buffer zone, the contents and patterns of human activities are also controlled to some extent.

The delimitation of boundary for the Langshan Heritage site mainly accord to the integrity of Danxia landform development and distribution, forest ecosystem and habitat of rare and endangered species. Rivers and ridgelines are regarded as natural lines, and the delimitation of boundary tries to avoid enclosing the modern artificial construction, settlements, mining, and hydropower stations. The delimitation of boundary for buffer zone not only accord to with the integrality of Danxia landform development and distribution, but also accord with the factors of protection of the Heritage site, such as avoiding the influence of human activities. As well as Heritage site, rivers, roads, ridgelines and phase-changed lines are regarded as natural lines. The western boundary keeps coherence with the boundary of Langshan Scenic Spot. The southern boundary keeps coherence with the common boundary of Guangxi Province.

14.1.3 Justification Statement of Outstanding Universal Value

(1) **Outstanding universal value of beauty**

The Heritage site is a typical representative region of peak forest with close and narrow valleys of Danxia landform at early mature stage in South China's humid region. The main body of Danxia landform in the heritage region is Danxia peak forest landscape; the keynote of it is "bare and red cliff." Langshan Mountain is a galaxy of series Danxia landforms with group structure, which displays the entire process of

H. Peng, *China Danxia*, Springer Geography,
https://doi.org/10.1007/978-981-13-5959-0_14

morphological formation, development and evolution from early rock sculpture, and segmentation to late corrosion morphology.

The Heritage site boasts thrilling and rugged positive landform as well as handsome and elegant negative landform, which forms various and unitized rhythm to combine inflexibility with yielding, to combine abundance with simplicity, to combine vividness with orderliness.

The bare and red cliff consisting of red sandy conglomerate is the most typical element of morphology in Langshan Danxia landform. The fixed shape and color of the heritage region may change from static state to dynamic state, from monotony to diversification, with the change of the environment. Blue and clear Fuyi River zigzag through this region, exhibiting the beauty of contrast as four seasons alternate. Beautiful flowers in spring, green bamboo sea in summer, blue water and red cliff in autumn, and agate color in winter, together form the rare natural beauty zone with distinct personality of special color beauty of Danxia landscape.

The heritage region retains farming activities followed several thousand years. The unbounded paddy fields show rural scenery changing with season, green in spring and yellow in autumn. Ancient folk houses with indigo roof and white wall, small bridges and flowing water are built along the mountains; castles, villages, and temples are hidden among green leaves in the mountains. Red cliffs, green mountains, ancient sites, and cottages are combined to integrity, a magnificent natural picture.

(2) **Outstanding values of earth history**

Langshan Mountain is located in the transition zone of Yangtze Plate and South China Plate, where is the just transition zone of China's second and third step. Zixin bed basin formed in Cretaceous period; Langshan Danxia landform formed in the late Neogene and the Quaternary. From the Cretaceous to the Quaternary, Chinese continental crust was compressed by the collision from the Indian Ocean plate and the Pacific Ocean Plate and was uplifted a lot. Especially ever since Quaternary, Tibetan Plateau known as the roof of the world rose, the lithosphere became mature and stable, and the distribution pattern of tectonic structure was set. Therefore, conclusion may be safely arrived at that Langshan Danxia landform marks the change of eco-environment caused by crust movement and climatic change in the latest geologic time; it is an outstanding representative of the main stage of earth evolution history; it is of great geoscientific significance! Especially the rise of the Tibetan Plateau known as the roof of the world is an important milestone of China's modern atmospheric circulation and terrain pattern. It is in this specific geological period that Langshan Danxia landform formed under the certain condition of crustal movement pattern and specific regional environment, climate changes, as a symbol of special ecological environment evolution.

Therefore, Langshan Danxia landform and the evolution of its climate and biomes, is a representative of the Earth evolution history in southeast Asia since the Cretaceous,and also represents the crust evolution process and the ancient environment change of southeast China in 100 million years. So, it is supposed to be the excellent example of the main stage of geoevolution history.

Danxia landform in the heritage region is a typical representative of early mature peak forest, peak cluster Danxia landform in southeast China humid area. Deep study on the Heritage Danxia landform will enrich, complete, and develop the Danxia landform theoretical system.

Danxia landform in the heritage region has obvious karst phenomenon, forming unique Danxia Karst landform such as funnel, depression, sinkhole, cave, and holes with calcium carbonate deposition landscape, which is a rare case of landform evolution process in China and the world, with high significance of stratigraphic correlation and special value of geoscientific research.

(3) **Outstanding universal values of biology and ecology**

Located in the subtropics moist monsoonal climate region, Langshan has been considered as a model region of "isolated island of ecology" and succession of Danxia vegetation, boasting original broadleaf evergreen forest on its top and ridge, organic combination of herbosa ecosystem and vine ecosystem on its rock wall, preserving endemic species which manifest narrow habitat of Langshan Mountain.

Langshan Mountain is the only representative for the South Mountain range among the Heritage sites, and it is the idea location for evergreen broadleaf forest in subtropics moist region, where mainly dominated by "isolated island of ecology" phenomena and narrow habitat. It is a region that has integrated phases of Danxia plant community. It is a distinctive habitat region to record the cooperated evolution relationship between angiosperm and animal (insects). It is a terrific model and experimental field to make comprehensive study on biologic diversity.

With a cover of vegetation rate of 85%, a forest greening rate of 75.9%, 9 vegetation types, and 71 plant formation, Langshan area is home to 1421 wild vascular plants and 150 large-scale fungus species including 75 species, among the plants, 21 species are listed in China Species Red List, 86 species are listed in IUCN Appendix List, 41 species are listed in CITES Appendix List, 23 species are listed in the state key protected plant species, in which three species are considered as I-level; Langshan area also possesses 26 mammal species, 94 birds species, 35 reptiles, 19 amphibians species, 36 fish species and 816 insects, including which are listed in China Species Red List (18 species), IUCN Appendix List (2), CITES appendix list (27 species) the state key protected plant species (18 species).

14.2 Description of Langshan Danxia

14.2.1 Physical Geography

(1) **Geology and Geomorphology**

The Heritage site is occurring in Cretaceous strip basin (Ziyuan—Xinning basin), which formed on the basis of Caledonian and Indo-Chinese folds in central south Hunan and controlled by NE and NNE directed fractures, with a material basis of continental facies, alluvial–diluvial facies, fluvial facies red conglomerate, and gravel of Cretaceous lower series Lanlong Group.

With large topographical unit, the Heritage site lies in the transition zone between China's second and third step. Located between Yuecheng Ridge and Miaoer Mountain, this mountain is high in the south and low in the north, boasting a highest peak with altitude of 818 m, and a lowest Fuyi River with altitude of 302 m in the north. The main landscape in the region is peak cluster Danxia landform, as well as karst landform.

(2) **Climate**

The heritage region belongs to subtropical moist monsoon climate zone.

Its annual average temperature is 15.5 °C. The hottest month is July with the average temperature of 26 °C; the coldest month is January with the mean temperature of 4 °C. The extreme maximum temperature arrives 37.2 °C, and the extreme minimum temperature is −6.8 °C. Langshan Mountain has typical characteristics of mountain climate in south with annual average rainfall 1450 mm, April to June occupies 45%, May has the most—an average of 213 mm, and December has the least—an average of 47 mm. Annual average sunshine duration is 1495 h, snowfall time is 18 days (10 days covered with snow), and frost-free period 291 days.

(3) **Hydrology and Water Resources**

The Heritage region is in the upper reaches of Zi River, which is called as Fuyi River flows in Heritage site. It runs through Langshan Danxia landform region from south to north, which is 24 km long, with annual average flow 78.5 m^3/s and the minimum flow 13.2 m^3/s. Lang Stream, Peng Stream, Li Stream, Qixing River, and Sanyuan River cross this heritage region, among which Lang Stream is 6 km long in the region with annual average flow 10.4 m^3/s and the minimum flow 1.4 m^3/s. Rainwater is the main source of surface water, and it is abundant in spring and summer and is relatively exhausted in autumn and winter.

(4) **Soil and Vegetation**

Soil types in the heritage region mainly include paddy soil, fluvo-aquic soil, red soil, purple soil, and so on. Thick soil layer has good capability in water conservation and fertilizer conservation, which is not easy to cause soil and water loss.

As located in the interjunctional transitional belt of various animal area and flora region, the Heritage site locates belongs to the south subzone of mid-subtropical evergreen broadleaved forest and is a typical subtropical hilly-type forest ecosystem in south China. Langshan Mountain is green in the four seasons with forest cover rate of 78.8%.

14.2.2 Geological Structure

14.2.2.1 Strata Lithology

The strata, from the Proterozoic Banxi Group to Quaternary System with absence of Middle and Upper Silurian, Jurassic, and Tertiary Systems, are exposed in the Heritage site and surrounding area. The Ziyuan–Xinning Basin, where the Heritage site is located, is 47 km long in N-S direction and 3–9 km wide in E-W direction. The Lanlong Group of Lower Cretaceous (K_1l) extending in NNE-NS direction consists of mauve continental molasse clastic formation unconformably overlaying pre-Cretaceous strata and granite. The lithology of the Cretaceous strata, with a total thickness of 200–2320 m, is described as follows:

- **The Upper part**: presenting morphologically as red hills in the Xinning–Yixinqiao area, is composed of thickly stratified mauve pebbled sandstone and pebbled argillaceous siltstone intercalated with mudstone and siltstone with gravels of sandstone, limestone, granite, etc., indicating the lacustrine facies.
- **The Lower part**: distributing in Langshanjie and Meixi–Ziyuan area, is a multi-component continental clastic formation composed of pebble, gravel, sand, silt, clay, etc., constituting thick to hugely thick bed of conglomerate, sandy conglomerate intercalated with pebbled sandstone and minor pebbled argillaceous siltstone and arkose sandstone, in which angular to subangular gravels of various sizes, are mixed together. The thickness of the strata decreases from 2320 m in Ziyuan and Meixi area in the South to about 150 m in Xinning in the North. The gravel components are also different from place to place, being mainly composed of granitic fragments (rock fragments, quartz, feldspar, and biotite) and minor siliceous rock debris, meta-sandstone and slate, etc., in Ziyuan and Meixi area in the south, but mainly of limestone, sandstone, and minor amount of granite in Langshanjie and Xinning area in the north. The formation belongs to piedmont pluvial and alluvial facies, as evidenced by the decrease of gravel size from posterior edge (mountain collado) to the front edge of the pluvial and alluvial fan, following the sequence of conglomerate-pebbled sandstone-sandstone-siltstone, and accordingly by the variation of psephicity from sharply angular, subangular to subrounded. The well-developed cross-beddings may be attributed to the sediments of short and torrential streams deposited according to topographical change of the palaeo-basin and the change of flowing direction after the mountain collado, or to the mutual transposition of several rivers of different flowing directions.

This set of coarse-grained red rocks of piedmont pluvial and alluvial facies constitute the material basis for forming the landscapes of Danxia landform in the Langshan Heritage site. Because of the low textural maturity and component maturity of the strata, as well as the material sources mainly from nearby granite and limestone, the landscape-shaping strata commonly contain unstable minerals with feldspar up to above 30% (Deng and Wang 1996), which could be easily weathered into kaolinite and sericite, causing the rocks to be readily disintegrated and rounded as accompanied by the decrease of resistance of the rocks to weathering.

On the other hand, the landscape-constituting strata usually contain considerable amounts of CaO, 7.28%–7.56% (7.24% in average) and $CaCO_3$ in cements, 2.00%–25% (8.75% in average) (Xiao et al. 1998). Therefore, the spectacular Karst-mixed landscapes, unique to the Heritage site, were formed in the landscape-constituting red beds (Table 14.1).

Table 14.1 Stratigraphic division and comparison of Cretaceous strata in the Heritage site

Geologic age		Central Hunan stratigraphic subdivision (VI_5^3)		
		Xinning–Ziyuan Basin	Hengyang Basin	Fossil feature
Tertiary	Eocene		Chashan-ao Group (E*c*)	Palaeo-vertebrate: Propachynolophus, Matutinia, etc.
	Paleocene		Zaoshi Group (E*z*)	Palaeo-vertebrate: Bemalambda, Hypsilolambda, etc.
Cretaceous System	Later Epoch		Chejiang Group (K*c*)	Dinosaur egg
			Daijiaping Group (K*dj*)	Dinosaur egg
	Earlier Epoch	Lanlong Group (K_1)	Shenhuangshan Group (K*sh*)	Ostracode: Cypridea (Pseudocypridina)–Eucypris–Cyprinotus Plant::Manica, Pagiophyllum, Brachyphyllum, etc.
			Dongjing Group (K*d*)	Bivale: Trigonioides–Nippononaia–Plicatounio combination Ostracode: Cypridea–Darwinula combination

Data from Hunan Bureau of Geology and Mineral Resources 1997, Lithology and Stratigraphic Divisions of Hunan, China Geologic University Press, 1997, Beijing

Cretaceous Amaranth Clasolite Formation

Alluvial–diluvial facies

Fluvial facies

Cross-bedding and erosion surface (white line)

Disintegration and sphering of rocks

14.2.2.2 The lithological effect on the development of landform

The Lower Cretaceous (K_1l)mauve clastic rocks constitute the material basis to form Danxia landform. It has some features as follows:

- The rocks, being thickly- and hugely-thick stratified, are hard and rich in calcareous matters, which constitutes the lithological basis to form Danxia landform. In dried condition, Langshan glutenite possess a compression strength of 831–1011 kg/cm^2. In water-saturated condition, conglomerate and sandstone have a compression strength of 441–546 kg/cm^2; Theoretically, for homogeneous rocks, the precipice height may be up to 3246–3949 m for dry rocks and 1683–2083 m for water-saturated.. However, due to the development of cleavages and cracks in rocks and to weathering process of rocks, the actual height is only tens to hundreds metre in maximum. The red glutenite was cut by networked structural cracks to form the morphological pattern of interlacing valleys and stone peaks as a result of water erosion and gravity devolution, in which the boundary faces between stone peaks and valleys are usually present as high and steep cliffs, namely *Chibi Danya*.
- Vertical cleavages are well developed in the strata with good perviousness for water. As such, precipitation water readily seeped downward with little surface runoff left and gullies were not developed on slope surfaces, which scarcely receded because of little erosion from flowing water. This is the main reason for the slopes to keep as steep cliffs.

Red cliff and wall

Red cliff and wall

Joint fissures of Yingzui rock

Joint fissures

- The strata are gently flat with vertical cleavages, which were expanded to form loaded cleavages due to progressive erosion and corrosion of seeping water. More vertical cleavages were formed from the stress release of opened rocks by loaded cleavages, directly causing the precipice wall to recede after collapsing. Meanwhile, the top parts of rocks were little eroded. The dilapidated rocks piled up at the foot of mountains. Thus, the pattern of flat top and steep body was developed.

14.2.2.3 The effect of geological structures on the development of Danxia landform

Langshan Mountain is located in the transition zone of south-central China's Yangtze Plate (northwestern Hunan in situ terrane) and South China Plate, southwest to central Hunan terrane (member of south Yangtze River block in the western margin of South China Plate) and east to Xuefengshan suture zone.

(1) **Features of main structures**

Fractures: The NNE-striking Gongtian–Ningxiang–Xinning–Ziyuan fracture belt extends more than 400 km with a strike of 20°–30° and dips NW with a dipping angle of 30°–45°, diagonally traversing the Heritage site. This regional fracture not only controls the distribution range and scale of the red beds in Langshan, but also determines the source materials and textures of red beds as well as the general distribution scale of Danxia landform.

Cleavages corresponding to the main fractures in the Heritage site are divided into four groups as follows:

- NNE-striking cleavages, dominating the Heritage site, strike 20°–30°, dip northwestward with dipping angle of 70°–80°, obviously having shearing property.

- NNW-striking cleavages, almost vertical, have tensile property with a strike of 320°–340°.

Besides, there are two other groups of compression-shearing cleavages called NE-striking cleavages and Near E-W cleavages.

Monoclinal structures: The Cretaceous red beds of sandstone and conglomerate occur as monoclines, dipping mainly SE with dipping angle of 12°–20°, occasionally up to 35°–45°, which is attributed to uneven uplifts in this area where the uplifting speed in the northwest is higher than that in the southeast, forming the monoclinal hills in this area.

(2) **Geologic structure control the development of landform**

The regional NE-striking Gongtian–Ningxiang–Xinning-–Ziyuan fracture zone and corresponding four groups of cleavages prepared the spatial conditions for forming various landscapes of Danxia landform in nature. The number and density of landscape scenes depend on structural cleavages and display alternating dense and sparse regularity.

The control of the four groups of cleavages on the formation of Danxia landscapes can be seen quite easily in the Heritage site. The Valley Cluster in Bajiaozhai and the Danxia Peaks in Jinyunaohai, as well Najiao Peak and Candle Peak, are controlled by two groups of cleavages, NNW-striking (310°–330°) and NNE-striking (310°–330°) cleavages. The block morphology in Baimianzhai is controlled by NE-striking (50°) and NNW-striking (330°) cleavages, in which the precipice wall is just the main cleavage face of NNW-striking cleavages. The General's Stone was developed along near EW-striking (80°) and NNW-striking (350°) cleavages. The Natural Stone Bridge in Tanjiaba is related to the NNW-striking (315°) cleavage and gravity dilapidation. The stretching direction of Tianyixiang in Niubizai is 3400, of Yixiatian in Yuquanshan is 550. Anyway, Danxia landscapes reveal the trace of the four cleavage groups in various form.

14.2.3 Important Geological Events

(1) **Pre-cretaceous Geological Event**

The 1000 Ma's stratigraphic record in the region surrounding the Heritage site indicates that collision and collage between South China Plate and Yangtze Passive Continent took place in late Neo-Proterozoic, causing the formation of a union terrane from the Northwest Hunan terrane and the Central Hunan terrane, which became the basements of the passive continental shelf of Yangtze in late Sinian and early Paleozoic, accompanied by intrusions of collision-type (S-type) granite. Later, A-type subduction was introduced, causing the mélange of structures and rocks in the suture zone between the early Paleozoic union terrane and the South Hunan terrane and the final collage between South China Plate and Yangtze Plate was accomplished,

System of cleavages and cracks in the landscape-constituting strata of Danxia landform

accompanied by large scale of intrusions of collision-type granite (Caledonian Movement).

In Mesozoic, the subduction of the Pacific Plate into the East Asia continent changed the Paleozoic structural pattern and caused the formation of continental red bed basins and several NE-NNE-directed tectono-magmatic zones. The Indosinian orogeny made the D_2-T_1 sedimentary covers to be detached on the basement, accompanied by the formation of thin-skinned tectonics and granitic intrusions (along the margin of the suture zone), as well as the entire recession of sea water.

The Yanshan Movement was featured by the activity of large scale NE-NNE tensile fractures, accompanied by the formation of continental faulted basins in Cretaceous and Paleogene. In East Hunan, rift basins and alkaline flooding basalt appeared (142–81 Ma). In Shaoyang–Xinning area to the west of Hengyang basin, a series of faulted basins of various scales were developed along the fracture zone, including the Ziyuan–Xinning red bed basin.

(2) **Cenozoic Crustal Uplift**

Due to the collision from the India Plate, the Tibetan Plateau was wholly uplifted since Paleogene, especially in Pliocene, which caused the formation of the three-step topography of China from west to east, including the I-level Tibetan Plateau (average altitude 4000 m), the II-level Yunan–Guizhou Plateau (including Xuefengshan) (average altitude 2000 m) and the III-level low hills and wide plains (below 1000 m).

The Heritage site is located in the low hills area of Central Hunan in the transition zone between the I-level and II-level steps of China. After undergoing the intermittent uplift in the Himalayan and Neo-Tectonic movements, three levels of planation surface lowering northward were formed in the Heritage site, i.e., 800–700 m (I, late Miocene), 600–500 m(II, Pliocene-early Pleistocene) and 400–300 m(III, middle Pleistocene). The Heritage site is still rising due to the effect of the uplifting movement of Tibetan Plateau, as evidenced by the valley constructional terrace in the Wuzhudong river bed, where the dualistic structure surface of valley constructional terrace is 8 ± m above modern river bed.

(3) **Floral evolution and breeding of new species**

The Heritage site is endowed with superior ecosystem environment for a unique species named *Chirita Langshanica* which is wildly grown in the red beds. The plant has such features as being drought-resistant, high temperature-resistant, good healing function, and strong vitality. The plant may rejuvenate and germinate in wet sand in a week even 45–50 days after drying up.

14.2.4 Types and Features of Danxia Landform

14.2.4.1 Types of Danxia Landform

According to the petrological characteristics and detrital composition, Langshan Mountain belongs to glutenite Danxia landform;

According to the geomorphic shape, Langshan Mountain belongs to peak forest and peak cluster Danxia landform.

According to the process of development, the main body of Langshan belongs to Danxia at mature stage, while its surrounding landform tends to be old.

According to monomer topographic form, Danxia landform in the heritage region can be classified as positive landform and negative landform. Positive landform can be divided into stone village, stone wall, stone column, stone peak, peak cluster, peak forest, colluvial rock, and so on; negative landform include linear valley, roadway valley, gorge, cave, and so on.

Monomer Topographic Forms of Danxia Landform in the Heritage region

Type		Quantity	Representative landscape
Positive Landforms	Danxia cliff	25	Longtouxiang, Red cliff stone, Baizhang Cliff
	Danxia stone fort	7	Bajiaozhai Village, Niubizai Village, Hongwazai Village
	Danxia stone wall	20	Hongwa Bare Cliff, Stonewall of Linjiazai
	Danxia stone beam	2	Warship stone, peckerwood stone
	Danxia column	5	General stone, muzhishi, peckerwood stone
	Danxia peak	225	Candle peak, Pepper peak, Doulifeng
	cuesta	65	Jingyunaohai
	Sculpting landform	22	General stone, Dog-head Stone, Pepper peak
	Colluvial block	12	Frog Stone, Shidao Stone, Luanshijian
Negative landform	Linear ravine	15	Tiyixiang Lane, Shuibo Lane
	Danxia valley	33	Bajiaozhai valley, Wuyunxia
	Natural bridge	3	Tangjiaba Natural bridge, Moon Natural bridge
	Cave	2	Xielin cave, Shuiboxia cave
	Flat cave	10	Shuangyanhuilang, Qicenglou, Louwailou
	Rock groove	4	Wanjingcao groove
	Beehive cave	4	Honghuachibi, Linjiazhai
	Cavern	4	Wuzhuyan Cvern, eilian Cvern
	Waterworn cave	2	Tuanyusdhi
	Certical cave	1	Poposhi
	Accumulated cave	5	Xxangquandong, Yeniudong
	Collapsed cave	2	Xianglongdong, Yinzhudong
	Natural wall painting	2	Qingfengxiang, Longdanbi

14.2.4.2 Characters of Danxia Landform

(1) **The Individual Characteristics: Diverse Forms and Majestic Landscapes**

There are diverse monomer morphological types in the Heritage Langshan Mountain. Landscapes there are high, steep, deep, and long in the form and majestic, odd, quiet, and beautiful to one's feeling. We can see magnificent Danxia mesa, soaring stone forest and peak cluster, precipitous bare and red cliffs, narrow and zigzag rock lanes and stone cisterns, as well as arched stone bridges, cliff caves, and incomparable pictographic stones like person, bird, beast, or others, all of which are outstanding for its unique and natural beauty.

Bajiaozhai Village Lumpish landform

Linjiazhai Village column landform

Danxia natural bridge

Danxia isolated peak

Bajiaozhai Village Cliff

Pagoda Peak

Trunk Split Stone

(2) **The Group Characteristics: Grand Scale and Magnificent Landscapes**

Langshan Mountain is a galaxy of series Danxia landforms with group structure, which mainly displays the dense peak cluster peak forest landform. Macroscopically, folds of mountains and peaks are magnificent and massy.

Pepper Peak Groups

Peak cluster–peak forest

(3) **Typical representative of Dense Peak Forest and Peak Cluster Danxia Landform at Mature Stage**

Langshan Danxia landform is typical for the early mature peak cluster Danxia landform with close and narrow valleys, as well as the old Danxia landform distributed along both sides of Fuyi River.

Danxia landform at mature stage

Danxia landform at mature stage

(4) **Unique Danxia Karst Landform**

The amaranthine sandy conglomerate constituting Danxia landform in the heritage region is generally in the form of cement or gravel with high $CaCO_3$ content and has obvious karstification. The main phenomena of this landform are:

(i) corroded funnel and corroded depression, a total of 30 have been found in Heritage site, the biggest one is named as Wanjing Trough, which is a roadway-shaped dissolution of 260 m in length, 40 m in depth, and 5–10 m in width;

(ii) underground river, a dozen meters long underground stream reach, is found near Baimianzhai Village of Yaoshi Town and near Datangchong and Yangjia Village south or north of Penholder Mountain.

(iii) two kinds of caves, one is the karst cave formed by erosion and gravity collapse, such as Five Column Cave of Yaoshi Town. It is a complex cave system developed in the calcareous limestone and gravel debris of red bed, 5 m high and 40 m wide, with five columns about 1 m from the ground to the roof, very exotic; the other is the cave formed by combined action of erosion and corrosion in the red layer of calcareous conglomerate and often effected by fracture.

(iv) The upper Cretaceous red bed conglomerate developed into Danxia landform, the lower limestone developed into karst landform, which is a prominent character of the heritage region in its landform development. Feilian Cave, for example, in devonian limestone under red bed conglomerate, is a complex underground river cave system.

Tianyi Alley

Shuibo Alley

(5) The Grand Scale of Roadway Valley, Linear Valley, and Natural Bridge Is Rare in Similar Danxia Landforms

"Deep slit" landscape which is rare in Danxia landform's formation has been discovered in Langshan Mountain for more than 10 places, among which the NNW-trending "Tianyi Alley" is 238.8 m long, 0.8–0.33 m wide, 0.5 m for average, and with rock wall on both sides of 80–120 m. When someone is walking down the bending lane, he could only see linear blue sky over head and bare cliffs on both sides, not finding the origin and the termination of the lane.

Yuxian Alley

Three Danxia natural bridges have been discovered in the heritage region, which are a type of landform with high ornamental and scientific value. Tangjiaba natural bridge has the span of 64 m, width of 14 m, height of 20 m and rock layer thickness in the central arch of 5 m. The second natural bridge called Moon bridge, which looks like a crescent, of 45 m in length, 20 m in height, 10 m in width, and its rock layer thickness is 7 m.

14.2.5 Natural Landscape

14.2.5.1 Danxia Landscape

The nature is the unique engraver, who makes use of the rising new formation as the elevator. And the fissuring and the layers are the lines which can be engraved. After many years' hard work, the engraver introduces to the nature a fabulous artwork.

The main landscape of Langshan Heritage site

No.	Name	Scenic spot	Characteristics
1	Tianyi Valley	Scenic spot of Tianyi Valley	It is 80–120 m high, 0.85–1.70 m deep, 0.8 and 0.33 m wide in widest and narrowest parts, respectively, with an average width of 0.5 m. Located at its bottom, linear sky on your head and bare cliffs on both sides
2	Pepper Peak	Scenic spot of Pepper Peak	Danxia huge stone with hight of 120 m, its top is bigger than bottom, the perimeter of top is 100 m, of bottom is more than 40 m. It is red-colored, prism-like, like a large red pepper
3	Natural Bridge	Scenic spot of Natural Bridge	A haplopore stone bridge originated from collapsed Danxia boulder, having spaning length of 64 m, arch hight of 20 m, surface width of 14 m, and thickness of 5 m. There are no concave and convex on interior arch
4	Whale peak cluster	Scenic spot of Bajiaozhai Village	It is consisted of hundreds of Danxia peaks lying tens of square kilometer, like a group of whales is playing in the sea
5	General stone	Scenic spot of Fuyi River	It is Danxia column at old age with height of 75 m and perimeter of 40 m. It is located at the east bank of Fuyi River, standing like a general
6	Camel peak	Scenic spot of Pepper Peak	Surrounded by steep cliffs on three sides, with two hollows left at its top, so It is looked like camel

(continued)

No.	Name	Scenic spot	Characteristics
7	Bajiaozhai Village	Scenic spot of Bajiaozhai Village	It located on the culmination of Mt Langshan, with an elevation of 818 m, a relatively hight of 502 m. It is surround by steep cliffs, and has eight horns pointing at various direction
8	Hongwa Bare cliff	Scenic spot of Zixiadong	It is a bare and red stone wall surrounded by mountains; it has a length of 700 m and a mean height of 120 m. The wall is fullfilled with various caves. It is also a fabulous echo wall
9	Bajiaozhai ravine	Scenic spot of Bajiaozhai Village	Five ravines under the Bajia stockaded village, having length of 3–5 km, width of 1–100 m. Boulders are scattered in the valleys, some have piled into caves. Valleys together with bare cliffs, boulders, green mountains and brooks that has staged natural galleries
10	Candle peak	Scenic spot of Pepper Peak	It is has elevation of 674.4 m, with the cliff hight of 99 m. It is cylindrical, candle-shaped. The eastside cliff is flat and bear, called unlettered tablet
11	Yuxian Valley	Scenic spot of Tianyi Valley	The alley is 170 m long, 0.85–1.70 m wide, 85–105 m high from bottom to peak, NNW-trending. It is a straight and even entry leading to Niubi Stockaded village
12	Yixiantian valley of Linjiazai Village	Scenic spot of Pepper Peak	Yixian Valley is NE-trending, surrounded by cliffs, and 60 m long, 40 m high and 0.5–2 m wide. It is only way to Linjia Stockaded Village
13	Baimianzai Village	Scenic spot of Natural Bridge	It has elevation of 510 m, cliff hight of 80 m, width of 100 m. Surrounded by steep cliffs, only a trail in the southeast could lead to the top. Four doors of Ming and Qing Dynasty, and Neolithic relics were reserved here
14	two ledges corridor	Scenic spot of Fuyi River	It is a NE-trending alley composed of C-type groove, which is divided into up and down layers with a distance of 5 m. Each gallery is 330 m long
15	Woodpecke Stone	Scenic spot of Fuyi River	It is located at the bank of Fuyi River, composed of a column cliff which is 90 m high. Its head part characterized by a inclined stone beam which is just like a peak sloping down. The holes on its top looks like eyes of woodpecker
16	Warship stone	Scenic spot of Fuyi River	It is composed of three giant rocks spanning from east to west and has an average length of 300 m and a mean height of 150 m around. The front of the giant rocks raises highly and the end part tilts in arc shape, it just looks like a warship
17	Conch peaks	Scenic spot of Bajiaozhai Village	Taking the lamination of red bed forms as its spiral striation, the stones look like thress conches and a big shell. The conch peaks are 100 m, 120 m, and 150 m high, respectively, and the shell peak is about 300 m high

Natural landscape of Langshan

Dense Danxia peak cluster

Langshan peak forest

Langshan Danxia peak forest

Langshan Danxia peak forest

Langshan cuesta cluster

Difference of occurrence and landform between two sides of the large rupture

Linjiazai Village

Danxia isolated peak

Langshan stone arch (natural bridge)

Camel peak and Candle peak

General Stone

Pepper peak

Goghead Stone

Conch Peak

14.2.5.2 Water Sight

Fuyi River without any water pollution runs along the Langshan Mountain, adding to the beautiful scenery. Waterscape and hill scenes reflect each other, benefiting by association together, that makeup an unusual region with natural beauty.

14.2.5.3 Biological Landscapes

Different biologic species result in diversity of biologic landscape, various plant communities make up Danxia polychrome and colorful. The forest communities together with the Danxia landscape have made compound landscapes. They are or green spots sprinkling on the red cliffs, or red patches prominent on the green vegetation, or green shades covering the mountain tops, or trees green and shady at the foot of mountains, or orchids blooming, vines climbing, that have contributed to this poetic and picturesque land.

14.2.6 Biodiversity (Biology and Ecology)

14.2.6.1 Ecoregion

According to the world classification of Udvardy's (1975) biogeographical provinces, nominating site belongs to subtropical and temperate rain forests or woodlands of Chinese subtropical forest in the Palaearctic Realm. Within the WWF's Global 200 ecoregions, it is included in the ecoregion of Southeast China-Hainan Moist Forests in tropical and subtropical moist broadleaf forest. The climate here is mild. Precipitation patterns are monsoon-dominated with summer precipitation and hot weather during the spring. The highest peak of hills is elevation of 818 m. The soil types mainly include red soil and purple soil types. The bulk of the vegetation in the wilderness is moist mid-subtropical forest vegetation.

The Langshan area was formed during the Cretaceous period that was contemporary with the origin and evolution of the advanced gymnosperms and the angiosperms. This contributes to rich ancient originated plant species. For example, there are four of six in China, seven in world conifer families in the nominating site. Some ancient angiosperms also have been preserved here, including the genera of *Liriodendron*, *Mangolia*, *Michelia*, *Altingia*, *Tapiscia*, and some families of Theaceae, Caylycanthaceae, Lauraceae, Bretschneideraceae, and Sargentodoxaceae.

The Heritage site covers 13 subecosystems of terrestrial, freshwater, and combined ecosystems, mainly dominated by evergreen broadleaf forests in terrestrial ecosystem.

14.2.6.2 Species

(1) **Communities**

The Heritage site supports rich communities, including nine vegetation types and 53 plant communities, of which woody communities are 38, herbaceous communities 11, and woody-vine communities four, but dominated by 18 communities of evergreen broadleaf forests. On the top and ridge of hills, there are dry-endurance evergreen broadleaf forests. Forms *Cyclobalanopsis chungii* and *Lithocargus hancei*, *Ternstroemia gymnanthera,* and *Symplocos ernestii* are representative. These forests are isolated small fragmentary association with green color all the year round. The communities are small in size with small or middle size of coreaceous thick leaf, branching, and curving trunk. Generally, there are tree layer, shrubs layer and herbs layer, or tree layer and herb layer within the community. The typical evergreen broadleaf forests grow in the valley, dominated by *Castanopsis tibetana* form and *Cinnamomum camphora* form. These forests are mosaic communities with relative large community size, small or middle size coreaceous leaf, and high height branches. Three layers including tree, shrub, and herb layers could be divided in these forests. Vine plants climbing on the cliffs are common in the nominating site, of which *Pileostegia viburnoides* and *Parthenocissus tricuspidata* are representative. The herbaceous communities are formed by some drought-tolerant species, also indicating the drought vulnerability of the Langshan Danxia region. There are 11 formations, including Form. *Chirita xinningensis*, Form. *Didymocarpus villosus*, Form. *Ranunculus xinningensis* and *Pogonatherum crinitum*, Form. *Lycoris aurea*.

(2) **Species**

Langshan area is home to 1421 plant species including 75 species of fern, 9 conifer species, and 1337 species of flowering plants; 150 species of macrofungi; 25 mammals, 94 birds, 35 reptiles, 18 amphibians, 37 fish species, and 816 insects.

Among the plants, 102 species are listed in the state key protected plant species (23 species), China Species Red List (88 species), and CITES Appendix List (41 species). The first-class state key protected plants include *Taxus wallichiana* var. *mairei*, *Amentotaxus argotaenia,* and *Bretschneidera sinensis*. For vertebrate, 61 species are listed in the state key protected plant species (18 species), China Species Red List (31 species), and CITES Appendix List (27 species). One bird, *Tragopan caboti*, is listed as the first-class state key protected animals.

Ranunculus xinningensis

Chirita langshanica

(3) Endemic species

Two species, *Ranunculus xinningensis* and *Chirita langshanica*, are endemic to Langshan Danxia area. They are found only growing on the cliffs of danxia hills within the Langshan area after more than twenty years of collection in Langshan area and around regions. Therefore, we can consider these two species as strictly environment limited endemic species.

Moreover, there are 1096 tree individuals aged more than one hundred years. They belong to 66 species in 50 genera of 34 families.

14.2.6.3 Biodiversity

Diverse and complex ecosystem For a single danxia hill, on the cliffs the communities of lichens and mosses as well as of mosses, herbs and vines have developed; on tops and ridges, the dry-bearable evergreen broadleaf communities have been conserved; in valleys, the typical evergreen broadleaf communities have existed. However, these communities are generally small in scale and are isolated from each other.

The features of this flora are represented by this evergreen broad-leaved forests of subtropical China consist mainly by the plants that shared by China and Japan, the endemic plants of China as well as some species of East Asia and North America. Meanwhile, the typical species of the humid evergreen broad-leaved forests of middle subtropical China can be found here.

Anyway, the ecological characteristics of their vegetation types are special and outstanding of the danxia vegetation, thus providing important models for studying conservation of primary communities and recovery of secondary communities in central subtropical danxia regions.

Comparing with the other series nominating danxia sites and other sites around the Lanshan areas, the unique biodiversity features showed in following two aspects:

(i) There are some endemic communities. There are three endemic communites, *Chirita xinningensis*, *Ranunculus xinningensis*, and *Ranunculus xinningensis* and *Pogonatherum crinitum*, dominated by two endemic species.

(ii) The "sugarloaf hills" and similar hills with sharp cliffs cause the outstanding ecological isolation effects. Within four 400 m^2 plots on four different tops of danxia hills, totally 71 plant species were found. Among those 71 plant species, the species that have not shared by any two plots are 59.15% of the total species, and the species that appeared in three plots only reached 7.04% of the total species. Within four 400 m^2 plots on the different positions of same danxia hill, there are 111 plant species. Among these plant species, 70.02% of them are not shared by any two plots and only 0.9% of them are shared similarly by three plots. Above results showed that the species similarity among different tops of danxia hills and different positions of same danxia hill is very low, and that indicated the apparent ecological isolation effects.

Progress of the primary succession is shown on the surface of the Danxia cliffs and tops and ridges of danxia hills in nominating site. On the cliffs of danxia hills with thin and shallow soils, there are high diversity herbaceous communities, mainly dominated by some lithophilous orchids, *Chirita* spp. and *Didymocarpus* spp. of Gesneriaceae and *Lycoris* spp. of Amaryllidaceae. The communities growing on the top or ridge of danxia hills are shrubs, evergreen broad-leaved forests, and coniferous and broad-leaved mixed forest. In valleys of the Langshan danxia region, there are pioneer forests (deciduous broad-leaved forests and *Pinus nassoniana* forests), evergreen and deciduous broad-leaved forests, and evergreen broad-leaved forests. The high diversity communities of the nominating site supply a natural laboratory for the study of the progresses of community successions.

The Heritage site is the especial place where the coevolution between the plants of Schisanderaceae (one of the plant group considering as being the basal angiosperm) and pollinators (insects) were recorded. Here, a high especially specialized deception pollination system is reported in *Kadsura longipeducu-lata*, one member of the family Schisanderaceae.

14.3 History and Development

14.3.1 Evolutionary Process of Langshan Danxia

(1) Evolution of Basin Base And Formed Stage of Basin

A collision and collage between South China Plate and Yangtze Passive Continent took place in late Neo-Proterozoic, causing the formation of a union terrane from the Northwest Hunan terrane and the Central Hunan terrane, which became the basements of the passive continental shelf of Yangtze in late Sinian and early Paleozoic, accompanied by intrusions of collision-type (S-type) granite. Later, A-type subduction was introduced, causing the mélange of structures and rocks in the suture zone between the early Paleozoic union terrane and the South Hunan terrane and the final collage between South China Plate and Yangtze Plate was accomplished, accompanied by large scale of intrusions of collision-type granite (Caledonian Movement).

In Mesozoic, the subduction of the Pacific Plate into the East Asia continent changed the Paleozoic structural pattern and caused the formation of continental red bed basins and several NE-NNE-directed tectono-magmatic zones, including Langshan Heritage Site. The Indosinian orogeny made the D_2-T_1 sedimentary covers to be detached on the basement, accompanied by the formation of thin-skinned tectonics and granitic intrusions (along the margin of the suture zone), as well as the entire recession of sea water.

From Jurassic period to Cretaceous period, Yanshanian Movement was featured by the activity of large-scale NE-NNE tensile fractures, accompanied by the formation of continental faulted basins in Cretaceous and Paleogene. In east Hunan, rift basins and alkaline flooding basalt appeared (142–81 Ma). In Shaoyang–Xinning area to the west of Hengyang basin, a series of faulted basins of various scales were developed along the fracture zone, including the Ziyuan–Xinning red bed basin.

(2) **Accumulative Phase of Red Beds**

The Xinning–Ziyuan Basin, where the Heritage site is located, is 47 km long in N-S direction and 3–9 km wide in E-W direction. The north Xinning County non-conformitily overlay on the south side of D_2-T_1 complex syncline and the south Xinning County non-conformitily overlay on Z-O and Caledon granite. The NNE-striking Gongtian–Ningxiang–Xinning–Ziyuan fracture belt extends more than 400 km with a strike of 20°–30° and dips NW with a dipping angle of 300–450, diagonally traversing the Heritage site. This regional fracture controls the Cretaceous sediment and results in unsymmetrical half graben-like basin. In Cretaceous period, a great deal of argillaceous silt and stone block carried by torrential flood rapidly sediment in the intermountain graben basin, accumulated thick amaranth conglomerate and glutenite characterized by alluvial–diluvial facies, fluvial facies, and low maturity, which were the material basis of the Danxia landscape.

(3) **Crust Uplifting and Developmental Stage of Danxia**

Because of the subduction and compression from the Pacific and Indian plates, the acting,Cretaceous red beds changed into monoclinal structure with four groups of derived structural fracture or cleavages. In early Neogene period, due to the Himalayan orogeny, Langshan region were uplifted,dissected valleys were gradually developed, Fuyi River incised into amaranth glutenite while it was running through the basin and carrying with silts, which cause the appearance of valleys and cliffs, that is Danxia landscape. After that, Danxia landscape has been denuded to flat until to last Neogene when new tectonic movement brought three intermittent uplifts to Mt. Langshan.

Thanks to the downcutting function of rejuvenated streams on netty vertical cleavages, a large area of Danxia landscape were formed, dominated by monoclinal peak forest. Meanwhile, third class valley terrace and alluvial flat plain were formed in Fuyi valley. Geological actions such as fluvial erosion, gravity action, concentric weathering, and biologic weathering also made great contribution to the modern Danxia landform.

14.3.2 History and Actual State

Mt. Langshan has been a place of graceful landscape and developed civilization. It was said that the human activity dating back to 4000 year ago was originated at Yaoshi, which was famous for earthenware in Neolithic period. It has become tourist attraction since from Song Dynasty and has inspired many great classics of ancient writers and calligraphers. Academician of Chen Guoda, the founder of Danxia landform, made field survey in 1992 and gave praise of "It is national treasure, the soul of Danxia landscape."

Mt. Langshan was approved of municipal scenic spot in 1992, of provincial scenic spot in 1993, and Management Office of Langshan Scenic Spot was found in 1995. Langshan was granted as National Geopark by former Ministry of Land and Resource of the People's Republic of China in December 2001, as the National Key Scenic Spot by State Council of the People's Republic of China in May, 2002. Xinning County was approved of Nationally Designated Eco-Demonstration Region in 2003. *Regulations on the Protection of Langshan Scenic Spots* was set down in October 2004 and was brought into effect in January 2005, which meant the protection of Langshan Scenic Spot was stepping into legislation and normalization.

The Heritage site has a total area of 66 km^2, including eight administrative villages of Langshan Town. And it has a population of 4090 with a population density of 62 persons/km^2, consisting of 3700 original residents, 90 managerial personnel, and 300 exotic service population.

Since 1992, five scenic regions with a total area of 3.2 km^2 have been developed. Agriculture and tourism are the main industry in Heritage site. Adhering to the development strategy of "tourism drive development," the local people has been presently alleviating poverty and becoming prosperous at.

14.3.3 Human Activities and Cultural Landscape

The Heritage site owns a long history and luxuriant cultural heritage. The new Neolithic sites in Zhoujiashan Mountain and Baimianzhai, in the southern part of Langshan, artefacts such as stone axes, cutting tools, scraping stones, stone chisels, stone spades, stone balls, stone clusters, and grinding stones, have been excavated. There are also some bone artefacts such as needles, hairpins, and crockery, including axes, cans, pots, and other pottery pieces. Research has shown that hunting and fishing, supplemented by gathering, were the main human activities at that time in Langshan.

Due to its precipitousness terrain, the Heritage site were the military position and stockaded villages for ancients, such as the relics of Yiyong village and Bajiao village. Additionally, Mt. Langshan embodies the ethnic civilization with natural landscape, is famous for Bajiaozhai village, Zixiadong, Holy Land of Taoism and Buddhism, and temples.

Agricultural civilization and traditional farming activities have existed here for thousands of years, their influence on the natural environments of the Danxia landscape has been minimal. The residential population in the neighboring valleys also remains low. There are no industrial activities in the heritage area at present nor any large-scale human disturbance. The Heritage sites are, therefore, largely preserved in their natural state. Over a very long period concepts of harmony in Chinese traditional culture and the implementation of laws in recent years have had a strong influence in the protection of environmental resources in the Heritage. For the most part, local residents have maintained a tradition of living in harmony with their natural surroundings and respecting nature.

Stone tools excavated in Baimianzai

Stoneware excavated in Baimianzai

Yijunzai Village

Door of Bajiaozai Village

Folk Custom

bamboo pole dance

countryside scene

spring scenery of camel peak

14.4 Physical Features and Its Value of Danxia Heritage

(i) Langshan Mountain is located in the transition zone of Yangtze Plate and South China Plate where is also the transition zone of China's second and third step. It is supposed to be the epitome of evolution of China's third step.
(ii) Peak cluster and peak forest of Danxia landform at mature stage is most typical, and the Danxia landform at outlying zone tends to be old, constituting a complete system with other nominations in development stage.
(iii) With Characteristics of Dense Peak Forest, Peak Cluster Danxia Landform, as well as Roadway Valley, Linear Valley and Natural Bridge.
(iv) Unique Danxia Karst landform such as funnel, depression, sinkhole, cave, and holes with calcium carbonate deposition landscape as well as compounding topography of both Danxia landform and karst, which is a rare case of landform evolution process in China and the world, with high significance of stratigraphic correlation and special value of geoscientific research.
(v) So many peculiar plants, especially *Ranunculus xinningcnsis*, *Chirita langshanica* are found in Langshan, sufficiently indicating its important position in the biodiversity distribution center in the southern mountainous regions in China, offering test area for investigation of species fractionation and geographical relationship.
(vi) Langshan Mountain is the idea location for evergreen broadleaf forest in subtropics moist region, where mainly dominated by "isolated island of ecology" phenomena and narrow habitat. It is a region that has integrated phases of Danxia plant community. It is a distinctive habitat region to record the cooperated evolution relationship between angiosperm and animal (insects). It is a terrific model and experimental field to make comprehensive study on biologic diversity.
(vii) Mt. Langshan landscape boasts prominent esthetics, combining its shape and color with temperament to constitute a piece of florid natural painting.

14.5 Coincidental Standards of World Heritage

According to the "Convention Concerning the Protection of the World Cultural and Natural Heritage," Langshan Danxia Heritage site qualifies three standards.

Criterion (vii): to contain superlative natural phenomena or areas of exceptional natural beauty and aesthetic importance.

The Heritage site is a typical representative region of peak forest with close and narrow valleys of Danxia landform at early mature stage in South China's humid region. The main body of Danxia landform in the heritage region is Danxia peak forest landscape; the keynote of it is "bare and red cliff." Langshan Mountain is a galaxy of series Danxia landforms with group structure, which displays the entire process of morphological formation, development and evolution from early rock sculpture, and segmentation to late corrosion morphology.

The Heritage site boasts thrilling and rugged positive landform as well as handsome and elegant negative landform, which forms various and unitized rhythm to combine inflexibility with yielding, to combine abundance with simplicity, and to combine vividness with orderliness.

The bare and red cliff consisting of red sandy conglomerate is the most typical element of morphology in Langshan Danxia landform. The fixed shape and color of the heritage region may change from static state to dynamic state, from monotony to diversification, with the change of the environment. Blue and clear Fuyi River zigzag through this region, exhibiting the beauty of contrast as four seasons alternate. Beautiful flowers in spring, green bamboo sea in summer, blue water and red cliff in autumn, and agate color in winter, together form the rare natural beauty zone with distinct personality of special color beauty of Danxia landform.

The heritage region retains farming activities followed several thousand years. The unbounded paddy fields show a rural scenery changing with season, green in spring and yellow in autumn. Ancient folk houses with indigo roof and white wall, small bridges, and flowing water are built along the mountains; castles, villages, and temples are hidden among green leaves in the mountains. Red cliffs, green mountains, ancient sites, and cottages are combined to integrity, a magnificent natural picture.

Criterion (viii): be outstanding examples representing major stages of earth's history, including the record of life, significant ongoing geological processes in the development of landforms, or significant geomorphic or physiographic features.

It is an outstanding example for ongoing ecological and physiological evolution process of continental, fluvio-terrestrial, coastal and marine ecosystem, and animal and plant community.

Langshan Mountain is located in the transition zone of Yangtze Plate and South China Plate, where is the just transition zone of China's second and third step. Zixin bed basin is formed in Cretaceous period; Langshan Danxia landform is formed in the late Neogene and the Quaternary. From the Cretaceous to the Quaternary, Chinese continental crust was compressed by the collision from the Indian Ocean plate and the Pacific Ocean plate and was uplifted a lot. Especially ever since Quaternary, Tibetan Plateau known as the roof of the world rose, the lithosphere became mature and stable, and the distribution pattern of tectonic structure was set. Therefore, conclusion may be safely arrived at that Langshan Danxia landform marks the change of eco-environment caused by crust movement and climatic change in the latest geologic time; it is an outstanding representative of the main stage of earth evolution history;—it is of great geoscientific significance! Especially the rise of the Tibetan Plateau known as the roof of the world is an important milestone of China's modern atmospheric circulation and terrain pattern. It is in this specific geological period that Langshan Danxia landform formed under the certain condition of crustal movement pattern and specific regional environment, climate changes, as a symbol of special ecological environment evolution.

Therefore, Langshan Danxia landform and the evolution of its climate and biomes, is a representative of the Earth evolution history in southeast Asia since the Cretaceous, and also represents the crust evolution process and the ancient environment change of southeast China in 100 million years. So, it is supposed to be the excellent example of the main stage of geoevolution history.

Danxia landform in the heritage region is a typical representative of early mature peak forest, peak cluster Danxia landform in southeast China humid area. Deep study on the Heritage Danxia landform will enrich, complete, and develop the Danxia landform theoretical system.

Danxia landform in the heritage region has obvious karst phenomenon, forming unique Danxia karst landform such as funnel, depression, sinkhole, cave, and holes with calcium carbonate deposition landscape, which is a rare case of landform evolution process in China and the world, with high significance of stratigraphic correlation and special value of geoscientific research.

Criterion (ix): be outstanding examples representing significant ongoing ecological and biological processes in the evolution and development of terrestrial, freshwater, coastal and marine ecosystems, and communities of plants and animals.

The Heritage site belongs to ancient north realm of biogeographic zoning (Udvardy 1975), called Chinese subtropical forest biogeography province, where possesses the largest and integrated broad-leaved evergreen forest in the world. However, most other regions at the same latitude as Langshan are grassland and desert.

Primary succession and secondary succession of plant communities are presented in Langshan.

Progress of the primary succession is shown by medium-term undershrubbery and late arbour evolving from the early moss and herbosa, by the ecosystem succession from inferiority to advanced class on the surface of abandoned cliff and colluvial block. The secondary succession took place in the secondary forests in valleys, which are affected by human activities. In valleys of the Langshan Danxia region, pioneer forests, transition forests, and basic forests coexist and the direction of succession is going forward.

This special Danxia landscape leads to terrain ecological effect, that is gulch effect and hilltop effect. The gulch effect causes plant shift in horizontal distribution and geo-climax community. While hilltop effect give birth to the hydrothermal distributional characteristics, which differ from common humiture difference. And moreover, it causes the significant biologic difference between mountain top and bottom.

15 Danxiashan, Guangdong

15.1 Introduction to the Heritage

15.1.1 Summary

Name of Property: Danxiashan.
County (City), Province: Shaoguan City, Guangdong Province.
Geographical coordinates (Central): 24°57′55″ N, 113°43′12″E.
Area of the Property:
Heritage Site (ha): 16,800.
Buffer Zone (ha): 12,400.

15.1.2 Boundaries of the Heritage Site and Buffer Zone

Danxiashan, an integral continuous region with 16,800 ha, is located in Guangdong areas. It consists of typical areas of Danxia landform and its natural geographic elements.

Almost all borderline of candidate site of Danxiashan was according ridge and valley and rivers as natural boundary. Such lines superpose themselves to the boundary of core area of the national park; the boundaries of the buffer zone is superpose with the park, basically taking natural lines above as boundaries, partly roads and land type lines as boundaries. Boundary surveying and strict protection have been completed in the place whose boundaries clearly defined are consistent with territory lines of candidate site and buffer zone.

Candidate site is the continuous natural area with typical Danxia landform as the main part, maintaining the integrality of Danxia landform, forest ecosystem and ecological environment for the rare and endangered species.

15.1.3 Introduction to Outstanding Universal Values

(1) **Outstanding Universal Values of Aesthetics**

Danxiashan, taking clustered, middle and late mature Danxia landform as the main part, features diversity of single landform type and rareness of geomorphic landscape. It represents one particular type of scenery landform in the world. Red mountains and green water reflect each other along Jinjiang River and Zhenjiang River and form a sort of beautiful landscape. It is one of the most outstanding Heritage sites where broad-leaved evergreen forests of subtropical regions are perfectly preserved. Red clustered mountains seem as if thousands of red diamonds scattered in the green sea, which forms Danxia landscape system with highly appreciated aesthetic value.

From aesthetics of form, Danxiashan could be described as the beauty of shape of Danxia rocks, of structure of orderly combined hills and rocks, of rhythm of picturesque disordered blocks, of color of red mountain, green tree, blue water, clear sky, and white cloud. From aesthetics of

H. Peng, *China Danxia*, Springer Geography,
https://doi.org/10.1007/978-981-13-5959-0_15

artistic conception, red cliff features majesty, ruggedness, marvel spectacle, and elegance, all of which compose the extraordinary beauty of Danxiashan. It is creditably the most distinguished mountain of Danxia landform in the world.

(2) **Outstanding Universal Values of Geo-science**

Danxiashan is the place where China Danxia was named, the most important basic theory, type, and feature constructed. Also, it is the representative of peak cluster of subtropical humid zone and the most outstanding example of China Danxia and red bed landform in the world. Both international promotion and comparison research are of great significance.

Danxia Landform is a symbol for a special phase in the development of continental crust. Danxia Basin cultivated at the central part of Nanling Folds Series of South China Plate is considered as the epitome of regional crust evolution, reflecting the particular process during which the crust of South China transforms from active zone to stable zone, and again to be activated. It has outstanding universal values of geo-science and plays an irreplaceable role in constructing integral evolution series of China Danxia.

The duo-peak volcanic activities of Danxia Basin before the large-scale descent in the middle Cretaceous period reflect the effect posed by declining zone of plate margin on inner continental lava activities, showing the huge difference between basins in special rifting-subsidence back-arc mode and marginal back-arc stretched basins and rifting-continental basins.

Danxiashan is in the mature stage of geomorphic development as a whole, possessing multi-phrased feature of geomorphic development. Since late Tertiary, multi-phrased uplifts make basins reserve landforms of different evolution stages. Danxia landform is still in the process of uplifting. The geological and geomorphic process is so clear-cut that would be regarded as museum of Danxia geomorphic evolution.

Danxiashan is the essence and representative of Danxia in humid zone, containing all the principal types and important features of Danxia in low-altitude humid zone. Danxia ecosystem and biodiversity form a unique natural geographic features and distinguished characters.

(3) **Outstanding Universal value of biology and ecology**

The area of Danxiashan has preserved the integrality and uniqueness of natural ecological environment and nurtured particular biodiversity and unique species. Besides, it is the biotope for masses of rare and endangered species and the typical region of ecological succession in the late geological ages. Among the listed sites, Danxiashan is the only candidate region in the southern side of Nanling. In many tropical species elements, the features of ravine rainforest are the most outstanding. Due to its complexity and spatial differentiation, biotope in mountains is considered as the research model place of Danxia landform eco-differentiation, biological lineages, effect of isolated islands and hot island. It provides researches concerning interrelation between the diversity of ecosystem and species with valuable references which are of great research significance for ecosystem administration.

15.2 Description of Danxiashan

15.2.1 Introduction to Natural Geography

(1) **Topography**

Danxiashan has been developed in a tectonic basin—Danxia Basin, which is located at inner Nanling Folds Series. A layer of red beds consisted of both coarse and fine grains were deposited in middle and Late Cretaceous. Along with violent topographic change, developing Danxia landform features clustered peak-forest and peak-group. Bazhai, the highest peak, has an altitude of 625 m and the interval of altitude is from 65 to 625 m.

(2) **Climate Features**

Danxiashan stands in the intermountain basin in the middle of Nanling Mountain Chain. Therefore, it has the climate feature of monsoon and rain characteristics of the transition from the middle subtropical zone to southern subtropical zone. The average air temperature in January is 9.3 °C, its average air temperature in July is 28.4 °C, its extreme minimum air temperature is −5.4 °C, and extreme maximum air temperature is 38.5 °C. The precipitation here is rich: the average precipitation in many years is 1715 mm, but the distribution is uneven. March–August accounts for 75%. The general climate feature can be summarized as: its summer is long while winter is short; its summer is hot while winter is cool; it is rainy in spring and summer while dry and crisp in autumn.

(3) **Hydrology and Water Resources**

All the rivers in the Geo-park are in the Zhenjiang River water system, mostly being in the Jinjiang River water system of first-class distributaries. The Jinjiang River cuts into Danxia Basin from the northeast in the form of an entrenched meander, about 34 km in this region, flowing into the Zhenjiang River in the south. The water flow of the Jinjiang River is rich, the long-time average annual flow being 45.8 cubic meters per second, and the runoff amount being as much as 1.44 billion cube meter. Due to the proper preservation of vegetation in upriver areas and Danxia Mountains, there hardly exists soil and water loss and the water quality keep beyond second-class of surface water.

(4) **Soil and Species**

The main soil type in Danxiashan is forest red soil; forest yellow soil, meadow red soil, and rhogosol red soil are distributed on partial peaks and slopes. Danxiashan nurtures lush subtropical evergreen broad-leaved forest, and in the valley, there are areas of quasi South-Asian tropical forests, single peaks preserve natural secondary forest. 1757 kinds of vascular plants have been found. Among them, two is the endemic species of Danxianshan and nine are under special protection of nation. The physiognomic type, structure, and diversity of vegetation communities have provided a very advantageous place for inbiotope and multiplication for all kinds of wildlife. Wildlife resource here is rich.

15.2.2 Geologic Background

15.2.2.1 The Geotectonic Background

Danxia Basin is located in the south of watershed of the middle part of Nanling Folds Series in South China Plate and it is the basin with Cretaceous fault-depression structure. Before the early Paleozoic Era, the Danxia Basin was a part of Nanling Folds Series and has the deposition of Sinian System and Lower Paleozoic Erathem. At the end of early Paleozoic Era, the Caledonian Movement had a great impact on Lingnan area and formed Nanling Folds Series. A stable developmental stage came into being. At the end of late Paleozoic Era, South China Plate experienced a large scale of sedimentation, thus Danxia Basin began to be accumulated by the coverage deposition of Devonian period–Triassic period. The crustal movement of Haixi-Yinzhi tectonic period makes the strata of the late Paleozoic Era and the early Triassic Period developed into folds.

Since the Jurassic Period, Danxia Basin turned into inland lacustrine deposition. From the late Jurassic Period to the early Cretaceous Period, the Yanshanian Movement resulted in folds and faults again in the Jurassic System in and around the basin, accompanied by violent volcanic activities and deposited by the volcanic clastic rock of Sandong Formation. Afterward, Danxia Basin began to sink to a great extent, deposited by the red beds of Changba and Danxia Formations.

15.2.2.2 Lithologic Features of Basin Strata

The outside strata of Danxia Basin layout generally in circles. The strata get older from inside to outside. Except a lack of strata belonged to the period of early Yanshanina Movement, Yinzhi Movement, and Caledonian Movement, strata of any period from Cambrian to Jurassic were distributed in Danxia Basin.

The inner Danxia Basin was mainly composed of Cretaceous strata. The lower Sandong Formation (K_1s) emerges at the edge of the basin with a total thickness of 700–1100 m, composed of duo-series of volcanic rocks, intercalated with mudstone, periodic siltstone, and mudstone and periodic siltstone of Maziping Formation (K_1c). The upper part of inner basin consists of alluvial phase red clastic rock of the Changba Formation ($K_{1-2}c$) and fluvial phase red clastic rock of the Danxia Formation (K_2d), both of which are main strata forming Danxianshan.

Changba Formation (K_2d), with the total thickness of more than 2400 m, is composed of fluvial-lacustrine phase mud-siltstone, fine-grained sands, and a little conglomerate.

Danxia Formation (K_2d) is a set of alluvial red bed, chiefly of fluvial phase with the total thickness of about 1300 m, mainly composed of mauve conglomerate, gritstone, fresh-colored feldspar–quartz sandstone and a little siltstone. The lithology is hard and it is the main strata where Danxia landform, such as cliffs, stone peaks, stone forts, and stone columns, has been formed and developed.

15.2.2.3 Tectonic Features

In Himalayan movement, geotectonic effect functioned as rigid block movement and formed faults and joints, while flexible change is inconspicuous. The Shaoguan-Renhua fault of NNE direction is the primary faulting structure, which develops many faults paralleled with Shao-Ren fault as the uplifts of basins. All of these faults are crucial tectonic lines controlling geomorphic patterns.

Crucial tectonic lines in Danxia basin

There are three groups of joints in Danxia Basin. The first group trends 10°–35°, the dig angle is almost uprightness. This sort of joints is sparseness and the trending length is long and of compressive-distorted property. The second trending 80°–110° develops stretching-distorted joint. Together with NNE group, these two are distributed in diamond form and considered as the important tectonic lines controlling block distribution. There is a third group of joints trending 320–340°. All of these three have their tectonic origins, they are characterized by their distinct direction and incising all the strata. Furthermore, there are some joints, which have no tectonic origins, formed by the outside force.

15.2.2.4 Influence of Strata and Structure on the Formation of Danxia

The development of Danxia is dominated by lithology of strata and geological structure.

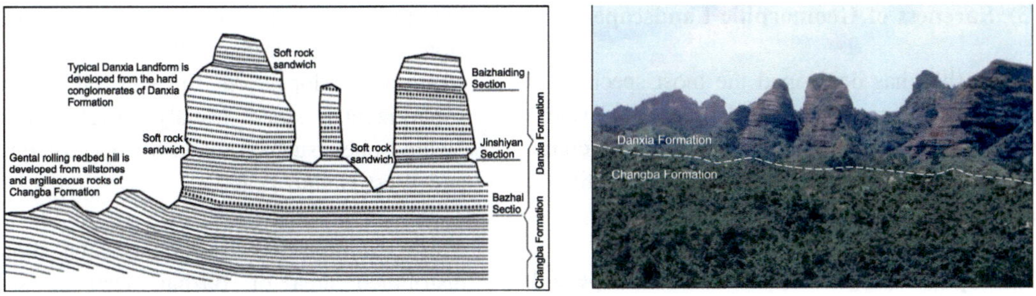

Fig. 15.1 Comparison in geomorphology of Danxia and Changba Formations

(1) **The strata with the development of Danxia**

The typical Danxia is developed in the Danxia Formation. Conglomerate and gritstone of Danxia Formation are gritty, silicic, and ferruginous cementation. The lithology is hard and vertical joints are developed. On the contrary, Changba Formation is mainly composed of soft siltstone and pelite, intercalated with marlite and gypsum bed. Except some steep-faced slopes, Changba Formation is mainly developed into low hills (Fig. 15.1).

(2) **Tectonic Lines and Geomorphic Spatial Pattern**

Generally, the scenery landform can be divided into four scenes with NNE trending, mainly dominated by faults of NNE direction. Under the tectonism of the secondary fault and big joint, the combination of block, wall, line, and point hills has been formed, presenting the orderliness of spatial pattern of Danxia.

15.2.3 Features and Types of Landform

Danxiashan mainly consists of subhorizontal strata and it is the representative of advanced clustered Danxia landform.

15.2.3.1 Main Features of Landform

All basic types Danxia in China subtropical humid zone develop well in Danxiashan. The summary of China Danxia types mainly comes from the classification research of Danxiashan. Therefore, Danxiashan is the most important typical model of China Danxia.

(1) **Discreteness, Orderliness, and Multi-layered Features of Geomorphic Combination**

Isolated by valleys, hills cut by faults form discrete fort; wall; column and prick isolated blocks. The arrangement of peak cluster is controlled by faulting blocks, while single landform in various places may continue to form landscapes of various sizes and densities. In the uplifts with intermittence and difference, multi-class even plane has been formed. On single hills, soft-hard alternated subhorizontal strata usually develop sharp-gentle alternated layered cliff slopes, forming vertical multi-layered features.

(2) **Representative of Clustered Danxia Peak-groups**

The spatial combination of discreteness, orderliness, and multi-layered features forms rich horizontal and vertical landscape layers and makes Danxiashan be the representative of clustered Danxia peak-groups.

(3) **Rareness of Geomorphic Landscape**

Danxiashan has developed the most spectacular red cliffs, the most mysterious maze-like vales, the densest potholes, the most peculiar structural landform and the greatest number and varieties of stone arches and perforated caves.

(4) **Representative of Activated Old-period Danxia**

Several single peaks with a relative height of 200 m, which dispersedly erect on denudation-planation surface of Bazhai scenery area with the altitude of 400 m, become the representative of activated old-period hills.

(5) **Representative of Mature Advanced Danxia**

Peak-groups characterized by sparse-dense alternated, orderly combined clusters are the representative of Danxia landform areas in the mature advanced period.

15.2.3.2 Main Types and Landscapes of Landform

(1) **Geomorphic Types of Single**

Daxiashan single landform contains almost all types of subtropical humid zone.

Main Positive Landform: Danxia cliff; mesa; stone peak; cuesta; stone wall; stone column; hill; single peak; single stone; colluvial pile and colluvial rock block.

- **Danxia Cliff**: Red cliff is the most distinctive landscape of Danxiashan. Large-scale cliffs, such as Jinshiyan Cliff and Shaoshiding Cliff, are more than 200 m high and extend more than 1000 m, forming natural strata section. Due to the difference of weathering and erosion resulted from different lithology, steep-faced slopes on soft-hard alternated subhorizontal strata usually develop into layered ones, such as the two slopes of East and West Hailuo Peak.

- **Mesa (Stone Fort)**: In subhorizontal strata distributed area of Bazhai, Danxia, and Shaoshi scenes, former deposited top or planation surface are cut into stone forts which feature flat top and vertical cliffs. The most famous is Bazhai stone fort, the main peak of Danxiashan, standing on denudation-planation surface with the altitude of 380–400 m. It is around 500 m long, nearly 300 m wide and more than 230 m high, and regarded as a typical single peak which is uplifted again after the old period.
- **Stone Wall**: Huge stone walls are scattered in every scenery area of Danxiashan. Among them, the most spectacular is the landscape of "Elephants Coming Out of the Mountains" composed of eight huge stone walls of Yangyuanshan. Stone walls of different sizes and heights form a lovely scene like a family of elephants walking toward Jinjiang River.
- **Stone column**: Isolated stone columns of various shapes are distributed in all main scenery areas, of which Danxia scenery area is mostly concentrated. Among them, the most peculiar one is Yangyuan Rock. The relative height of Candle Rock is 35 m, but the diameter of the thinnest part of the base is less than 5 m. It is the most slender Danxia stone column (height: diameter = 7:1). The relative height of Guanyin Rock is 143 m measured from the right point where the rock separates from Guanyin. It is the column whose relative height is greatest. Teapot Peak is surrounded by five columns whose heights vary from 50 to 100 m.

Great cliff of Jinshi Rock, the hugest cliff is 250 m high and 2.1 km long

Shaoshiding Cliff, the hugest cliff is 225 m high stretching 2.3 km

Cluster of stone forts of Datangzhai

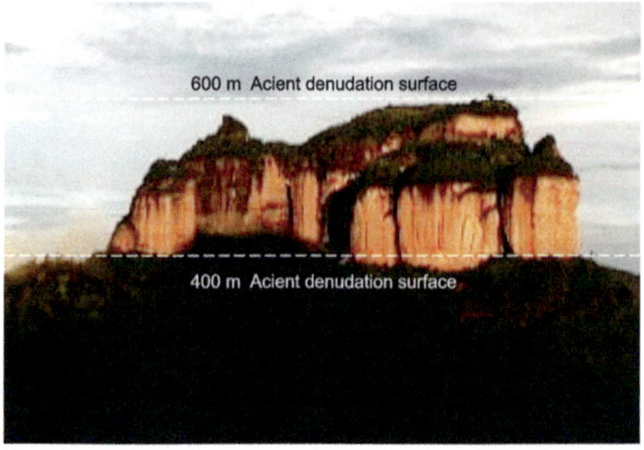

Great stone fort of Bazhai (High Single Peak)

Bazhai Great Stone Fort of Bazhai Scenery Area of the west

Wangujincheng Great Stone Wall

Part of stone wall cluster of "Elephants Coming Out of the Moutains", Yangyuanshan

Wall ridge with great thickness is called stone beam

Landform of stone column, Guanyin Rock, Wangfu Rock, and Candel Rock

Stone peak and cluster of stone columns around at the top of Teapot Peak, Bazhai Scenery Area

Main Negative Landform: Negative landform of Danxiashan forms an integral system, including valley; bedding notch; cave; perforated cave; stone arch and pothole.

- **Danxia Valley**: Danxia Valley includes wide valley; incised meanders; canyon and lane valley/line valley. Most of the main channels flowing through Danxia Mountains, such as valleys of Jinjiang River and Zhenjiang River, are wide valleys formed by incision along fragmentized belt of structure and original bottomland. Part maintains meandering incision and forms canyon-shaped incisive meanders.

The longest and deepest lane valley of Danxiashan is Shaoshiding Lane Valley with the depth of around 200 m, length of around 800 m, and width of 0.5–2 m. It is the largest Danxia Lane Valley that ever be found in China. The most peculiar lane valley is Sisters Peak Lane Valley Cluster. More than 20 lane valleys interweave and the whole hill block has been cut into fragments. Most lane valleys are directly linked together while parts are connected by perforated cave at the bottom, which seems like mazes. Inside such connected lane valleys, there could be found collapse; dislocation; cave; stalactite; ancient mountain villages and paleoanthropic relics.

Maze-like lane valley/line valley cluster inside Sister Peak

Canyon reaches, Jinjiang River in Danxia Scenery Area, and downstream of Jinjiang River

Wide valley, Yangyuan reach of Jinjiang River

Wide valley, Xiafu Reach of Jinjiang River

Perforated cave; stone arch and natural bridge are considered as a sort of significant characteristic of Danxiashan. There are more than 60 sites of perforated caves and stone arches that ever be found.

The span of incisive stone arch is 38 m, Tongtai Bridge, Yangyuanshan.

Aizhai Weathering Stone Arch—Triumphal Arch (The whole hill is integrally an arch with the span of around 30 m. However, the height of inner arch is more than 20 m)

- **Danxia Pothole**: On almost every site of riverbed, the current carries pebbles and whirls on its way, forming small mouth-large bottom potholes deep into rocks of the riverbed, and even expanding and enlarging constantly to generate big ponds. On the base rock valleys of Feihuashui and Wangshankeng, bead-shaped potholes are developed.

Cluster of potholes at the valley bottom of cluster of Feihuashui Stream and valley-in-valley formed by the linked potholes in Miaozhai Hole

Faveolate grotto of Danxiashan is the place where such mini-type landform has been named. There are lots of sites that have been ever found. The most typical case is the dragon-squama rock inside the Jinshiyan Grotto: on the surface of the sandstone of the inner wall, there is a zone of faveolate grottos about 1 m wide and 10 m long, crossing the whole rear grotto wall.

- **Modeling Landform**: Danxia is famous for abundant modeling. Particularly, Danxiashan is characterized by miraculousness of mountains; stones; caves and valleys. They are so marvelous that it is unbelievable that they are natural creatures. Instead, they are easily mistaken for immoral work or sculptor's masterpieces. Among them, Yangyuan(male) stone is called "the most miraculous stone in the world". Together with Yinyuan(female) stone, Dragon-squama stone and Guanyin stone, they are called "Four Miracles of Danxia". Yangyuan(male) stone, together with Yinyuan(female) stone and Shuangru(-double breast) stone forms the combination of "Three Romantic stones". Therefore, Danxiashan has been named "Nude Natural Park".

Dragon-squama rock on rear wall of Grotto Dragon-squama temple in Jinshiyan part of the cave

Fancy modeling landform: three romantic stones

(2) **Types of Colony Landform and Landscape**
Danxia Mountains are rich in types of geomorphic combination, generally forming typical area of clustered Danxia groups.

Cluster of peak forest-stone column

Dense peak clusters: sun rising from clouds, Sengmao Peak

High cluster of single peaks, Bazhai Scenery Area

Scattered column-like cluster of peak forest, Shaoshi Scenery Area

Scattered column-like and tower-like peak forest, Huangsha Hole

15.2.4 Biology and Ecological System

(1) **Bio-Geographic Zone (Province)**

According to Udvardy's (1975) biogeographic system, Danxiashan belongs to biotope in mid-southern region of bio-geographic province of China subtropical forest of Old-north zone, where the extent to biodiversity is rich. It locates "China Southeast tropical forest" zone of "India–Malaysia" region of "200 biological zones of World Wildlife Fund". It generally exhibits characteristics of mid-subtropical eastern humid zone evergreen broad-leaved forests, featuring transitional property of south subtropical and mid-subtropical zone. Complex biotope types of

Danxiashan have developed abundant and various ecosystems. Especially, ecosystem in small scale features diversity and complexity. On the same hill block, the ecosystem of evergreen broad-leaved and coniferous mixed forest exists at the top of hill; the ecosystem of arid cliff consists of low-plant and liana; while in the valleys, there exists the ecosystem of typical evergreen broad-leaved forest and ravine rainforest.

There are 7 orders, 24 families, 58 genera, and 88 species of mammals; 17 orders, 44 families, and 156 species of birds; 3 orders, 11 families, and 41 species of reptiles; 1 orders, 6 families, and 37 species (or subspecies) of amphibians; 6 orders, 20 families, 69 families, and 100 species (or subspecies) of fishes and 16 orders, 176 families, 783 genera, and 1023 species of insects. 59 species of animals have been listed in Red List of China Species, 73 in Red List of IUCN, 66 in CITES, and 13 of which have been listed in Appendix I. There also lists 54 species of animals under key protection of China, among which there are 7 species under protection of Class I and 47 species of Class II.

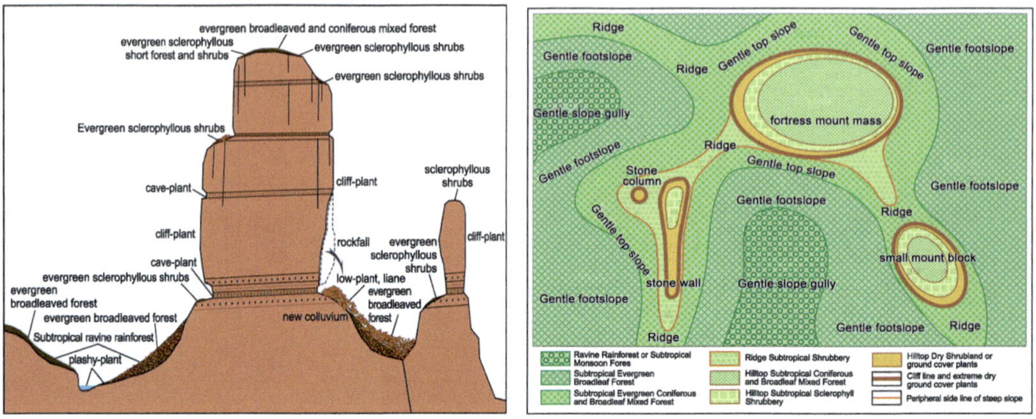

Profile and planar of differentiation of Danxiashan vegetation as the change of geomorphic location

(2) **Species of Vegetation**

Danxiashan possesses 210 families, 755 genera, and 1757 species of vascular plants, including 36 families, 61 genera, and 102 species of fern plants; 7 families, 12 genera, and 15 species of gymnosperms; 174 families, 714 genera, and 1520 species of angiosperms. There are 20 species listed in the China Species Red List; 10 listed in IUCN Red List and 39 listed in CITES. There are 11 species of plants under key protection of China, among which there are 2 species of plants under protection of Class I. There also lives two unique type specimen of Danxia: *Firmiana danxiaensis* and Danxia: *Chritopsis subulata* W.T. Wang var.*danxiaensis*.

The vegetation coverage rate of Danxiashan is above 90%. There are 11 types, 27 formations, and 48 associations.

The diversity of Danxia landform types, structures, and vegetation colonies generates complex and various environmental conditions and maintains favorable ecological environment for the living of wildlife. According to the diversity of biotopes, species colonies can be classified into mountain forest, grotto, cultivation zone, community of aquatic animals, and so forth.

(3) **Ecological Process**

Southeast China possesses the largest and most typical area of evergreen broad-leaved forest, which has global significance for the researches of forest vegetation colony. Candidate region preserves typical virgin and virgin secondary evergreen broad-leaved forest, including typical type of evergreen broad-leaved forest distributed in valleys and morphological alienated typical evergreen broad-leaved forest distributed on the tops and cliffs of Danxia landform area. There exists unique dynamic process of succession of colonies of plants. On the surfaces of fresh collapse piles and colluvial rocks, especially, there preserve the primary ecological succession of different stages and starting from different biotopes, which has formed an integral succession series in various space. It has provided the researches of vegetation colony with a hard-won perfect place and reference system of ecological restoration.

(4) **Ecological Features**

Danxiashan locates in mid-subtropical zone and original vegetation has been characterized by transitional features of mid-subtropical evergreen broad-leaved forest and lower-subtropical monsoon evergreen broad-leaved forest. There are abundant species of tropical and lower-subtropical plants, which has accounted for 70%. The resource of woody lianas is rich, which accounts for 14%. Especially, there exists the phenomenon of strangler, cauliflory, and epiphyte, together with the wide distribution of plants, such as Cyatheaeae, Angiopteridaceae, Dicksoniaceae, Annonaceae, Saurauiaceae, Moraceae, Palmae, Musaceae, Araceae, Bambusoideae, and liana bamboo of bamboo subfamily, which proves that

Appearance of evergreen broad-leaved forest of eastern part of Danxiashan

Dense landscape of valley rain forest

Unique species of Danxiashan: colony of *Firmiana Danxiaensis* and flowers

Danxia landform has exerted significant influence on plant biotope, such as partial little climate, water, soil, and so forth.

Danxia landform generates "isolated island effect" affecting regional (partial) climate, which makes it be richer in diversity of ecosystem compared with neighborhoods. This is one of great values of Danxia landform as natural heritage.

15.2.5 Natural Landscape and Beauty

(1) **Natural Landscape**

Natural landscape of Danxiashan can be summarized as plentiful landscape of rocks, orderly landscape of mountain group, colorful combined landscape of red mountain, green tree, blue water, clear sky, and white cloud, all of which together forms landscape system of aesthetic value.

- **Geomorphic Landscape**: It mainly consists of Danxia cliff; peak-fort-wall-column; combination of peak group; modeling landform; Danxia valley, and so forth, featuring diversity of geomorphic types and miraculousness of geomorphic landscape, exhibited especially by spectacular red cliffs, towering high single peak, mysterious maze of lane valley, dense potholes, various-shaped rock arches and perforated holes, peculiar modeling landform, and so forth. It has become the most outstanding representative of China Danxia and red beds landform in the world, exhibiting the unique natural beauty.
- **Danxia Water Landscape**: There are red group of mountains in Danxia Basin, among which Jinjiang River and Zhenjiang River meander through. Along the way, Red Mountains and Blue River reflect each other, together forming a beautiful mountain–river landscape. Besides, there are numerous streams and splendid currents, waterfalls, deep pools, and potholes in Danxiashan.
- **Biological Landscape**: Danxiashan is one of the most outstanding Heritage sites where subtropical evergreen broad-leaved forest is well preserved. Red clustered mountains seem as if thousands of red diamonds scattered in the green sea, interspersed by

vegetation and layered cliff vegetation belt, which makes landscapes even more abundant and colorful.

In the low-flat valleys of Danxiashan, there preserves areas of lower-subtropical monsoon rainforest and ravine rainforest. There exists the phenomenon of strangler, cauliflory, epiphyte, buttress root, large woody liana, and huge number of typical tropical species. The forests are characterized by clusters upon clusters of trees of all kinds, ancient lianas and wild bines entangling each other, which exhibit a scene of tropical rainforest. There preserves plenty of old and famous trees. Huge trees standing in the valleys and on the single block and the twittering of birds on trees, all of which forms a scene of vitality. It could be frequently seen that some animals of ornamental values, such as rhesus macaque, small Indian civet, sambar, Eurasian otter, and so on and that rare species of bird, such as yellow-bellied tragopan and a mass of raptors. In addition, there are silver pheasant, Chinese frencolin, bamboo partridge, and little egret and so on living in the forest.

(2) **Importance of Aesthetics Values**

Danxiashan is at mature advanced stage and clustered Danxia landform, featuring dispersed blocks, the general layout of stones and peaks, high canyons and deep valleys and various types and shapes of landform. It has been evaluated as "the most beautiful Danxia of China" by Chinese National Geography in 2005.

From aesthetics of form, Danxiashan could be described as the beauty of shape of Danxia rocks, of structure of orderly combined hills and rocks, of rhythm of picturesque disordered blocks, of color of red mountain, green tree, blue water, clear sky, and white cloud. From aesthetics of artistic conception, red cliff features majesty, ruggedness, marvel spectacle, and elegance, all of which compose the extraordinary beauty of Danxiashan. It is of extremely high ornamental values.

15.3 History and Development

15.3.1 Evolution of Danxia Basin

Danxia Basin is located in the early Paleozoic Nanling Active Belt of South China Plate with late Proterozoic base and formed Nanling Folds Series during Caledonian Movement. At the end of late Paleozoic Era, South China Plate experienced a large scale of sedimentation, thus Danxia Basin began to be accumulated by the coverage deposition of Devonian period–Triassic period. During Haixi-Yinzhi tectonic period, folds, and uplifts of the crustal layer took place. Since the Jurassic Period, Danxia Basin turned into inland lacustrine deposition. From the late Jurassic Period to the early Cretaceous Period, the Yanshanian Movement resulted in folds and faults again in the Jurassic System in and around the basin, accompanied by violent volcanic activities and deposited by the volcanic clastic rock of Sandong Formation. Afterward, Danxia Basin began to sink to a great extent, deposited by the red beds of Changba and Danxia Formations.

Orogeny from the late of late Cretaceous Era to the early of early Tertiary ended deposited environment of internal-flow basin and transformed it into external-flow eroded area. Himalayan movement in the early of late Tertiary made it uplifted the area in the west of the main fault of Danxia Basin, that is, Danxia Formation strata of nowadays Bazhai and Feihuashui scenery area suffered from erosion. Until the late of late Tertiary, this area had been eroded into peneplain, on which single peak, such as Bazhai and Yanyan with the relative height of around 200 m, had been erected.

From the late Late Tertiary to Quaternary Period, this area had integrally experienced many intermittent uplifts, meanwhile, Jinjiang River and Zhenjiang River kept meandering incising. Such uplifts were divided by intervals of peaceful time, during which corresponding planation surface was formed, which made this area form multi-stage planation surface and river terraces

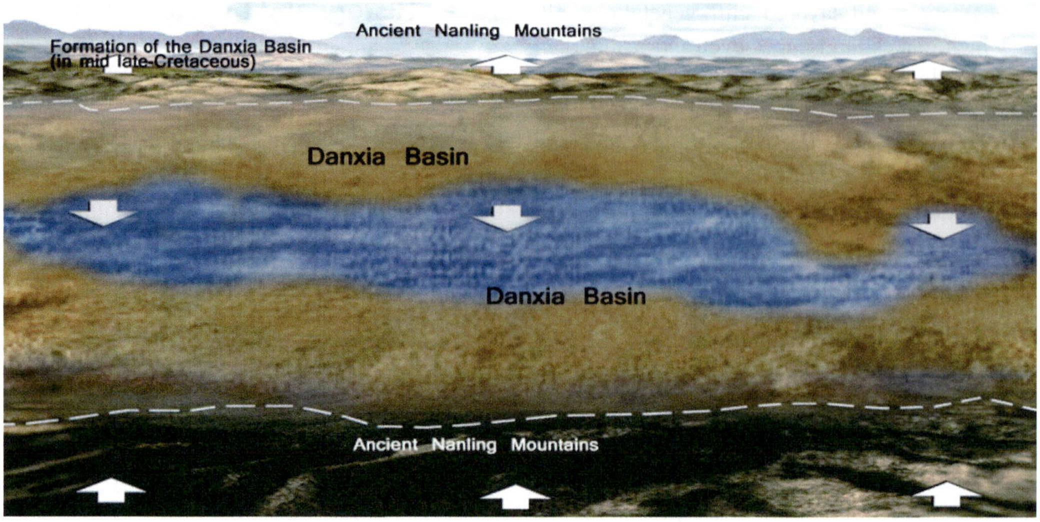

Nanling Mountains had a significant uplift during the Yanshanian movement. During this period Danxia basin experienced extensional rift process and intense volcanic activity, forming the Cretaceous volcanic Sandong Formation (K_1) and the upper Cretaceous volcaniclastic Maziping Formation (K_2) . In the mid-Cretaceous, Danxia Basin started to sink substantially.

Danxia Basin sunk and began to be accumulated by the coverage deposition

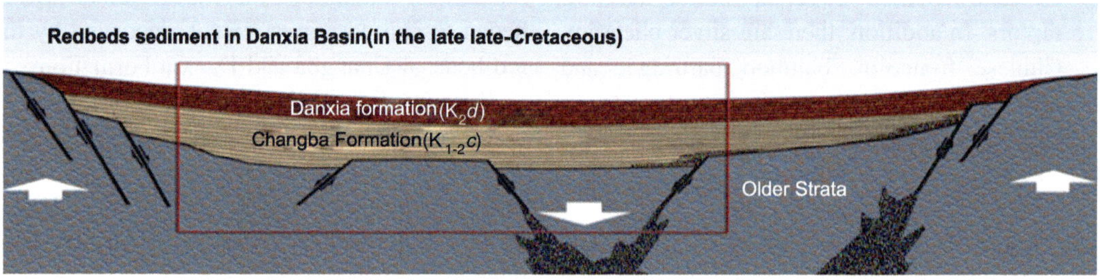

Danxia Basin substantially sank from the mid-Cretaceous and deposited a 3700m-thick red terrestrial clastic rock series of Changba Formation ($K_{1-2}c$) and Danxia Formation (K_2d) in the late Cretaceous Period.

Danxia Basin was deposited by red beds of Changba and Danxia Formations during advanced Cretaceous Period

with the altitude of 400, 300, 200 m, and so on. According to the isotopic calculation of river terrace sediment, this area has uplifted by 0.87 m per year in the late 500,000 years. Thus, it is supposed that it has experienced 4–5 million years since the erosion of planation surface. Also, it is the right period in which modern landform has been formed.

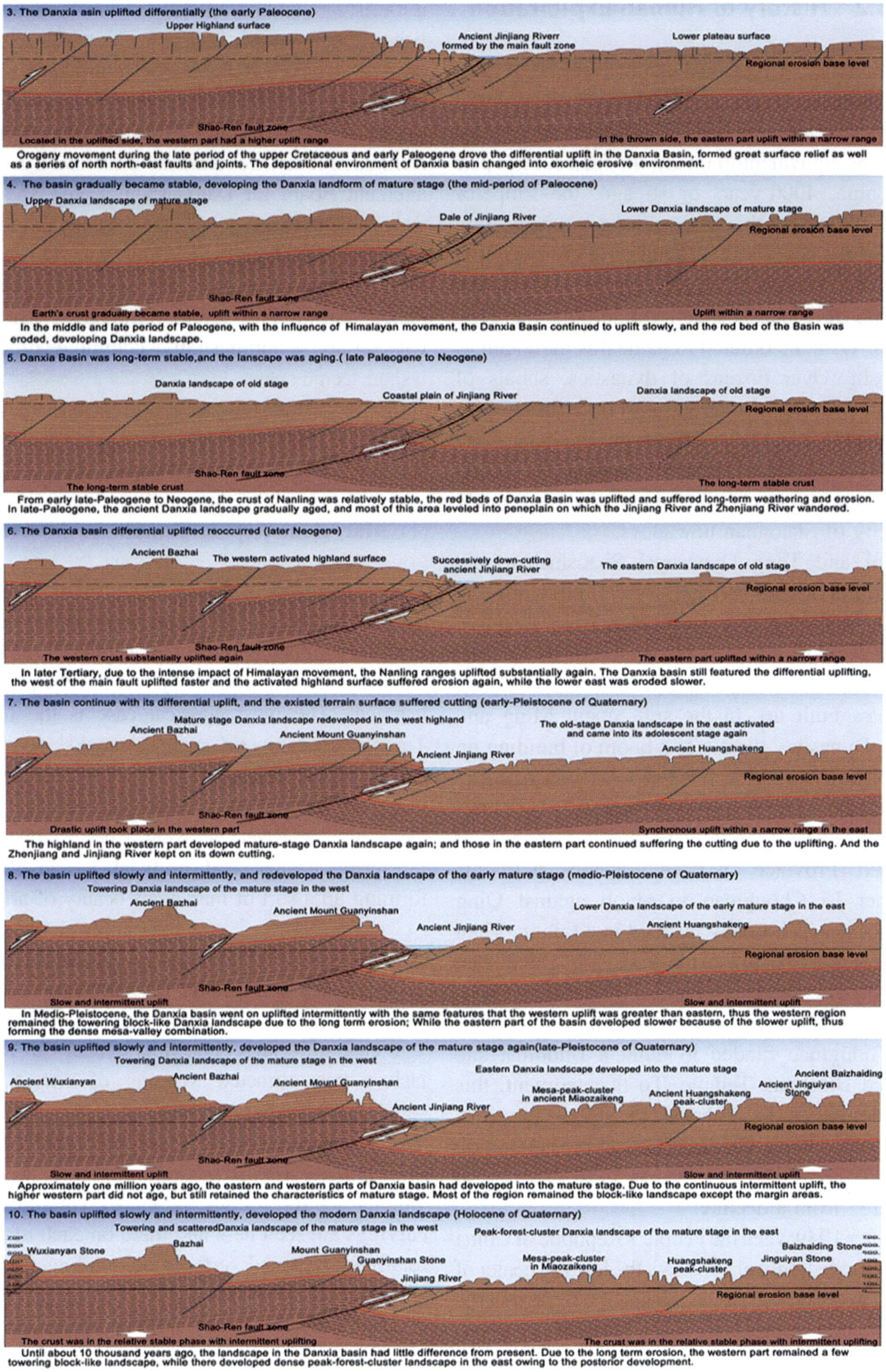

Illustration of evolution process of Danxiashan

15.3.2 History of Human Exploitation

As the relics of Nianyuzhuan of Neolithic Era in the southeastern part have indicated, the cultural affairs development level in this area 5000 years ago was equivalent to that in the then central part of China. 4000 years or so ago, the Emperor Shun passed by this place during his inspection around the south part of China. He climbed up a hill and played Shao-music, the sound and rhythm of which sounded extraordinary euphonious. Then, he ordered to name this mountain as Shaoshi. Over dozens of dynasties, shoals of literati visiting the south part of China were attracted to take the lead to see around here. After Sui Dynasty, the government renamed the former Donghengzhou to be Shaozhou, which is namely the city of Shaoguan nowadays.

Sui and Tang Dynasties, Shaoshishan and Danxiashan have become a great attraction due to its picturesque scenery; also from then on, some monks and nuns had begun to make their debut into the mountain and then had Buddhist temples built up. In the subsequent Ming and Qing Dynasties, there was a boom of building up temples in the mountain. Later in 1644 (the late Chongzhen Times in Ming Dynasty), Li Yongmao, as the Imperial Inspector at Ganzhou of Jiangxi Province, worked together with his brother Li Chongmao to revolt against Qing Dynasty, which finally resulted in a failure. Then they went to Zhanglaozhai Village in Danxishan and led a secluded life there. Till the first year of Kangxi Emperor Reign in Qing Dynasty (1662), Li Chongmao funded to build a Buddhist site named Biechuan Temple. To that moment, this temple began to embrace shoals of pilgrims. From late Qing Dynasty to 1949, Danxiashan experienced several catastrophes and the religion suffered from a decline.

After 1949 when the People's Republic of China was founded, July of 1963 saw the forestry center of the Danxiashan set up, in a main aim to protect the mountain forest and landscape resources in Danxiashan. In 1976, Mt. Danxiashan Management Office was established and organized the erection of main scenic spots. In 1980, Guangdong Province Government announced the Danxiashan to become a tourist site and be opened to tourists and established the Danxiashan Scenic Spot Management Bureau, which are liable for refurbishment and development of the Changlao Peak Scenic Spot and the riparian lands along Jinjiang River. In 1993, Renhua County Government established a Management Board for Danxiashan Scenic Spot and developed two new scenic sites: Yangyuan Scenic Spot and Xianglong Lake Scenic Spot. Up to now, the total area developed and utilized has reached about 6 km^2. In 2005, the government of Shaoguan City set up the administrative committee of Danxiashan scenic spots.

15.3.3 Human Landscape

(1) **Marvelous Arts of Temple Architecture**

The development of Danxiashan Scenery in history adhered to the principle conforming to the natural landforms. The temple architectures exhibit the harmony of natural landscape and artificial establishment. The most classical case is the Grotto Temple at Jinshiyan Cave, the monks' dormitory and Buddhist halls are all erected on steep cliffs or in caves. The mouth of each of these cave holes faces Jinjiang River, showing a perfect fusion of the great natures and artificial establishments, forming up a sort of marvelous beauty of art.

(2) **Magnificent Carvings on Standing Stone Surfaces and Tablet Inscriptions**

The elegant, unique, and hilly scenery in Danxiashan has attracted swarms of literati and scholars, who have composed verses and left superscriptions, stone carvings, and tablet inscriptions. These works have now become valuable cultural relics. Around Zhanglao Peak, carvings are seen here and there on each of those relatively flattened surfaces of standing stones.

(3) **Splendid Ancient Fortified Mountain Villages and In-cave Temples**

Danxiashan region has a saying which goes as, "there is always a village on each hill, each village has a gate of its own, and each gate is erected on a steep site". All of these villages are built conforming to natural terrain, being adjacent to abyss or gully. Usually, a hill is furnished with three gates, forming up a reliable defense layout. The prank way is narrow and steep. Most paths leading into the mountain are steep and rugged. Villages are furnished with abundant water sources, and the foodstuffs restoration is available, providing fundamental necessities for human existence and living.

Strategic place of mountain villages

Gate of village and relic of rock Temple

As the findings of a preliminary survey have showed, there are over 40 ancient temples built on ancient caves in Danxiashan. Just like ancient fortified mountain villages, these ancient temples are distributed inside cave cavities which lie in the middle of the mountain or under a cliff and furnished with a tranquil, attractive, and elegant environment. They fit well into the inside environment of these cavities and natural environment in the peripheries, looking like a fairy land to live in.

(4) **Mysterious Cliff Tomb**

In Danxiashan, there are a great number of steeply cliffs and hidden caves, which provides much desired natural conditions for cliff tombs or hung coffins. According to a survey, cliff tombs have been found in many places of Shaoshi, Dashi, and Danxiashan. As such mysterious cliff tombs are normally scattered inside remote mountains and on cliffs/bluffs, being have to see not to speak of touching; they also represent the ancient local burial custom, appearing rather mysterious and stimulating tourists' curiosity. Finally, they have provided hard-won references for the researches concerning about history, culture, and last folk custom of Danxiashan.

15.4 Natural Features and Its Values in Series of Danxia

15.4.1 Natural Features

Danxiashan has been developed in a tectonic basin which locates in the central part of Nanling Folds Series. Basins and strata and structure around have documented development history and crucial geological cases of Nanling Folds Series. Therefore, no other place in this series of heritages is more appropriate than Danxiashan to reflect evolution history of Nanling. Diluvial and alluvial red bed of Danxia Formation is the main stratum developing Danxia landform, which is also the standard stratum of Upper Cretaceous Danxia Formation in Lingnan.

Danxiashan is the model place of basic Danxia type in China subtropical humid zone, whose landform combination appears violent discreteness, orderliness, and multi-layered features. It is considered as the representative of advanced clustered Danxia landform.

Danxiashan possesses the diversity of geomorphic types and marvelousness of geomorphic landscapes. Especially, there appears spectacular red cliff, protruded high single peak, mysterious maze of lane valley, and vivid biological scenery, all of which together promote Danxiashan to be the most outstanding representative of China Danxia and World Red Bed Landform.

Among series of candidates, Danxiashan is located in the most further south position and regarded as the transitional humid monsoon-climate zone from mid-subtropical zone to lower-subtropical zone, nurturing, and preserving typical ecosystem of subtropical evergreen broad-leaved forest. There preserves areas of lower-subtropical ravine forest in closed valleys and natural secondary forest on the top of single peak. It is the model region of Danxia vegetation lineage and ecological isolated island effect.

15.4.2 Value and Status in Series of Heritages

Danxiashan has the same or similar geological tectonic background, evolution history, geomorphic, and natural geographic features with each of series of Heritages, satisfying the qualification of World Natural Heritage (III.C, No. 137). It is playing an irreplaceable role in this series and its outstanding values showed as below:

- Danxiashan is the place where China Danxia was named. With regard to Danxia, it is also the model place where principle basic theories, types, and features are formed. Moreover, Danxiashan has been regarded as the model region of Danxia comparing researches in China.
- Danxiashan is the typical representative of advanced clustered Danxia and the key link of great significance in series of candidates.

- Danxia Basin cultivated at the central part of Nanling Folds Series of South China Plate is considered as the epitome of regional crust evolution, documenting development history, and crucial geological cases of Nanling Folds Series. It plays an irreplaceable role in constructing integral evolution series of South China Red Bed Basin.
- Danxia Formation strata of Danxia Basin are regional standard strata model place of Upper Cretaceous Danxia Formation.
- Among the listed sites, Danxiashan is the only candidate region in the southern side of Nanling, featuring the transition from mid-subtropical zone to lower-subtropical zone. It is also the candidate possessing the greatest number of species elements and the most outstanding ravine rainforest features. Moreover, it has been considered as the research model place of Danxia landform eco-differentiation, biological lineages, effect of isolated islands, and hot island.
- In general, Danxiashan is at the mature stage of geomorphic development, however, possessing multi-phrased feature of geomorphic development. Since late Tertiary, multi-phrased uplifts make basins reserve landforms of different evolution stages and modern geomorphic process is clear-cut.
- Danxiashan is one of most beautiful China Danxia landforms, featuring diversity, miraculousness, rareness, and naturalness of Danxia landform. It is the representative of one particular scenic landform in the world.

16 Longhushan, Jiangxi

16.1 Introduction to the Heritage

16.1.1 Summary

Name of Property: Longhushan
County (City), Province:
Longhushan: Yingtan City, Jiangxi Province.
Guifeng: Shangrao City, Jiangxi Province.
Geographical Coordinates (Central):
Longhushan: 28°04′15″N, 116°59′05″E.
Guifeng: 28°19′03″N, 117°28′10″E.
Area of Property:
Heritage Sites (ha): 16950 (Longhushan), 2740 (Guifeng).
Buffer Zones (ha): 59820.

Longhushan Heritage site is a broad valley-type Danxia landform area with broadening and evacuated peak forests composed of the Late Cretaceous red clastic rock developed along the margin of the Mesozoic continental basin, containing the highest density and finest pictographic stones, peak clusters, and caves of well-formed Danxia landform known anywhere in the world, and featuring the diversities of geology, geomorphology, ecology, and cultural interest, which constitute a natural landscape system bearing an excellent scientific value and artistic value. Thanks to its splendid and unique landscapes, it was designated a national park by the State Council of the People's Republic of China in 1988. In 2005, it was chosen as "the most beautiful Danxia landscape in China" by the China National Geography. In 2006, it was inscribed on the tentative list of national natural and cultural heritage by the Ministry of Construction, and in 2007 it officially jointed the UNESCO Global Geoparks Network as a member.

16.1.2 Physiographic Features

(1) **Geology and Geomorphology**

The Xinjiang Mesozoic red basin where Longhushan is located lies in the east section of suture zone between Yangtze Paleoplate and Cathaysia Paleoplate, close to Mount Wuyi Range in the south and to the Xinjiang valley in the north. The Upper Cretaceous Hekou Formation (k_2h) and Tangbian Formation (k_2t) of Xinjiang Basin provided materials for the development of Danxia landform. They are mainly distributed as peneplanation low hills, with scattered residual tolts (or isolated peaks) or lonestones along both sides of Xinjiang valley, while the geomorphologic landscape such as the peak clusters, hoodoos, isolated peaks, and Unakas (kopjes) only can be seen in Longhushan, Guifeng, etc., which just like two great tray landscapes (bonsai) stand erect in the south edge of Xinjiang Basin with typicality of peneplanation. The general relief is high in the south and low in the north. The highest part reaches an altitude of 401.1 m a.s.l. at the Paidao Peak while the lowest 48 m a.s.l. with the relative height of 353.1 m.

H. Peng, *China Danxia*, Springer Geography,
https://doi.org/10.1007/978-981-13-5959-0_16

(2) **Climate**

The climate in the Heritage site belongs to the subtropical moist monsoon continental type. By turning of water bodies and dense vegetation, there is seldom extreme weather in summer or winter, and seasonal difference is obvious, but comfortable. The annual average temperature is 18.0 °C, and the average temperature in January and July is 5.5 °C and 29.7 °C, respectively. The maximum temperature is 40.7 °C while the minimum temperature 8.6 °C. There is plenty of rainfall within the area, with annual precipitation of 1878 mm and the annual average evaporation of 1648.4 mm.

(3) **Hydrology**

The Heritage site is located at the south of middle section of Xinjiang Basin, the drainage of Poyang Lake in the middle reaches of the Yangtze River. The main tributary through the area is Luxi River, 43 km long from southeast to northwest within Longhushan area. The surface water resource is rich with better quality of water, meeting the national II-grade of *Criteria on Environmental Quality of Ground Water* (GB3838-2002). The underground water is also abundant within the area, suitable for drinking or other purposes.

(4) **Soil**

This mainly occurs as red earth, mountain yellow soil, and terra roxa within the area, with wide distribution of red earth and mountain yellow soil. And the acid terra roxa and red earth are distributed on the low hill country with an altitude of 50–500 m a.s.l.

(5) **Vegetation and wildlife**

Based on the Udvardy classification system, the organisms in the Heritage site belong to the Sino-Japanese forest flora domain of the Palaearctic Realm. There is still preserved primary vegetation in the valleys, especially on the brow and scarp-foot, with development of precious mid-subtropical evergreen broad-leaved forest in the humid area with low altitude. Identified so far within the Heritage site are 262 families, 938 genera, and 1626 species of higher plant. According to the classification of zoogeographic zoning in China, the Heritage site is located in the Oriental Realm (subtropics). However, in terms of composition, it shows distinct characteristics of the Oriental Realm and is mixed with some characteristics of the Palaearctic Realm. The latest survey shows that there are 33 orders, 101 families, and 387 species of aquatic and terrestrial vertebrate, with typicality of extremely rich wildlife resources.

16.2 Description of Longhushan Danxia

16.2.1 Physical Geology

(1) **Regional Geology**

Longhushan Heritage site is located at the southern edge of the middle section of Xinjiang Mesozoic fault basin, the suture zone between the Yangtze and Cathaysia paleoplates, the southeastern part of Eurasian continent. The Heritage site and its surroundings contain a continuous stratigraphical record from the Meso-Proterozoic to the Quaternary period. The Caledonian Movement, which gave rise to collision and amalgamation of the two tectonic units in south and north, greatly impacted on the basement structure and the upper cover layer as well as the formation and evolution of the Meso-Cenozoic basin.

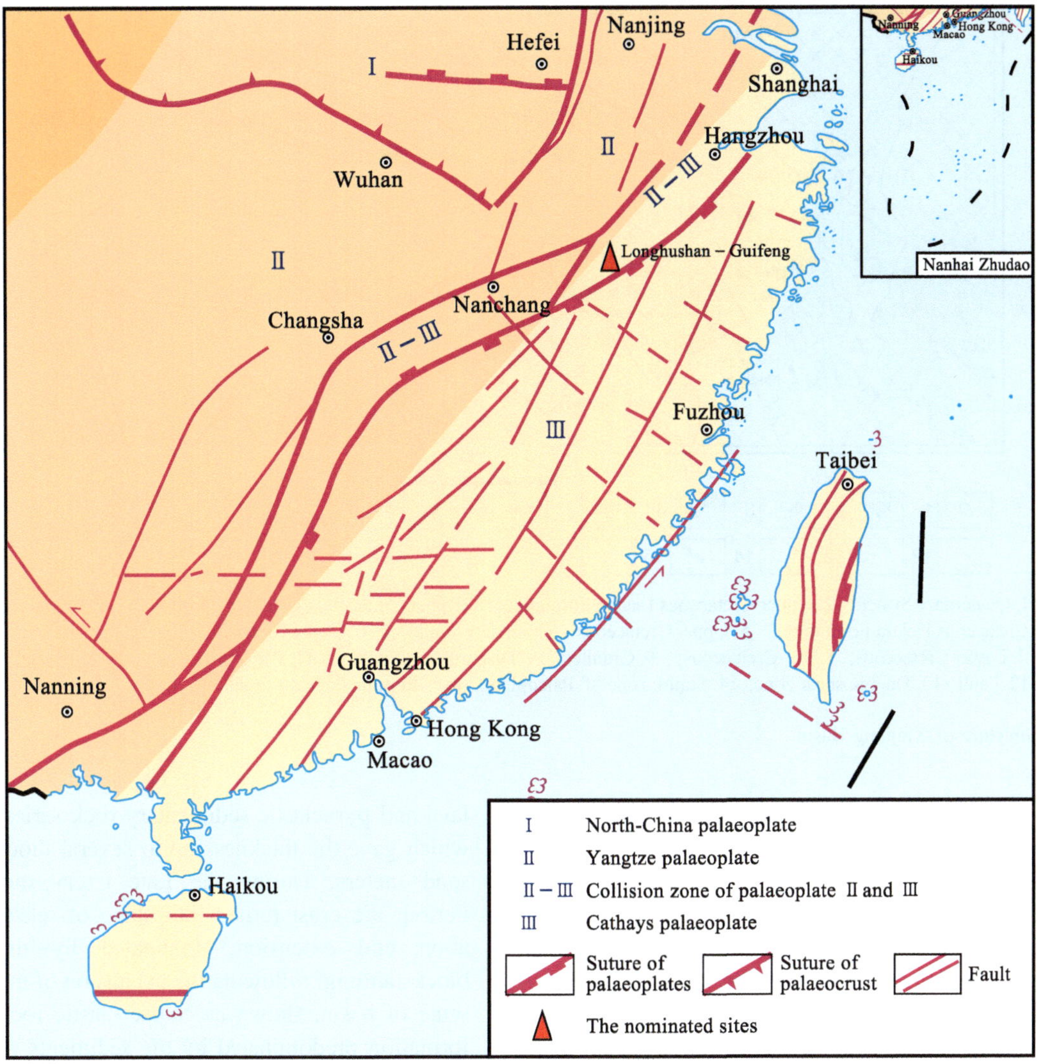

Tectonic setting of the Heritage sites

Late Triassic–Early Jurassic intermountain depression-type basin formed as a result of the Indosinian orogenesis constitutes the superimposed basin, together with the Early Cretaceous continental volcanic basin and the Late Cretaceous red clastic basin by the end of Yanshanian Movement.

Not only was the development of the Late Cretaceous Xinjiang Basin controlled by nearly EW-trending fault, but also by a pair of NNE- and NW-trending X-shaped faults, which made the lithology, lithofacies, sedimentary cycles, and sedimentary structures present different characteristics in different parts within the basin.

(2) **Geologic characteristics of the basin**

- **Types of the basin**. By the end of Indosinian Movement, this area entered into the mobile stage of the circum-Pacific continental margin, generally controlled by Beihai–Shaoxing fault zone, forming a composite structural basin distributed as nearly EW strike, with typicality of the compound structure of "gray on the top and red at the bottom."

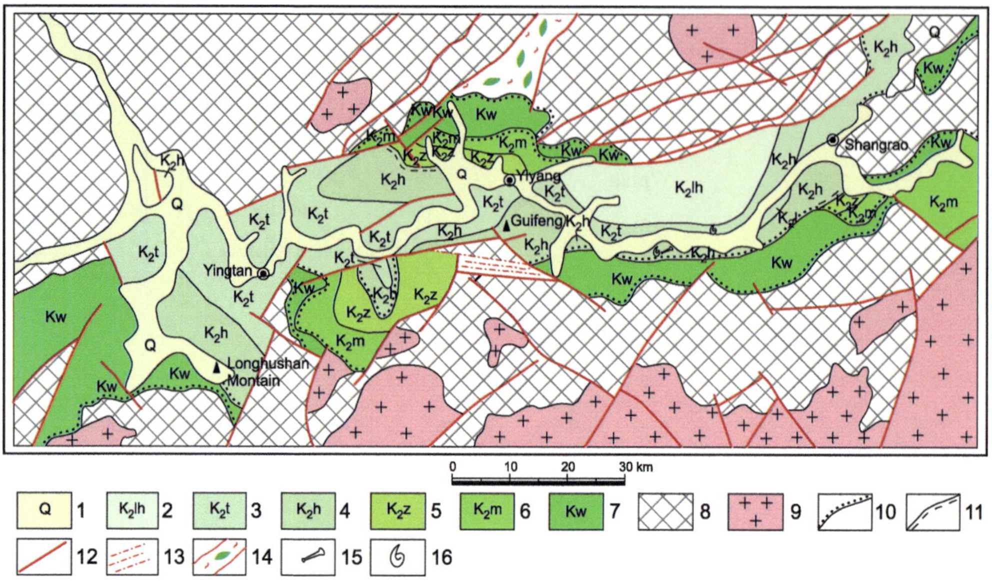

1. Quaternary System; 2. Upper Cretaceous Lianhe Formation; 3. Upper Cretaceous Tangbian Formation; 4. Upper Cretaceous Hekou Formation; 5. Upper Cretaceous Zhoutian Formation; 6. Upper Cretaceous Maodian Formation; 7. Lower Cretaceous; 8. Pre-Cretaceous; 9. Granite; 10. Discordant boundary; 11. Parallel-discordant boundary; 12. Fault; 13. Ductile shear zone; 14. Suture zone of Paleoplate; 15. Dinosaur skeleton fossils; 16. Dinosaur egg fossils

Structure of Xinjiang Basin

- **Depression(sag) basin**. During the Late Triassic–Middle Jurassic Periods, when most parts of the area became the terrigenous denuded area with undulating mountains, a few, including Yiyang and Shangrao, were transformed into depression–subsidence zone, and piedmont or intermountain depression-type fluviolacustrine coal-bearing basin after the receding of widespread marine transgression; thus, an embryonic form of Xinjiang Basin came into being.
- **Fault basin**. In the Early Cretaceous, a new fault basin was developed on the base of the original one and expanded. Lithology and lithofacies varied greatly. At that time, the tectonic activity was transformed from depression to faulting, and the volcanic basin rejuvenated based on the original basin, accompanying with accumulation of volcanic lava and pyroclastic sedimentary rock series, which gave the thickness up to several thousand meters. During the Late Cretaceous Period, the crust turned into stage of relaxation and extension, dominated by the block-faulting, following the expansion of the scale of basin, shown as coarse clastic rock formation predominated by the sediments of piedmont, pluvial, and alluvial facies with huge thickness, intercalated with sediments of lake basin facies bearing gypsum salt under the conditions of extremely arid climate.
- **Texture and formation of the Cretaceous basin**. The Cretaceous stratum was developed completely in the Xinjiang Basin. And the unique superimposed basin is composed of the Early Cretaceous continental volcanic basin, together with the Late Cretaceous continental red clastic basin.

Cretaceous stratigraphy of Longhushan Heritage site

Geological age		Lithostratigraphy			Lithology	Tectonic environment	Sedimentary environment	Geomorphologic features
Cretaceous	Late	Guifeng group	Lianhe Formation	K_2lh	Purplish red conglomerate, sandy conglomerate, fine sandstone, siltstone, thickness>1600 m	Extensional depression	Stream	Red bed hills
			Tangbian Formation	K_2t	Calcareous fine sandstone, siltstone, fossils such as *Elongatoolithus elongathus and Magaloolithus* sp. 462 m in thickness		Eolian sand dune	Danxia landform
			Hekou Formation	K_2h	Purplish red conglomerate, sandy conglomerate, pebbly sandstone, intercalated with sandstone, siltstone. The fossils such as Spheroolithus minor, *Ovaloolithus, Paraspheroolithus, Macroolithus rugustus, and Coelurosauria Fam.gen et.indet*, 687 m in thickness		Pluvial–fluvial fan stream	
		Ganzhou group	Zhoutian Formation	K_2z	Purplish red calcareous sandstone, siltstone, gypsiferous, calcareous, and mirabilite. Plant and *Ostracoda* fossils, etc., 650 m in thickness	Collision and compression	Littoral–shallow lake	Red bed hills
			Maodian Formation	K_2m	Purplish red conglomerate, sandy conglomerate, pebbly sandstone, siltstone, locally intercalated with basalt, plant, silicified wood, etc., fossils, 830 m in thickness		Pluvial–fluvial fan stream	
	Early	Huobashan group		K_1H	Dominated by gray and purple rhyolitic ignimbrite, volcanic breccia, andesitic tuff in the lower part, intercalated with sandstone, argillite, conglomerate, sandy conglomerate, sandstone at the bottom, variegated pebbly sandstone, or sandy conglomerate, tuffaceous sandstone, microsandstone, siltstone, argillite in the upper part. *Bivalvia*, plant fossil, *Estheria, Ostracoda* fossils, etc., thickness>2711 m		Eruption, effusion, overflow, and volcanic lake basin	Mid–low mountains and hills
		Wuyi group		K_1W	Rhyolitic ignimbrite, spherical rhyolite, andedite, dacite, agglomerate breccia, tuff and sandy conglomerate, pebbly sandstone, tuffaceous sandstone, fine sandstone–siltstone, etc., the age of the suite of volcanic rock ranging from 138 to 130 Ma, thickness>8785 m			

- **Early Cretaceous continental volcanic basin**. The volcanic rock, derived from an intensive volcanic eruption in the Heritage site, is mainly distributed in the North Wuyi mountain area, southern part of Xinjiang Basin. During the Early Cretaceous period, the volcanic eruption cycle was predominated by explosion of detrital flow rhyolitic ignimbrite or rhyolitic rubble-bearing ignimbrite, secondary effusion facies andesite characterized by the multi-phase eruption cycles, complete and complex volcanic textures, volcanic organs and eruptive materials with different compositions while different eruption centers as well as eruption materials were superimposed and penetrated with each other, forming different volcanic structures (depression). It appears as acid intermediate meta-alkalic rock series from early to late in time, while new evolutional tendency gradually. became new from west to east at the beginning of eruption in space. There exist certain differences between lithology and lithofacies as controlled by the fractures at different areas and different directions, generally distributed as nearly EW strike.
- **Late Cretaceous continental red clastic basin**. The Late Cretaceous red clastic rock series is a product resulted from further extension and depression of the continental activity margin. Maodian Formation and Zhoutian Formation in Ganzhou group are represented by a suite of red clastic sediments belonging to the piedmont pluvial–fluviolacustrine and alluvial–fluviolacustrine facies, showing inherited and continuous relationship between them. In the normal sedimentary sequence, Hekou Formation of Guifeng group generally overlays with parallel and unconformability on the underlying Zhoutian Formation while along the margin of the basin, the bottom of Hekou Formation often overlaps on the Early Cretaceous strata.
In summary, the red clastic rock series in Hekou Formation and Tangbian Formation of the Upper Cretaceous is a measure or carrier for building the landscape of Danxia landform.
- **Hekou Formation**. It is a suite of sedimentary assemblage of red clastic rock distributed along piedmont pluvial and alluvial fan-braided river. Longhushan belongs to the two typical pluvial and alluvial fan bodies developed at the south edge of the basin and overlapped on the underlying strata. In general, the material of the formation is made up of "lithic conglomerate," with the development of the basal scouring structures and cross-bedding and polygonal structures in the sandstone intercalation, containing the fossils such as dinosaur eggs, *Spheroolithus minor, Ovaloolithus, Paraspheroolithus, Macroolithus rugustus,* and *Coelurosauria* Fam.*gen* et.*indet.*

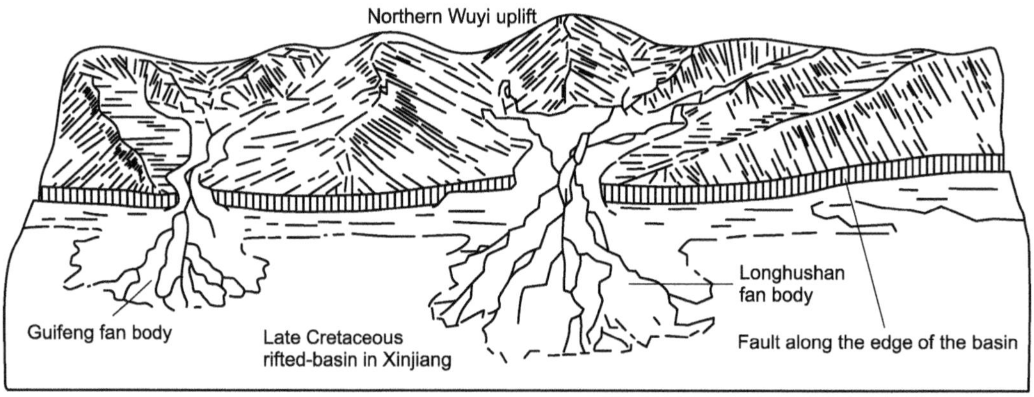

Evolution of Longhushan alluvial fan body

Record of gypsification sediments under the condition of evaporation environment and air hot climate

Hekou Formation

Hekou Formation

Hekou Formation

Dinosaur egg fossil in Tangbian Formation

Sandstone in Tangbian Formation (Nanyan in Yiyang)

Features of quartz sand particle of Tangbian Formation. **a** Grandiose pits; **b** butterfly pits; **c** enlarged butterfly pit partially

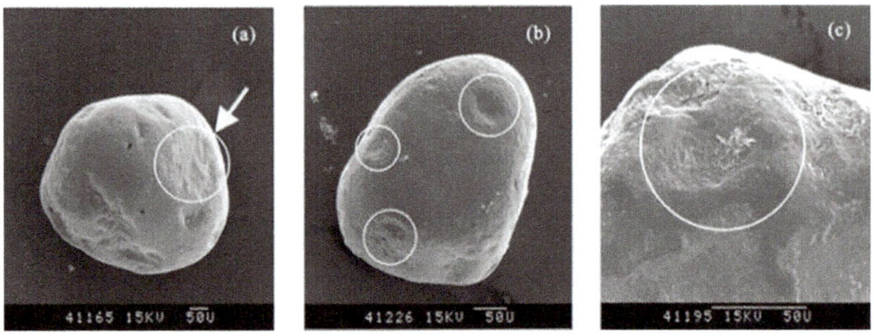

- **Tangbian Formation**. The host lithology is red massive (bulk) fine-grained lithic graywacke. The typical sedimentary structure is large-scale and huge tabular cross-bedding. The individual measures of bed ranging from 10 to 20 m in thickness are common, and some even attaint to 50 m in thickness, rarely seen elsewhere. It belongs to eolian sand dune sedimentation based on the study of sedimentary texture, structure and the surface features of quartz sand grain, bearing fossils such as *Elongatoolithus elongathus and Magaloolithus* sp.

(3) **Structural evolution and the significant geological processes**

The development of Xinjiang Basin is mainly controlled by the nearly EW-trending Beihai–Shaoxing fracture zone under the impacts of faults at lying wall and its blocks, experienced a series of transformation processes from depression (sag) basin into fault basin and from sequential basin to neogenic basin as well as extension faulting compression. Accordingly, it can be divided into three evolution stages.

- **Pre-cretaceous evolution stage**. The Yangtze Plate and Cathaysia landmass were welded into a united continental terrain by the Caledonian tectonic event. Indo-China compression orogenic event not only made the strata folded and uplifted before the Late Triassic, but also led to reallocating of the oceanic–continental terrain(framework). After the Indosinian Movement, the area entered into the mobile stage along the continental margin. At that time, it was characterized by the extension of uplift, contraction of depression, intensification of faulting, and frequent magmatism. The Xinjiang depression traversing the area was transformed into faulting. Meanwhile, the formation, sedimentary, volcanic activity, and magmatism of secondary basin were controlled by NNE- or NE-, and NW-trending deep fault zone through the area, which further reveals the formation processes and evolution history of synthems(structural layers) in different geological ages.

- **Cretaceous–Paleocene evolution stage**. It is mainly the formation stage of extension fault of post-intracontinental orogeny and the Xinjiang red bed basin.
- **Early–Early cretaceous period**. It is an intracontinental orogenic period. Its dynamic environment is mainly intracontinental contraction under the subduction of Pacific Paleoplate. Firstly, a suite of lacustrine facies clastic rock of pre-volcanic eruption developed within the depression area. As extruding stress increased in the directions of NWW-SEE, it entered a period of high-frequent intensive volcanic eruption, and the eruption debris flow facies and effusion facies continental eruptive rock were formed as long as the fractures controlling the volcanic eruption closed.
- **Late–Early cretaceous period**. It is an initial extension period of post-orogeny. The crust was transformed from drastic movement into relaxation stage by the intensive volcanic eruption, receiving lake basin facies sedimentation. The crustal oscillatory movement was frequent, accompanying with the volcanic eruption, but the volcanic activity trended toward decreasing due to the unstablility of the sedimentary environment at that time. Succeeding this, the tectonic movement became relatively stability gentle with the velocity speed of sedimentation outpacing that of subsidence. The climate was getting arid; finally, the basin was withered and subjected to denudate.
- **Early–Late cretaceous period**. It is an intensively fault basin-forming period. A new basin was developed again before complete extinction of the former basin and entered into expansion stage. The lithology and lithofacies varied greatly in different places, mainly representing fluviolacustrine sedimentary environment, locally intercalated with basalt and gypsum salt bed.
- **Late–Late cretaceous period**. The sedimentation of the basin began to wither resulting in the formation of the landscape-building strata, the basin subsidence and the sedimentary center as a whole was in general migrated from south to north, and the sedimentary sequence is dominated by the piedmont alluvial fan body, meanwhile the sedimentation predominated by the eolian sand dune was developed based on it, which indicates that the climate at that time became extremely arid and hot. At the same time, the dinosaur extincted, leaving a large amount of dinosaur eggs and dinosaur osteolith and branner algae skeleton fossils.
- **Paleocene**. Lying in the faulting area of withered (suppression) stage, the basin was migrated westward, following the uplifting of crust and the end of sedimentation.
- **Neogene–Quaternary evolution stage**. It is not only the continental uplifting stage but also the formation stage of Danxia landform within the basin. Under the impacts of neotectonism, the crust began to uplift since Neogene. The margin of Xinjiang Basin took place inhomogeneous uplift, and the whole body was tilted toward north. The reasons for its dynamic environment are as follows: Tibetan Plateau was uplifted rapidly as a result of the collision between Indian Plate and Eurosian Plate in southwest China, while the subduction of Philippine Sea plates led to the uplifting of South China continent, and the principal stress field trended toward NWW-SEE strike (Fig. 16.1).

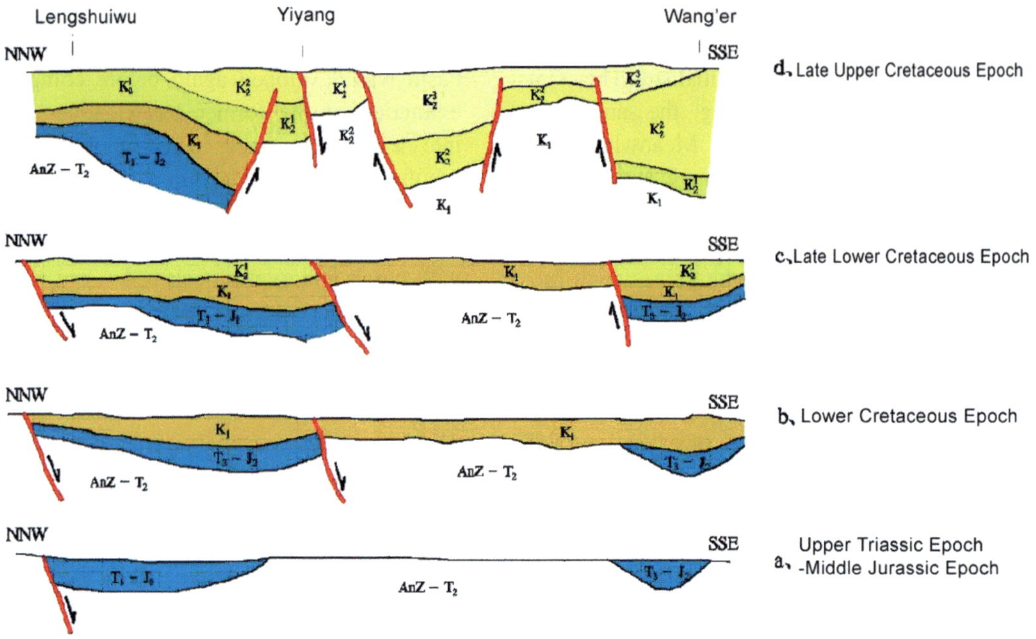

Fig. 16.1 Evolution model of Xinjiang Basin

(4) **Significant geologic events**

The Xinjiang Mesozoic continental basin where Longhushan is located contains rich paleogeographic and paleontologic information, displaying the outstanding evidences of major geological events since Mesozoic Era in the world.

- **Indosinian orogeny and basin-forming**. The features of molasses are common in the Upper Triassic strata of Xinjiang Basin, and the coarse clastic rock overlies unconformably on the underlying strata, which provides an evidence for Indosinian Movement. The sedimentation of Late Triassic–Early-Middle Jurassic laid a foundation for the formation of Xinjiang Basin after the event.
- **Early Cretaceous volcanism**. During the Early Cretaceous period, the area controlled by the nearly EW-trending Pingxiang–Guangfeng suture zone(Jiangxi section), the NNE-trending deep fracture zone of Wuyuan–Ningdu–Anyuan, etc., took place multi-phase intensive volcanic eruption represented by the Daguding Formation and Ehuling Formation, apart from the formation of remelting plutonic granite, forming South China volcanic rock series of Mesozoic circum-Pacific volcanic rock zone.
- **Late Cretaceous gypsum salt sedimentation**. Purplish microsandstone and calcareous siltstone are intercalated with the light gray and light gray green laminar or network gypsum bed in the Early–Late Cretaceous Zhoutian Formation, reflecting the extremely arid sedimentary environment at that time.
- **Late Cretaceous eolian sand accumulation (deposition)**. There was developed a large scale of high-angle tabular cross-bedding in the purplish red arkosic sandstone of Late Cretaceous Tangbian Formation. The individual measures have the huge thickness and quartz particle with excellent sphericity. And the special butterfly pits and granulose pits of eolian sand can be observed under the microscope, indicating that it belongs to the genetic type of eolian sand dune sedimentation.
- **Late Cretaceous dinosaur cataclysm**. Hekou Formation and Tangbian Formation of Late Cretaceous are the host bed of dinosaur fossils, indicating that the Late-Late Cretaceous

Xinjiang Basin used to be the major habitat of dinosaurs. The Dinosaur extincted from the end of the Cretaceous Period.

- **Uplifting of the basin in the end of Late Cretaceous**. By the end of Late Cretaceous, Xinjiang Basin took place inhomogeneous uplifting under the impacts of Himalayan orogeny, and the whole body was tilted toward north. Thereby, the sedimentation of Cretaceous red bed basin was getting to the terminal, and the basin became denuded area. Neotectonic movement not only made the primary fault structure re-revived but also produced a series of faults and vertical joints with different directions and different properties. Therefore, the red clastic rock shaped by the scouring and weathering created the beautiful Danxia landscapes of Longhushan and Guifeng Heritage sites.

16.2.2 Danxia Landform

(1) **Geomorphologic types**

- **Types of pattern**. The Heritage site contains the major types of individual Danxia landform and colonial Danxia landform in the subtropical humid area in China. And the individual scale of Danxia is big, while the colony level is clear.
- **Types of the individual pattern**. Individual Danxia landform in Longhushan almost covers all of the types in subtropical humid area, including Danxia cliff, rock, stone "village," stone wall, stone peak and stone pillar and Danxia cave, Danxia groove as well as unique and pecurious Danxia pictographic stone. Evidence of the formation processes and stages is extremely well preserved, which has enabled the construction of a typomorphology.
- **Types of the colonial pattern**. It is characterized by the combination of erosion residule flat and rounded peak cluster, peak forest (hoodoo) and isolated peak, and Unaka (kopje), being a model of evacuated Danxia peak forest (hoodoo) landform, of which those along the bank of Luxi River and Guifeng feature the peak forest-type Danxia landform; Paiya Peak marks the peak cluster-type Danxia landform, while Nanyan is typicality of the isolated peak-type and hill-type Danxia landform.

Types of Danxia landform element in the area of Longhushan and examples

Series	No.	Type	Name of example	Description
Danxia positive landform	1	Danxia cliff wall	Splended Silk Peak	The stone cliff stands on the bank of river, 300 m wide and 150 m high above river, symmetrical in appearance. There are waved vertical solution grooves and caves in cluster on the cliff wall, which formed by water dissolution. The cliff wall is patched with numerous nests and excrements of birds, multi-colored, and looks like a huge bolt of silk hanging down
	2	Danxia stone stockage	Celestial City	Original mountain body was cut and separated to form the stone stockade that has rounded form in plan and surrounded by steep stone cliffs in all sides, 180 m high, el. 244 m above sea level
	3	Danxia cuesta	Flying Flag Peak	The cuesta rises to 110 m, with a gentle slope in northwest side, 20°–25°, 300–500 m long, and a steep slope in southeast side, 75° and 20–30 m long
	4	Danxia stone wall	Sacrifice Cliff	With el. 289 m above sea level, relative height 175 m, and width 24 m, the cliff strikes in meridian direction. As cut by two sets of joint in NE and NW directions, it uprises from ground, sheer, and steep

(continued)

Series	No.	Type		Name of example	Description
	5	Danxia stone pillar		Gold Spear Peak	Running water washed and scoured the huge original mountain body along vertical joints to cause it collapse, leaving an isolated stone pillar, which uprises steeply to sky, 118 m above sea level
	6	Danxia peak		Celestial Peach Peak	El. 95 m above sea level, 45 m high from ground. Original stone peak was scoured by running water and collapsed, and as a result, the middle part appears protruded and the stone peak looks like a peach; however, the west part of the peach-like body was destroyed through later collapse
	7	Danxia tolt and lonestone		Mourning Person Stone	Original mountain was subjected to collapse leaving two residual masses, a butte (121 m high) and a lonestone (33 m high) with interval between them over 100 m
	8	Colluvial mass and boulder		Lotus Stone	Angular blocks caved from Celestial Peach Peak have piled in the Luxi River to form a block mound, which uprises above the river for 8–10 m and looks from far like a blooming lotus flower
Danxia negative landform	9	Danxia gorge	Linear	A Thread of Sky in Guifeng	The V-shaped valley is formed by two parallel peaks—threefold Turtle Peak and lying Cow Peak which are only 0.9–2.3 m apart. The valley, 100 m deep and 21 m long, straight and serene, sees only a ray of the sky. In some places, it only permits one person to get through at the same time
			Narrow	Celestial Stone	There is a large U-shaped valley that separates Celestial City Peak and Hetunpu Peak, 50 m wide and 200 m deep, and extends for 1000 m in NE 75° direction. There are numerous holes, grottos, and caves on cliff walls of the valley, spectacular and pleasing one's eyes
			Broad	Luxi River	Deeply dissected the mender structurally is controlled by a NW-trending fault and several sets of joint. Scouring and erosion of running water have "sculptured" the peak forest that comprising Danxia cliffs and stone stockades with rounded top on both sides of the river. The landscape resembles a vivid and charming Danxia mountain and water brush painting
	10	Bedding trough		Majishan	5–7 parallel layers of shallow troughs are developed along the soft rock beds, while the hard rock layers protrude upward
	11	Bedding stone trough		Divine Nunnery	There is a trough with flat floor, in which caves with flat arch-like roof are developed. Of them, the Divine Nunnery Cave is 40 m wide, 10 m high, and 40 m deep. In some of the caves are stone coffins which left from the Spring and Autumn period and the period of Warring States

(continued)

Series	No.	Type		Name of example	Description
	12	Danxia cave	Weathering	Celestial Stone	Distributed conformably with rock bed in moniliform arrangement, grottos and caves belong to three layers. The lowest layer is 30–40 m high from the river. A single cave usually 1–7 m long, 0.5–2 m high and 0.5–4.5 m deep, in elongated, oval, or ellipsoidal form. Some caves have treasured the coffins and funeral objects left from 2600 years ago, the Spring and Autumn period, and Warring States period
			Erosion	Nanyan Buddhism Cave	The stone gate is 70 m wide, 30 m high, and 30 m deep, appearing in semicircular form. The Buddhist grottos inside of the cave were dug in the Tang and the Song dynasties. In total, there are 40 stone notches and grottos, and over 10 engraved inscriptions reserved inside
			Collapse and erosion	Majishan	Some rock beds were scoured and eroded in the rock trough, and the roof without support collapsed freely to form a bigger space. The colluvial mass is piled on the bottom of trough
	13	Danxia perforate cave	Weather and erosion	Hetunpu	The cave, 34–42 m long, 8 m wide, and 4 m high with altitude 140 m, was formed by progressive weathering and dissolution from both sides of the stone wall along the same gently dipping soft rock layer, which perforated the stone wall
				Majishan	There is a huge perforate cave on the halfway of the Majishan stone beam in Mazuyan, which is 50 m long, 20 m wide, and 2–3 m high, crosscut the stone beam, and became a part of the intermountain footpath
			Colluvial mass	Entrance to Fairyland	Huge colluvial blocks support each other to form a perforate cave, while the small blocks and debris were scoured away by running water and the cave emptied. It is 30 m long and 1–2 m wide, permitting a person to get through with bended back
	14	Natural bridge		Celestial Bridge	The stone bridge is in sub-meridian direction, put up high in the mountain. The west part of bridge is steep with a 19 m high cliff, while the east slope is gentle, one may walk on. Bridge floor area is 20 m^2
	15	Honeycomb-like cave		Xiongbatianxia	The stone pillar, nearly 100 m high, is surrounded by vertical cliffs in all sides. Numerous caves, holes, and pits with diversified form are developed on the pillar and its fundament, resembling a honeycomb. The caves are commonly 0.2–1 m wide, 0.2–0.5 m high, and 0.1–0.5 m deep
	16	Notch-like cave cluster		Swan Lake	Caves are distributed in cluster, conformable with rock bed and connected with each other, and the large one overlaps the small one, all different in size, usually 1–4 m long, 0.5–2 m high, and 0.5–3.5 m deep, rounded, ellipsoidal, and irregular. They are put on coffins and funeral objects inside by ancient people of the ethnic group Yue in the Spring and Autumn and Warring States periods

(continued)

Series	No.	Type	Name of example	Description
	17	Danxia pot-shaped den	Celestial Pool	Seven deeply holes in distance of 20 m are distributed interruptedly on the bed rock of an upper reach tributary of the Longmen Lake. Diameter of opening 0.5–1 m, while hollow bigger
	18	Erect trough	Flying Flag Peak	Both sides of the 110 m high mountain are developed with erect troughs formed by rainfall erosion and Danxia caves formed by dissolution and weathering of the rock
	19	Erect cave	Fairy Lady Peak	Dropping water along a 330° trending extension fracture continuously scours and dissolutes the rock wall. For the water affected to the lower part more intensively than to the upper part, the potholing has made 95 m high an erect cave with a big lower part. In addition, there is a "人"-shaped hanging trough, and the cave resembles a female private part and bottom
Figurative landform	20	Figurative stone	Trunk Mountain	A longtime scouring by running water has made an erect cave at the bottom of stone beam. The rock above the cave collapsed along joints leaving the residual stone pillar that stands outside. The landscape of the huge stone elephant composed of the beam and pillar has been appreciated as "No.1 stone elephant in the world."
			Old Man Peak	The residual and isolated peak rises 51 m above ground. The summit and slope with caves and a clear contour make an old man's figure, which changes, if viewed from different angles. It is an elite microlandform among the Danxia geomorphologic landscapes

Danxia cliff and wall: Fairylady Sprankling Flowers

Mesa: Celestial City

Danxia cuesta: Flying Flag Peak

Stone wall: Sacrifice Cliff

Hoodoo column: Pillar of the Sky

Hoodoo column: Pillar of the Sky

Hoodoo column: Pillar of the Sky

Golden Bell Peak

Isolated peak: Mourning Person Stone

Isolated peak: Strongest Pear Peak

Danxia perforate

Natural Bridge

Narrow geoge: Carmel Peak

Danxia peak: Fairy Peach Peak

Danxia peak: Carmel Peak

Danxia rock trough: Celestial Rock

Cliff cave: Swan Lake

Water erosion cave: Spoon Cave

House-Head Peak

Haw-Peak

Seonior Peak

Warrior Stone

Fairy Lady Rock

Celestial Quest Stone

Drunk Monkey

Bajie Stone

Queen Stone

(2) Genetic types

On the basis of the formation mode of Danxia landform, Longhushan Danxia landform can be divided into five genetic types as follows:

- **Running water erosion type**: Permanent and seasonal running water scouring and erosion are the major reasons why Danxia landform was formed, including major exogenic processes which played an important role for the formation of Danxia landform of each stage, taking Luxi River as a typical example.
- **Rainwater erosion type**: It denotes that the rainfall leaches and dissolves Danxia slope face (the impacts of raindrops and scouring and sheet erosion) as well as the ephemeral runoff and erosion processes eroded the root of mountain (or hills). The rain erosion provides the exogenic process for sculpturing the various types of microlandform landscape with Danxia slope face. The Flying Flag Peak is the typical example of this kind of genetic types, which is dominated by the linear and laminar washing action, but the rainfall action is more intensive than the interstratal dissolution.
- **Collapse residual type**: Gravity process plays a vital role in the development of Danxia landform because the most characteristic geometric element of Danxia landform, i.e., the escarpment slope, is often formed by avalanche.
  The rock on the upper part lost the balance as long as the multiple vertical fissures in the red sandy conglomerate bed were incised, washed, or laterally eroded by the flowing water, leaving the Danxia landscapes such as castle-like, wall-like, and column-like.
- **Collapse accumulation type**: The collapsing blocks often formed the huge block accumulation at the piedmont, while some of them formed fantastic life-like landforms, such as Lotus Stone and Jade Comb Stone.
- **Solution cavity(cave) type**: The surface water or groundwater dissolved the solubility materials in the rocks, while the change of temperature gave rise to weathering and spalling of the rock, further expanding of the rock fissures, resulting in the formation of the concavo landform and

rounded convex landform with different sizes and shapes such as rock trough and solution cavity. The representative ones include Xianshuiyan cliff-comb cave group and Fairy Xiangu Temple trough-like cave.

(3) **Development conditions of Danxia landform**

The development of Danxia landform mainly depends on such prerequisites as stratigraphic lithology, structure, and exogenetic processes.

- **Stratigraphic lithology**: The Danxia landscape-building strata are mainly Late Cretaceous Hekou Formation and secondary Tangbian Formation. Conglomerate and sandy conglomerate in Hekou Formation are characterized by hardness and anti-weathered and anti-denuded, while the sandstone and siltstone intercalation vice versa, which the rock troughs and caves formed as a result of the erosion of the flowing water, constitute the various geomorphologic types and shaped peak (stones) together with the coarse-grained rocks. The microclastic rock formed by the eolian sand dune deposition of Tangbian Formation was easy to be weathered into the gentle low hills, locally steep cliffs, and red walls as well as various sizes of caves or cave groups such as Nanyan solution cavity (cave).
- **Tectonic conditions**: Gentle stratigraphic occurrence, fracture, and vertical joints (fissures) play a principal controlling role on the formation of Danxia geomorphologic landscape. The dip angle is generally 10°–25° where the gentle hilly top plane is common in the Heritage site. The larger dip angle is up to 20°–50° where the mountain body has characterized by cuesta such as Flying Flag Peak in Guifeng Scenic Spot. After the Late Cretaceous, deposition of Xinjiang Basin ceased, and the crust was uplifted into eroded area, accompanying with the formation of fault, joint structure, and the mutual dissection among the joints as well as differential uplift of block, so the size, direction, and silhouette of Danxia mountain body were controlled by nearly EW-NW-trending fault structures while the spreading directions of castles, peak clusters, peak forest(hoodoo), stone wall, stone peak, and stone pillar(column) were controlled by NE-NEE- and NW-trending joint zoning, orderly forming the spatial topography. Since Neogene, the crust was characterized by differential and intermittent uplift. The denuded leveling plane with the same height was formed as a result of each uplifting. In view of the above, the Heritage site can be divided into five leveling planes and three stream terraces, indicating that the neotectonism included eight rapidly uplifted stages, with the elevation of about 300 m.

Vestiges of crustal deformation and the red bed dissected by the tectonic movement (shown in white line)

- **Exogenous conditions**: There is plenty of rainfall in the Heritage site, which is in favor of the planar and linear water flow scouring and incision of deep valley. The red beds are being eroded and leached intensively by the paroxysmal rainfall continuously. So, the formation of Danxia landform benefited from the exogenic processes, that is, water flow erosion and infiltration.

(4) **Evolution stage of geomorphology**

Division of geomorphologic evolution stage. Longhushan Heritage site is a typical representation of the Late Mature–Early Old Stage in the geomorphological evolution of Danxia landscape, of which:

- **Late mature Danxia peak forest(or hoodoo) in Guifeng**: It features stone forest, stone, stone wall, and steep cliff. On both sides of cliff wall, there are well-developed rain-eroded-type longitudinal linear grooves with weak solution processes, and the landscape assemblage is typical of shaped stone forest and stone column.
- **Late mature peak wall-type Danxia peak cluster in Paiya Peak**: Paiya Peak in Longhushan lies to the east bank of Luxi River, being a large scale of peak wall-type Danxia peak cluster. Various stone peaks, stone walls, and stone pillars are in picturesque disorder and ordered assemblages with rhyme and gradation.
- **Early–old stage evacuated-type Danxia broad valley peak forest along the bank of Luxi River**: The Luxi River with green water and red stone scattering along it, including many stone mesa, stone column with flate or round peak and hoodoo cluster, is a perfect balance between dynamic and static.
- **Early–old stage Danxia landform in Mazuyan and Nanyan**: It is characterized by isolated peaks, kopjes, and castle-like peaks.

Major geomorphologic processes. Clean water and red cliff along the Luxi River including stone castles and flattop or rounded-top long pillars and short pillars as well as stone castle group, displaying a perfect combination of static landscape and dynamic landscape.

The crustal movement since Neogene, characterized by the stage and intermittent uplift, created multilayer and multi-type, step-type Danxia landscapes. The formation processes of leveling plan can represent the evolution processes of Danxia landform, which holds an important significance for studying the ongoing geological processes and eco-environment.

The basin was uplifted differentially on the edge and tilted toward north as a whole under the impacts of neotectonic movement since the Neogene.

- **Terrace vestige and its major characteristics**: It is mainly distributed along the bank of Luxi River, including three-grade stream terraces, representing the rhythm of the neotectonism. The Xinjiang river bed is 10–15 m a.s.l, and the valley was generally subsiding in north while uplifting in south.
- **I-grade terrace**: 10–20 m a.s.l. The stratigraphic unit corresponds to Lianyu Formation (Q*hl*) ranging from 3 to 10 m in thickness, which generally consists of I-grade basement of terrace. Some of them developed on the alluvial flat composed of a suite of binary alluvial facies.
- **II-grade terrace**: 10–25 m a.s.l. The stratigraphic unit is equivalent to Liantang Formation (Qp_3lt) ranging from 2 to 13.19 m in thickness, forming II-grade buried terrace, 3–8 m higher than that of I-grade terrace, which

appears as the typical binary texture, belonging to the alluvial facies.

- **III-grade terrace**: About 24–30 m a.s.l., the stratigraphic unit corresponds to Jinxian Formation(Qp_2jx) with thickness of 2–13 m, distributed on the basement of I-grade and II-grade with obvious binary texture, belonging to alluvial sedimentation.

Vestiges of the leveling plane and its major characteristics. From Guifeng or Longhushan to Xinjiang valley, the geomorphologic patterns can be divided into five leveling planes due to the pattern presenting multilayers and step subsidence by and large, which holds an evidence that there had been obvious crustal intermittent uplift for five times, and the interval between each uplift keeps a relative stable leveling stage (phase). The altitude of the same grade leveling plane in Longhushan and Guifeng is different from each other, which implies that there existed differential between uplifting amplitude and speed, as the uplifting speed in Guifeng was faster than that in Longhushan.

- **I-class leveling plane**: The distribution area of leveling plane is small due to the limited conservation. The height of Longhushan at an altitude of 240–280 m a.s.l. corresponds to the position where Paiya Peak, Songshan, Longhushan, and Shuiyan scenic spots are located. The height of Guifeng with 360–400 m a.s.l. is equivalent to the position where the Camel Peak (Paidao Stone) and Tianzhu Peak are located. The majority of their patterns are characterized by the stone peak and stone "village" with typicality of "flattop, steep cliff, and gentle piedmont."
- **II-class leveling plane**: The distribution area is next to that of I-class leveling plane. The height of Longhushan at an altitude of

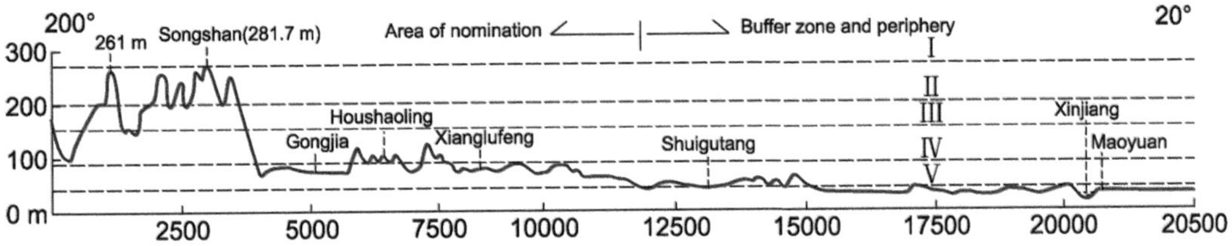

Map of planation surface of Danxia landform in Longhushan

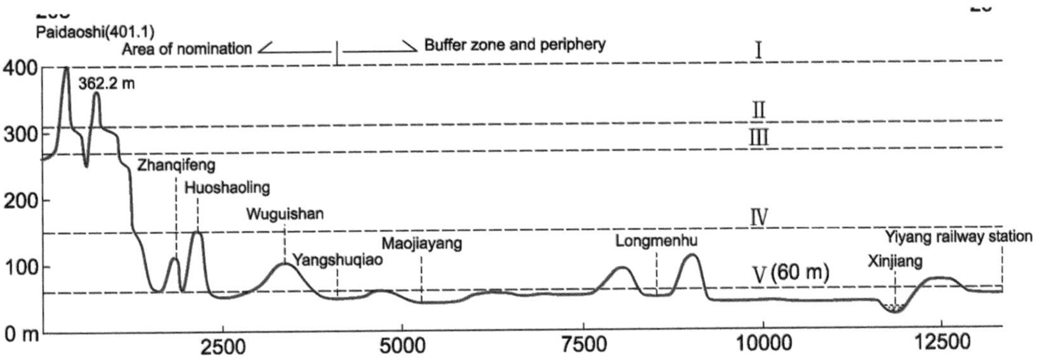

Map of the planation surface of Danxia Guifeng

200 m a.s.l. corresponds to those of Yunjin peak, Buddhist monks, and Sengni (Buddhist Nun) Peak in Xianshuiyan Scenic Spot, as well as Xianglu Peak and Golden Dragon Peak in Mazuyan Scenic Spot. Their patterns exhibit round top stone "village" and stone peak. The height of Guifeng at an altitude of 300–310 m is assigned to the position of first platform of Camel Peak (Paidao Stone) and Tianzhu Peak with gentle topography. The geomorphologic pattern mainly displays stone forest, stone peak, and gentle slope.

- **III-class leveling plane**: It is widely distributed. The height of Longhushan with an altitude of 120–155 m a.s.l corresponds to the height of Golden Bell Peak and Celestial Peach Stone in Longhushan Scenic Spot and Golden Spear Peak (Strongest Penis Peak) in Mazuyan Scenic Spot. The height of Guifeng at an altitude of 250–270 m a.s.l. is equivalent to the position of second platform, namely the saddle between Camel Peak and Tianzhu Peak, with gentle topography in Camel Peak(Paidao Peak), Tianzhu Peak, etc. The geomorphologic pattern is represented by stone peak and stone column.
- **IV-class leveling plane**: It is widely distributed. The height of Longhushan with an altitude of 80–90 m a.s.l. corresponds to that of Longhushan, Zweikanter in Longhushan Scenic Spot, and that of Mazuyan Cave in Mazuyan Scenic Spot. The height of Guifeng with an altitude of 40–60 m a.s.l. is equivalent to the height of Double Turtles, Flying Flag Peak, Mourning Person Stone. The geomorphologic patterns on the planes mostly display low-rounded Unakas(kopje), isolated peaks, and uneven hills.
- **V-class leveling plane**: It is the most extensively distributed. It is located at the position of Longhushan with an altitude of 30–45 m a.s.l, while Guifeng at an altitude of 40–60 m a.s.l, namely being presented in valley plain and peneplanation areas; meanwhile, the rice field and residential areas along Luxi River mostly lie on the leveling plane.

(5) **Development model of landform**

The bedrock consisting of Longhushan Danxia landform is a suite of red massive conglomerate assemblage of piedmont pluvial–alluvial fan. The rock bed trends toward nearly EW strike, dipping toward north, and its dip angle occurs as nearly horizontal or gently dipping. The neotectonic movement since Neogene caused the basin to be uplifted above the base level of erosion; thus, three sets of NE-, NW-, and NEE-trending faults and vertical joint were formed (newly born).

During the early stage of the development of Danxia landform, the primary rainwater and current eroded and incised along previously formed faults and vertical joints resulted in the formation of narrow deep gorge and "a-thread-of-sky"-type valley. The valleys were further expanded, following the ongoing incision and erosion of surface water as well as the lateral erosion.

The gravity collapse occurred along the vertical joint planes resulted from the overhanging of the valley wall; thereby, the Danxia geomorphologic patterns were gradually evolved into peak cluster, peak forest(hoodoo) (composed of stone "village," stone wall, stone peak, stone column, etc.), isolated peak and kopje, valley plain as well as equisite stones and various forms of pictographic stones with descending order.

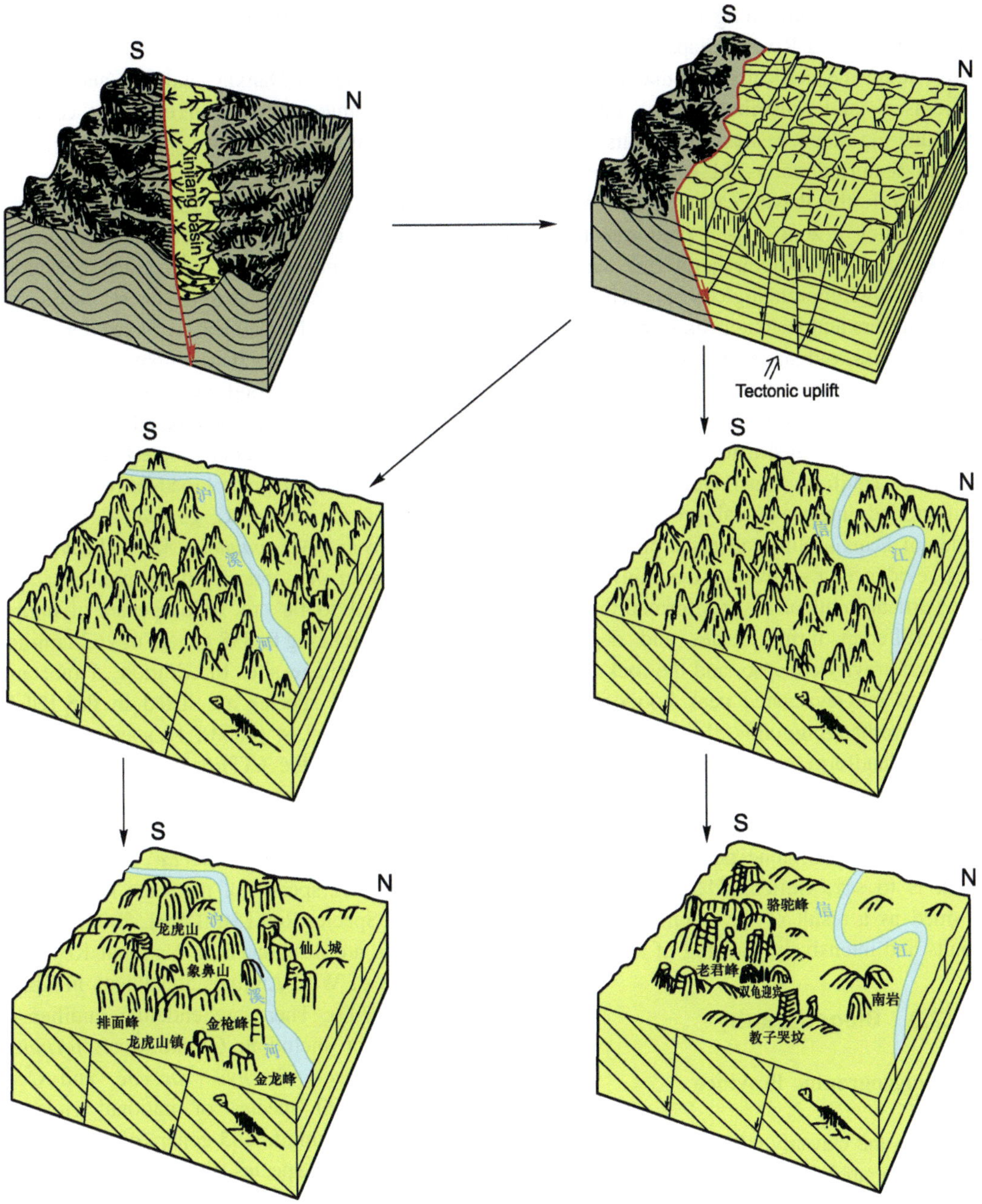

Evolution model of stone forest terrain in the Heritage sites. (Left: Evolution model of Longhushan stream-eroded Danxia hoodoo landform. Right: Evolution model of Guifeng rain-eroded hoodoo landform)

Luxi River of Longhushan is an inherent and permanent stream formed under the control of NW-trending fault and multiple joint, which exists from the periods of basin formation to geomorphologic formation even till now. Therefore, it not only provided a transportation trannel for the sediments of the Late Cretaceous piedmont pluvial–alluvial fan body and controlled the spatial distribution of the sedimentary fan body but also transformed (modified) the geomorphologic area continuously. So, the fluvial erosion process is most obvious. Impacted by it, the rock body along the valley margin of the bank of Luxi River collapsed to form steep cliffs along the vertical joints under the gravity processes and distributed as flattop and rounded-top stone castles, tall pillars and short pillars featuring early–old stage Danxia landform, while the Paiya Peak area far away from the valley zone is predominated by rain-eroded-type Danxia landform, with development of geomorphologic types such as Danxia peak clusters and Danxia peak forests (hoodoos). Nevertheless, old stage Danxia geomorphologic pattern is characterized by the Mazuyan in the lower reaches of Luxi River. The meteoric water provides a prerequisite for the development of Danxia landform in Guifeng area. The rain-eroded linear vertical grooves along both sides of Danxia cliff wall (escarpment) hold the best evidence; meanwhile, the rainwater made the vertical joints or fissures formed the grooves with different forms, scales and depths, and induced gravity collapse, and the rock body was denuded into evacuated peak hoodoo forest or steep overhanging cliffs, forming late mature Danxia landform characterized by stone "village," stone wall, stone cliff as well as the microlandform landscape featuring rare and pecurious shaped stone peak and stone pillars. That is to say, Guifeng Danxia landform was formed as a result of development of joints or faults and intensively leaching of rainwater.

16.2.3 Biogeography

(1) **Biogeographic Province**

Based on the Udvardy (1975) classification system of biogeographic province, the organisms in the Heritage site belong to the Southeast Coastal Sub-Unit of the China Subtropical Forest Province of the Palaearctic Realm. The natural ecosystems mainly are mid-subtropic low altitude evergreen broad-leaved forest and ecosystem of natural stream wetland, while the artificial ecosystem is predominated by wetland of rice and reservoir.

(2) **Diversity of Species**

The diversities of Danxia landform and vegetation community provide an excellent ecologic environment for the wildlife.

- **Flora**: There are nine vegetation types in the Heritage site, including evergreen broad-leaved forest, evergreen broad-leaved and deciduous and broad-leaved mixed forest, evergreen coniferous and broad-leaved mixed forest and warm temperate coniferous forest, mixed with warm temperate coniferous forest, and warm temperate coniferous and broad-leaved mixed forest as well as bamboo forest and coppice forest. In the evergreen broad-leaved forest, *C. tibetana, C. carlesii, C. glauca, L. glabe, M. leptophylla, M. pauhoi, M. velutina and P. sheareri* show primary property.
  Identified so far are 1626 species of higher plant, belonging to 838 genera of 262 families in the Heritage site. In addition, 5 rare and endangered species are listed in the IUCN Species Red List (2003), 18 species of plant are listed in the CITES Appendix I, II, III (2007) (Convention on International Trade in Endangered Species of Wild Fauna and Flora) mainly being *Orchidaceae* and *Euphorbiaceae*, 32 species of plant are listed in the China Species Red List (CSRL 2004), and 11 species are designated as National Key Protection of Wild Animals and Plants in China (first 1999). There are endemic families in East Asia, including *Cephalotaxus fortunei, Actinidiaceae, Stachyu- racceae,* and *Cornaceae*, 220 species of national endemic plant, belonging to 145 genera of 67 families. Among these species, there are 3 species of pteridophyta from 2 genera of 2 families; 13 species of gymnosperms from 11 genera of 6 families; and 204 species of Angiospermae from 132 genera of 59 families.
- **Fauna**: According to the classification of zoogeographic zoning in China, the Heritage site is located in the Oriental Realm. How-

ever, in terms of composition, it shows distinct characteristics of the Oriental Realm and is mixed with some characteristics of the Palaeoarctic Realm. It is identified so far that the vertebrates belong to 387 species of 101 families of rare and endangered animals from 33 orders, of which 16 species are listed in the IUCN Species Red List (2003), 55 species are inscribed in the CITES Appendix (2007), and 36 species of wild animals are under State-Level Key Protection (1989). The Heritage site is the habitat of 37 species endemic to China, including 6 species of mammals, 9 species of birds, 5 species of reptiles, 10 species of amphibians, and 2 species of insects.

(3) **Typicality of Vegetation**

The vegetation type of the Heritage site is diverse and shows a typical zoning vegetation character, belonging to the type of humid evergreen broad-leaved forest in the east of China mid-subtropic zone. There is preserved the complete ecosystem of evergreen broad-leaved forest of mid-subtropic zone with an area of 1500 hm^2 being a typical representative in northern section of Wuyi Mountain Range in China. The natural

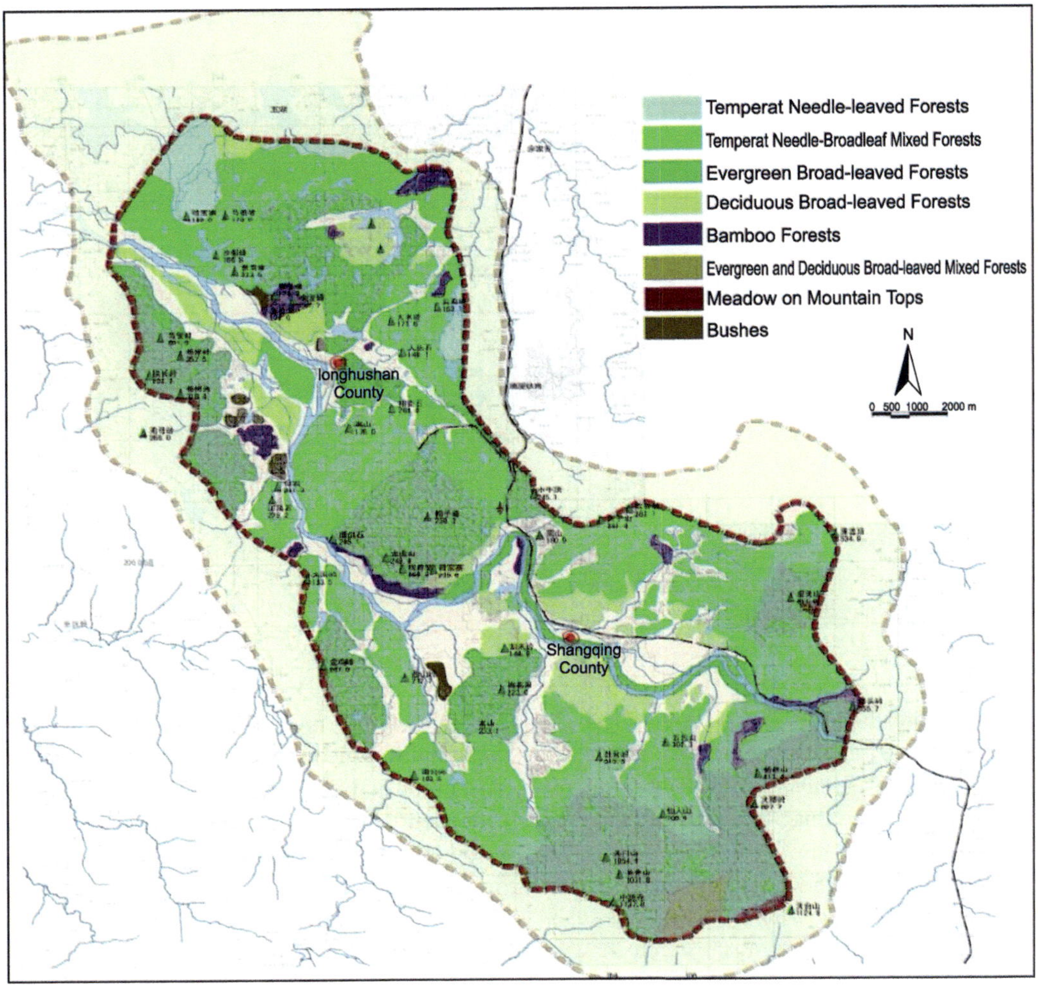

Vegetation distribution in Longhushan

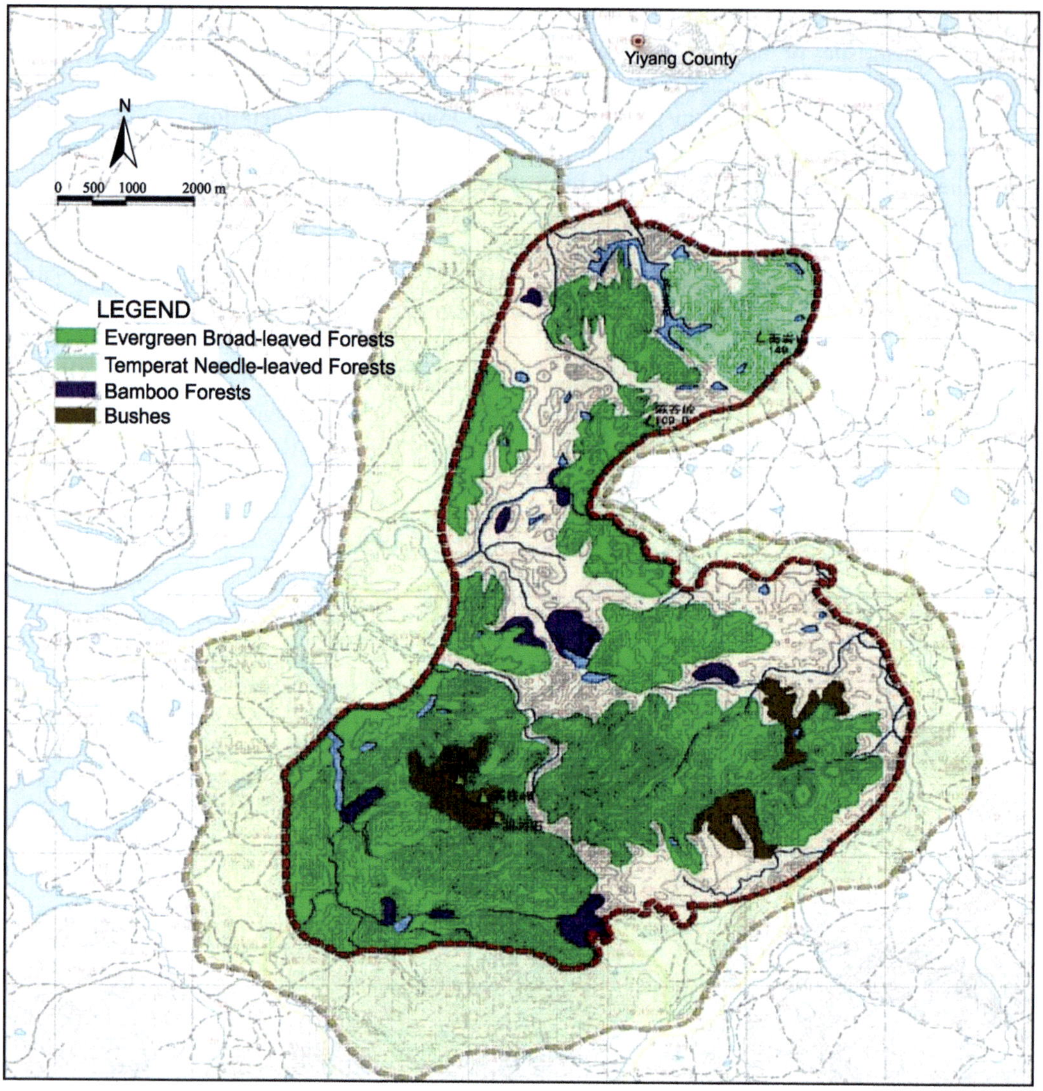

Vegetation distribution in Guifeng

evergreen broad-leaved forest sweeps all the types, including *Fagaceae*, *Lauraceae*, *Theaceae*, *Elaeocarpaceae*, *Symplocaceae*, and *Magnoliaceae*. At the same time, this type of evergreen broad-leaved forest also shows typical tropical composition. Since thousand million years, the communities breed by themselves and experienced the unique succession of biologic community, which provides the natural reference system for studying the recovery and rebuilds the ecosystem of the mid-subtropical evergreen broad-leaved forest, also a typical representative of evergreen broad-leaved forest with the global significance in China.

Typical evergreen and broad forest in Guifeng

(4) **Rarity of Species**

Surviving from the Quaternary Ice Age (glaciation), mergas squanatus (or Chinese merganser, the same as below) belonging to the northern species is listed as the endangered birds of the world in IUCN Species Red List. It is believed that the amount of population of mergas squanatus is less than 5000 in total all over the world according to IUCN Species Red List (2004). Furthermore, its population is decreasing gradually owing to its recession of habitat, which gains the global importance. In addition, the Wetland International pointed out that there were only 3600–4500 *mergas squanatus* all over the world in the 1990s. Meanwhile, the Wetlands International also holds that the Northeast Jiangxi area, including the Heritage site where the Chinese mergansers gather, should be paid great attention. Birdlife International issues four global scientific criteria for classifying the international important bird area: conserving international endangered birds in habitat; narrowly distributed birds; living in habitat particular ecologic area for bird communities; and major habitats for waterfowl, seabird, and other birds (raptors and cranes).

The section of Luxi River and Yiyang Qingshui Lake in the Heritage site is full of fish and shrimp with clean rushing and unceasing water. Along the bank, there are steep terrain, flourish forest vegetation, and towering ancient trees, suitable for the *Chinese mergansers* for wintering habitat. The amount of *Chinese mergansers* has maintained constantly for 150 in the past 10 years in the Heritage site, which is the major wintering habitat for *Chinese merganser*. It is reported that this is the biggest wintering population of Chinese merganser which is unwonted throughout the world (rarely seen in the world), covering about 3.7% of total amount of their entire migratory route, being one of the major wintering habitats. In terms of criteria for Wetlands of International Importance, Criterion II: the wetland should be regarded as international significance if it is the base of the vulnerable, endangered and critically endangered species, or threatened ecological community, and Criterion VI: the wetland should be considered as international significance if one of the waterfowl species or subspecies accounts for 1% of the total population. Consequently, the Heritage site where Chinese mergansers inhabit holds great scientific research value and important international significance as it meets the relative requirements of the IUCN Species Red List, Conservation International Biodiversity Hotspots, and Wetlands International.

Volitant living fossil *Chinese merganser*

16.2.4 Natural Landscape

(1) **Danxia landscape**

Danxia landform is a principal part and base of natural landscape in the Heritage site, containing rare and unique landscapes, with outstanding scientific and aesthetic value thanks to its Danxia peaks distributed densely and sparsely, in picturesque disorder, especially the fantastic shaped peaks and pictographic stones.

Types and characteristics of Longhushan Danxia morphologic landscapes

Location	Type of landscapes	Landscape features
Along the bank of Luxi River, Longhushan	Green water and red mountain, dispersed Danxia hoodoo	The peak forests (hoodoos) are distributed as strips along the middle reaches of Luxi River, stretching 20 km. Along the bank of the river, there are evacuated Danxia hoodoos featuring round stone peak, sparsely distributed and unconnected at the root, belonging to the evacuated Danxia peak forest (hoodoo), displaying the characteristics of early–old stage Danxia landscape which constitutes the fantastic Danxia landscape together with dales and meandering streams

(continued)

Types and characteristics of Longhushan Danxia morphologic landscapes

Location	Type of landscapes	Landscape features

Location	Type of landscapes	Landscape features
Guard Yanmen Peak, Longhushan	Green water and red mountain, large-scale Danxia Peak Cluster	Large-scale combination of landscape of Danxia peak walls and clusters is made up of Danxia Peak groups rising above hills. The highest peak Mount Song rises around 281.7 m a.s.l. stone peaks, walls, and stelae with various shapes demonstrating Danxia natural sceneries

(continued)

Types and characteristics of Longhushan Danxia morphologic landscapes

Location	Type of landscapes	Landscape features
Guifeng	Green water and red mountain, Miniascape-typed hoodoo	There are grid-like vertical joints in Danxia mountain. Due to the rainfall erosion and the interaction of gravity, original mountain has split into different shaped peaks, stelae, and strange stones. They are characteristics of flat summit and steep slope of cliffs. Guifeng is a miniature of Danxia topography with the recognition of "Danxia miniature garden."

Longhushan: the fairy water rocks	Combination of Danxia cliffs and cliff grave cluster	Peak groups and steep cliff tower are aloft along the banks of Luxi River; 10–60 m above the river level is the caves where coffins and burial objects in Spring and Autumn and Warring States periods of China (before 2600 A.D.) are placed. How did the ancestors put the coffins in the cave has become a mystery

(continued)

Types and characteristics of Longhushan Danxia morphologic landscapes		
Location	Type of landscapes	Landscape features
Longhushan: Splendid Silk peak	The combination of bare and red cliffs and bird ecology	It is the huge precipitous cliffs that stand out on the river. The mountain is toward northwest direction with a height of 204 m above the sea level. The cliff peak is 150 m high above the river and 300 m wide. Because of the longtime water corrosion, slopes of cliffs have formed wavelike vertical swallow holes and relatively intensive caves. Bird's nests and dung paint the colorful cliffs, which is an excellent representative of Danxia topography

Guifeng: Old Man Peak	Danxia view stone	Danxia massif has grid-like vertical joints. Due to the wind and rainwater erosion, as well as the interaction of gravity, the stone peak has been shaped like an old man. It is a wonder made by nature in Danxia

(continued)

Types and characteristics of Longhushan Danxia morphologic landscapes

Location	Type of landscapes	Landscape features
Guifeng: Flying Flag Peak	Rain erosion landscape	The Danxia cuesta is with a height of 110 m. The slow slope of the mountain is toward northwest and the steep slope for southeast direction with relatively flat surface. On both sides of the cliff slope, there are longitudinal slots caused by the linear rainfall erosion and slope surface erosion; cave groups are also formed due to erosion corrosion

Longhushan: strongest Penis Peak	Danxia isolated peak in senior period	It rises to a height of 118 m above sea level. Hoodoo column is 60 m high. The diameter of the broken section is less then 10 m. The abrupt column stands erectly among the isolated hills. It is the typical representative of Danxia topography in its senior period

(continued)

Types and characteristics of Longhushan Danxia morphologic landscapes		
Location	Type of landscapes	Landscape features
Longhushan: Elephant Trunk Hill	Danxia view stone (the combination of perforate and stone beams)	It rises to a height of around 100 m. Due to the water erosion from opposite directions, caves took shape in the stones. The huge stone carving, which is composed of stone beams, perforate, and stone column, is known as "the most significant signs of drought." The narrow-top perforate and wide-bottom perforate are 48 m long and 5–8 m wide

Guifeng	Danxia view stones (the combination of turtle-shaped peaks)	Guifeng wins over other Danxia topography of the same category due to its dense hoodoo. It is also famous for the turtle-shaped stones. Given a bird's-eye view from southwest direction, the entire Guifeng looks like a great turtle

(continued)

Types and characteristics of Longhushan Danxia morphologic landscapes

Location	Type of landscapes	Landscape features

(2) **Hydrologic landscape**

Luxi River within the Heritage site rises in the primeval (virgin forest) of Guangze, Fujian Province. Lakes, waterfalls, and springs in the Heritage site add more dynamic and interesting elements to the tranquil landscape of Danxia landform, displaying the characteristics of early–old stage stream-eroded Danxia peak forest landscape. The hoodoos are profiled against the river and create the unique landscape, just like a gallery with the water and red cliff, when you are drifting along the Luxi River, you will feel you were in the fantastic picture. Lakes, waterfalls, and springs in Longhushan, embellishing the hills and mountains, all highlight Danxia topographic characteristics, revealing a perfect combination of static and dynamic landscapes.

Landscape along Luxi River in Longhushan

Landscape around Longmen Lake in Guifeng

(3) **Ecological landscape**

The Heritage site has a superiority natural environment, creating rich and colorful ecological landscape.

- **Plant and seasonal landscape**: The Heritage site has very distinct four seasons, with corresponding beautiful sceneries: In spring, flowers flourish, rich in color; in summer, green plants are everywhere, refreshing, with birds singing and flowers blooming; in autumn, leaves seem to be dyed in red, colorful, and in winter, it is as cool as autumn, charming.
  Verdant sweet-scented osmanthus: There is an ancient osmanthus garden in Guifeng, which is a favorite place to enjoy the flowers. Since ancient times, scholars, poets, and writers had been attracted to the place by its reputation, lefting so many poetries and articles.
- ***Pinus taiwanensis***: *Pinus taiwanensis* community are growing on Tianmen Mountain and Yingtian Mountain with an altitude of 800 m a.s.l, which are hundred and thousand years old, covering with thick dark green moss. Some of them come from the cliff, some from stones, while the others are profiled against the sky, and all of them have imaginative shapes and forms.
- **Plank buttress landscape**: The plank buttress is a special ecological phenomenon in tropical rain forest. There is distributed a typical plank buttress in Guifeng, located in the northern part of mid-subtropic zone. Straight trees with shoot bearing plank buttress contribute to the wonderful landscape in the Heritage site.

Three folds waterfall, Mt Tianmen in Longhushan

Qingyun waterfall, Mt Tianmen in Longhushan

- **Migrant bird and wetland landscape**: Animals and birds make the natural scenery more vigorous and interesting. In October, wintering migrant birds, like *Mergus squamatus* and mandarik duck, will arrive here. Thousand of common cormorants stay all night on steep cliffs. They are flying in the morning and returning to the cliffs at night, hovering about the sky.

Thousands of common cormorant stay all night on steep cliffs

- **Ancient tree community**: The Heritage site abounds in ancient trees such as *Ginkgo biloba* and *Taxus mariei*, displaying a fantastic landscape.

16.3 History and Development

(1) **Evolution of the basin**

By the end of Indosinian Movement, this area entered into the mobile stage of the circum-Pacific continental margin, generally controlled by Beihai–Shaoxing fault zone, forming a composite structural basin distributed as nearly EW strike, with typicality of the compound structure of "gray on the top and red at the bottom." During the Early Cretaceous, the volcanic lava and pyroclastic sedimentary rock series were accumulated, which gave the thickness up to several thousand meters, while the formation of Late Cretaceous basin was characterized by the coarse clastic rock predominated by the sediments of piedmont, pluvial, and alluvial facies with huge thickness, intercalated with sediments of lake basin facies bearing gypsum salt under the conditions of extremely arid climate. About 65 Ma ago, in particular, the neotectonic movement not only made the primary fault structure re-revived but also produced a series of faults and vertical joints with different directions and different properties. Therefore, the red clastic rock shaped by the stream erosion, scouring, differential weathering, calc solution, gravity collapse, etc., created the beautiful Danxia landscapes of Longhushan Heritage site.

(2) **Human activities**

In Neolithic Age, the Heritage site was in sphere of influence of Miao Nationality. During the Xia–Shang–Zhou Dynasties, Baiyue Nationality was very active in Xinjiang area. In the Spring and Autumn Period (770–476 BC), the Gueyue Ethnicity lived in Longhushan area. The unique cliff grave culture became the museum for studying the history of the Guyue Ethnicity. The Daoism Leader Zhang Daoling in Eastern Han Dynasty (25–220 AD) found the Chinese Taoism in Longhushan, which has become the Ancestor Hall of Taoism. During the Tang Dynasty (618–907 AD), Mazu Daoyi was a prominent Taoism. Chan master had been preaching Chan doctrines here. At the same time, many Buddhist abbots across the country gathered here, and listened and studied the Buddhist sutra. Hence, the site was entitled as "the universal center of dhyanas (Chan)." The Xiangshan Academy founded by Chinese eminent idealism philosopher Lu Jiuyuan in Southern Song Dynasty has become one of four great academies in China.

There is no industrial activity within the Heritage site. After settling down in Xinjiang valley nearby, the local residents began farming life and collected the Chinese medicine since ancient times, which, to a certain extent, has contributed to its primitive natural condition. Meanwhile, due to the influence of Taoism and Guyue Nationality, the local residents embrace the principles of "harmony between people and nature." They respect the nature and protect the environment consciously, and all of these customs and ideas have kept its original look.

(3) **Protection history**

Generally, there are three stages to restore the protection history of the Heritage site:

- **Protection by the local residents**: As mentioned above, influenced by the religious culture and led by monks and priest, local residents have the tradition of protecting the history relics, respect the natural environment, and construct the green homeland on their own initiatives.
- **Protection by the village rules and regulations**: Village rules and regulations, as a sort of primitive law, provide the basis for a code of conduct by community residents. This code is accepted through common practice by villages, which formulate and observe the rules and regulations. The regulations, therefore, enjoy high authority. Village rules and regulations, together with forest and water source protection activities, are very efficient customary protective measures, especially the protection of heritage.

- **Protection and management by government**: The government has paid great attention to the environment protection of the nominated site, and established protective and preservative organizations and administrative departments due to its outstanding geologic scientific significance and topography values.

(4) **Sights of cultural interest**

Human civilization in the Heritage site was developed on the beautiful Danxia landform, leaving rich and varied historical relics, which has formed the Danxia cultural landscape integrated with cultures of the poetic literature, cliff grave funeral custom, Taoism, Buddhism, and stone inscription well known in the world.

- **Poem and landscape culture**: Since Tang Dynasty, Chinese prominent scholars and powerful ministers visited the nominated site and wrote poems that had been passed on till now. The wonderful scenery of "the peculiarity of the mountain peaks in Longhushan, Guifeng cannot be found in any other place" has been described in the *Jiangyou Dairy,* in the book of *Xiake' Travelogue* written by Xu Xiake. In the first chapter of the Chinese classic masterpiece *Outlaws of the Marshes,* it describes the natural beauty of Longhushan and tells the story of Heaven Master Zhang Daoling, the founder of Taoism.
- **The Cliff Grave**: The cliff graves in Longhushan area are scattered in groups or zones, and are mainly distributed in Danxia grottoes along the banks of Luxi River and Swan Lake. The cliff graves are all located in the precipices. Longhushan cliff graves can be dated back to Spring and Autumn period, 2600 years ago, holding great scientific and aesthetic values thanks to their the earliest history and the richest burial articles.
- **Taoist Culture**: Longhushan in the Heritage site is the birthplace of Chinese state religion Taoism. The second year of Yongyuan in Eastern Han Dynasty (25 A.D), Zhang Daoling preached and practiced alchemy here. After the fourth generation, Taoism saint Zhang Sheng moved to Guixi, and his offspring settled down for more than 1900 years up to now. The old and famous family is the only one in Chinese history

Cliff grave Danxia grottoes in Heritage sites

Sceneries in Buddhism Cave (Nanyan, Guifeng)

which can be mentioned in the same breath with Confucius's. The Heaven Master Masion and Shangqing Palace are sheltering under the cover of the Danxia pines, and it completely demonstrates the integration of Taoism culture and natural landscape, being of extremely great cultural value.

- **Buddhist Culture**: The Nanyan Grotto in the Heritage site is the representative of Buddhism culture. The figures on the stone were engraved during the Song Dynasty and were carved based on the natural Danxia Grottos, attending the cliff. There are still preserved 40 statues of Buddha inside the Nanyan Grotto with complete types, which is the biggest one in China.
- **Art Gallery of Calligraphy**: There have been preserved more than 200 rock inscriptions since Tang Dynasty, which embodies a wide range of Chinese calligraphy types, including Seal Script Calligraphy, Clerical Script Calligraphy, Standard Script Calligraphy, Running Script Calligraphy, Cursive Script Calligraphy. And the bigger ones can be up to several meters high. The carving skills are proficient, with the varied artistic styles. Therefore, it may be said that distinct calligrapher's assorted arts are all displayed here.
- **Red Mountain Academy**: There are some famous ancient academies in the Heritage site. The Xiangshan Academy in Longhushan has become one of the four greatest academies in China during the Southern Song Dynasty, while the Dieshan Academy in Yiyang was found by the eminent scholar Xie Dieshan in Yuan Dynasty.

16.4 Natural Features and Its Value in the Serial Heritage Sites

(1) **Natural features**

Having the unique geotectonic settings, Longhushan is just located in the suture zone between Yangtze and Cathaysia paleoplates, the southeastern part of Eurasian continent. The Xinjiang Basin where Longhushan is situated is a Mesozoic continental basin with complex texture and evolution history, and possessing evidence in its stratigraphy, lithology, paleontologic fossils, basin texture and structure, geology and geomorphology as well as modern biology of many of the world's great geologic events, especially the history of formation and evolution of South China continent of West Pacific. The heritage is a major component part which witnesses the history. The Heritage site therefore is a preeminent world location for the study of the history of the earth, in particular the history and geologic structure of the continental basin. However, impression as the geology is, the geomorphology, biodiversity, and aesthetic qualities of the Heritage site are of equal international importance. The incision, stream erosion, weathering, and denuding of sedimentary rock mass dominated by the red sandy conglomerate have sculpted a wealth of remarkable landforms at a variety of scales, being a natural museum for studying the geomorphology. In addition, the forest vegetation that clothes the mountain slopes and natural environment contains the world-ranking biodiversity and harbours internationally important rare and relict species. Together, these features form a stunning visual landscape. In summary, the Heritage site is a rare treasure in the world.

(2) **Status and value in the serial heritages**

Compared with other serial Heritage sites, Longhushan also meets the criteria of the serial nomination of the World Heritage because it not only has the same or similar geotectonic settings and evolution history, but also has the geomorphologic features and physiographic features, but the status and value of Longhushan in the serial Heritage sites are irreplaceable. Its outstanding value lies in as follows:

- Longhushan is an outstanding representative of Danxia peak forest geomorphologic landscape with blue water and red mountains, a natural model for artistic creation. The Danxia landscape integrated with the red mountains and blue water is superior to those of the serial Danxia Heritage sites in aspects such as diversity, uniqueness and peculiarity and natural form, holding the irreplaceable scientific value and aesthetic value.
- Danxia assemblage of stone peak and stone column in Longhushan is a typical example of rare microlandform landscapes in the world. There preserved the most densely distributed and the finest Danxia sculptured landscape with the highest quality.
- Longhushan Heritage site is a typical representation of the late mature–early old stage in the geomorphologic evolution of Danxia landscape, also a typical example displaying the features of rain-eroded Danxia peak forest landform and early–old stage stream erosion evacuated Danxia broadening valley peak forest landform, being an irreplaceable type in the serial Heritage sites, holding an outstanding geomorphologic value.
- Longhushan Heritage site is a refuge and habitat of rare and endangered species rarely seen at anywhere else in the world, bearing the major value of biodiversity. The Heritage site has best preserved the complete ecosystem of the low-altitude subtropical evergreen broad-leaved forest and the biggest wintering population of extremely endangered Chinese mergansers in the world, meeting the criteria of International Important Wetland in the International Wetland Convention.
- The Xinjiang Basin records the evolution history of the continental basin in the earth since Mesozoic Cretaceous, while the geoheritages hold six geologic events of the earth since Mesozoic, which is of typical importance of the model in the continental fault basins of the world.
- Longhushan Heritage site is a typical example of perfect combination of Danxia landscape, Danxia ecosystem, and Danxia culture as well as harmonious coexistence between people and nature. Archeological findings demonstrate that people inhabited and dwelled in the Heritage site for about 2600 years. Historical and cultural heritages are embraced in Danxia topography, and characterized by cliff burial culture, Taoism culture, and Buddha culture, which is an outstanding example of perfect harmony between people and nature.

17 Jianglangshan, Zhejiang

17.1 Introduction to the Heritage

17.1.1 Summary

Name of Property: Jianglangshan.
City, Province: Jiangshan City, Zhejiang Province.
Geographical Coordinates: 28°31′44″N, 118° 33′43″E.
Area of Property:
Heritage Site (ha): 610.
Buffer Zone (ha): 571.

17.1.2 Description of the Boundary of Heritage Site

As one of Heritage sites in Danxia Zhejiang area of China, Mt. Jianglangshan consists of typical Danxia landform and its physical geography elements with complete and continuous 708 ha area.

Mt. Jianglangshan Heritage site as a whole, borders as the ridge line, the valley line, rivers, and other natural lines; due to coincidence boundary between buffer zone and Mt. Jianglangshan scenic spots, so they basically border as the above-mentioned natural borderline and a part of them border as road and land classification. Strict protection shall be conducted in the scenic spots with a clear demarcation with the same boundary to Heritage site and the buffer zone.

Heritage site scope refers to connected natural areas centering on the typical Danxia landforms to maintain the integrity of Danxia landforms and forest ecosystems, as well as the integrity of rare and endangered species ecological environment.

17.1.3 Justification Statement of Outstanding Universal Value

(1) **Important scientific value**

During the Cretaceous Period, the most conspicuous geological event of the Southeast China was the volcanic events in Early Cretaceous, the faulting depression event in middle Cretaceous, and the uplifting event in the early stage of Late Cretaceous. The volcanic rock series, the gray lithic part, adding the lower class of molasse formation (faulting depression rock suite), and the upper class of molasse formation in the Xiakou Basin, bearing the typical lithologic features, widespreadness as well as the isochroneity, enable them to be the perfect mark layers for regional comparison.

Jianglangshan belongs to the old stage of Danxia landform, whose Three Pieces of Stone is still erupting, while the other red stones of

H. Peng, *China Danxia*, Springer Geography,
https://doi.org/10.1007/978-981-13-5959-0_17

Cretaceous in the surrounding areas had been eroded to lower land. The reason of this is closely related to the lithologic characters of the Three Pieces of Stone. After the Cretaceous sedimentary of the Fangyan Formation, the speed of accumulation decreased and the action of the deep-seated fault lasted long, with the upwelling of magma which intruded in the Fangyan Formation forming the pluton of dolerite, trachyte, andesite, and olive basalt in the form of vein. As a result, these rocks are mainly made up of volcanic debris, which experienced the volcanic process of high temperature as to be quenched, leading to the high resistance in erosion. On the other hand, the vein of dolerite, trachyte, andesite, and olivin-basalt bedding in the three-part stone, bear the unique ophitic-intergranular texture as well as the plagioclase and black minerals grouping, forming the rocks of high resistance ability in erosion, just as the effect of consolidating the steel in the cement concrete. The phenomenon is rare in other Danxia landform of the foreign countries, which deserves the research in lithology.

The reaction of the platform: on the basis of topography, the three-part stone of Jianglangshan is elevated about 800 m, while Danengxia and Xiaonengxia are about 500 m, showing two different planation surfaces at different elevation. Meanwhile, there are wide alluvial plains in the areas of Shimen, northwestern-wing of Mt. Jianglangshan, as well as Jianglangjie and Ya Peak, western-wing of Jianglangshan t, indicating that the crust of this region is relatively steady in the recent period. However, there are second-class river terraces at the relative altitudes of 4.8 and 8.0 m, indicating that there was still lifting action in recent period in the region of Jianglangshan, which perhaps belongs to the reaction of platform. It can be called that the Three Pieces of Stone is the witnesses of the "two-generation" or "three-generation", even "four-generation". As a result, the complexity and uniqueness of the development process of the Danxia landform in Jianglangshan deserves further research.

(2) **Unique value of landscape esthetics**

The most featured spots of Jianglang Danxia Landform are Three Stones Hill Group, Yixiantian Valley and Danxia Stone Wall made up of Three Pieces Stones. Jianglangshan is the top scenery in Danxia landform with profound science meaning and educational value in research and geology education. At the same time, the scenery spot is full of esthetic meaning and natural beauty.

Jianglangshan meets Chinese traditional concept of landscape, which stresses the harmony between human and natural. Jianglangshan bears a rare, beautiful "scenery resources."

The combination landscape elements of Jianglangshan are expressed by esthetic appreciation. The most important one of them is amazement.

The composing elements of landscape are diversity from macro to micro, from material background, and stone characteristics. Though the styles are changeable, they meet principle of beauty in form. All landscape of Jianglangshan is united under the theme of formation of Danxia landform and Three Pieces of Stone.

17.2 Description of Jianglangshan Danxia

17.2.1 Natural Environment

(1) **Climate**

Jianglangshan belongs to the humid subtropical monsoon climate area. It has a rich quantity of heat and enough sunlight. Its range of the mean annual sunlight is 2063.3 h. The range of the mean annual rainfall across the sites is from 1650 to 2200 mm. In the end of spring, plum rain is easily to be seen. In July and August, the range of the mean annual rainfall is only 218.3 mm, which is less than half of evaporation and often causes drought. In winter, north wind dominates this area and its wind speed is the highest comparing with other seasons. In spring, the wind is diversity and weak. In summer, south wind is common. Autumn is cool and dry.

(2) **Hydrology**

The site has lake, reservoir, brook, and fall. Xunv Lake is the representative of this landscape.

Xunv Lake, which also is an important reservoir, situates at the Yujiawu Village. It is positioned on the upper reaches of the Jianglang River. The total area of the catchment is about 1.05 km^2. Its capacity is 392,500 m^3. The area of the water surface is about 44,000 m^2. The drainage area of Xunv Lake is a mountain area. It has rich precipitation rain fall and thus, helps to keep a well natural environment. The vegetation coverage is 95% above.

(3) **Soil**

Due to the diversity of terran of Jianglangshan and wet climate, Jianglangshan has a variety of soil and vegetation. The red clay is mainly distributed at the broad hills of eastern and western part; The yellow soil is mainly distributed at the mountain area of eastern and southern part; The moisture soil and paddy soil are distributed at the intermountain basin and valley.

17.2.2 Geology

17.2.2.1 Features of Geotectonic Background

Jianglangshan belongs to Xiakou Basin between the Jiangshan–Shaoxing deep fault and Bao'an-Xiakou-Zhangcun fault zone. Jiangshan–Shaoxing deep fault is the fault that passes through salsphere and reaches siferna. It controls the development of geologic structure. Bao'an--Xiakou-Zhangcun fault zone belongs to the subfault of Jiangshan–Shaoxing deep fault. In the early Yanshan Movement period, the southeast of this area shows different types of fault tectonic basin (volcano tectonic basin and rift tectonic basin). The northwest shows flexible fold and contains one small tectonic basin. Jianglangshan is located in the Xiakou Basin which states in the east of Jiangshan–Shaoxing deep fault. Its southeast side is Bao'an-Xiakou-Zhangcun fault zone. The layer in the bottom of basin is Malm layer. In the late period of Late Jurassic epoch, due to the acceleration of speed of conflict between Pacific Plate and Eurasian Plate, the deep fault continued moving and magma movement reached to the florescence. It formed a continuous and large distribution of pan volcanic eruption and deposits with other places in the east of Zhejiang Province.

In the early Cretaceous, the tension rift of the two faults mentioned above led to the formation of Xiakou Basin. Also, the basin changed to an asymmetric pan-shaped downfaulted basin. In consequence the red glutenite, fluvial–lacustrine deposit pied glutenite Guantou Formation (K_1g), talus fluvial facies, Chaochuan Formation (K_1c), and Fangyan Formation (K_1f) became the sediment in the basin. In the Late Cretaceous, the movement of magmatite still existed, though structural feature was hard to be found in this area.

When coming to Cainozoic Era, the fault-block movement is weaker than before. The dominants are differentiation and oscillation of up-and-down movement. Subsequently, this area stepped into a new period. The well-known Danxia landform gradually formed in the stratum. In the Late Cretaceous Period, the two faults formed an intense compressional movement and forced the rise of Xiakou Basin. Since Cainozoic Era, incision of rock mass, and the sequent collapse and erosion process led to the growth of Danxia landform.

17.2.2.2 Strata and Lithology

Seen from the 1:50,000 geological map, Jianglangshan is located at Changtai amplitude (sheet designation: H50E021019). The map shows that most strata are volcanics of Mesozoic Upper Jurassic and sedimentary basin rock of Cretaceous. A small part is Triassic series Wudu Formation and Middle Jurassic Majian Formation. According to the definition of lithostratigraphic unit, it can divide into 15 lithostratigraphic units.

Strata of Jianglangshan

Era	Period	Epoch	Stratigraphic Province		Southeast Stratigraphic Province		
Cenozoic	Quaternary	Holocene	Yinjiangqiao Formation Qh*y*				
		Pleistocene	Lianhua Formation (Qp*l*)				
Mesozoic	Cretaceous	Lower Cretaceous	Qujiang Formation	Zhongdai Formation (K_1z)	Yongkang Group	Fangyan Formation (K_1f)	
						Chaochuan Formation (K_1c)	
						Guantou Formation (K_1g)	Member III(K_1g^3)
							Member II (K_1g^2)
							Member I (K_1g^1)
	Jurassic	Upper Jurassic			Moshishan Group	Jiuliping Formation (J_3j)	
						Xishantou Formation (J_3x)	
						Gaowu Formation (J_3g)	
						Dashuang Formation (J_3d)	
		Middle Jurassic	Majian Formation (J_2m)				
	Triassic	Upper Triassic		Wuzao (T_3w)			
Neoproterozoic	Sinian		Xiuning Formation (Z_1x)				
			Shangshu (Pt_3s)				
Middle Proterozoic				Chencai (Pt_2C)			

Generally, Danxia landform at Jianglangshan is a continental formation system that deposits in the synclis. It has eruptional lava, rhyolite, tuff, pyroclastic rock, and fluvial–lacustrine deposit strata that regard running water and lake water as force.

(1) **Features of Jurassic strata**

- **Majian Formation (J_2m) of Middle Jurassic Series**. Majian Formation is located at the northwest of Xiaoqing Lake. Its outcropping area is about 0.05 km^2. Lithology mainly is quartzose sandstone and siltstone. Shed coal is also found in some places. According to the mapsheet information, early basis of well logs of Majian Formation shows the fluvial face feature and the clast has a relative high maturity. The late one shows that it has lacustrine and swamp deposit, and clast are mostly siltstone and silty mudstones with instable shed coal interbedded in.
- **Upper Triassic Moshishan Group**. Upper Triassic volcanic series are mainly distributed at Tangyuankou and Lishendu. Some is also found at The N-E Shishangdong–Hemu fault. Its outcropping area is about 70 km^2. The majority of volcanite strata are Gaowu Formation and the minority ones are Dashuang Formation, Xishantou Formation, and Jiuliping Formation.

(i) **Gaowu Formation (J_3g)** Gaowu Formation covers the largest areas (about 50 km^2). It appears at Tang Yuan Kou and Li Shen Du. Its lithology composition is relative simple. The majority is french gray, dark gray rhyolitic crystallinoclastic shard tuff and some cinerite and rubble conglomerate tuff interbedded in it. It features with crassitude of crystallinoclastic particle size and high mineral. The main part of crystallinoclastic formation is quartz and feldspar. Plagioclase and mica group follow. Their proportions are about 40–50%.

(ii) **Dashuang Formation (J_3d)** Dashuang Formation is scatteredly distributed at the southeast of Xiaoqing Lake. The area is about 2 km^2. It is connected with Chencai Group metamorphic rock and Wudu Formation. They belong to the part of Shishangcun–Hemu fault.

(iii) **Xishantou Formation (J_3g)** Xishantou Formation is mainly distributed at Wanqingshan and Changqiu. Its outcropping area is about 15 km^2. According to the section information, its lithological characters composition is rhyolitic shard crystallinoclastic welded tuff, rhyolitic crystallinoclastic shard welded tuff, and rhyolitic shard welded tuff. The crystallinoclastic is about 15–40%. Its particle size is relative small and the major composition is feldspar and quartz. At Wanqingshan the Xishantou Formation distributed narrowly. It is controlled by the N-E Shishangcun–Hemu fault.

(iv) **Jiuliping Formation (J_3j)** Jiuliping Formation scatteredly distributed at Zhangcun and Tangyuan. Besides, it also has a 4 km^2 distribution at Shimenshan. It is composed of lavenderblush rhyolite, rhyolite–porphyry interlining multi-layer tuff, tuffaceous sandstone, and acid pyroclastic rock. The volcano movement of Jiuliping can be at least divided to four flow units. The sediment comes up with volcanic spurting. Jiuliping Formation (J_3j) of upper Jurassic Moshishan Formation is composed of pinky rhyolite, rhyolite–porphyry interlining multi-layer tuff, tuffaceous sandstone, and pyroclastic rock. It distributes widely and is the preexistence of Guantou Formation and Chaochuan Formation.

(2) **Features of Cretaceous Layer**

The main revealing part in this zone is the Late Cretaceous Yongkang Group. The Late Cretaceous Yongkang Group distributes central and southeastern area, which is component of Xiakou Basin, which is composed with Guantou Formation, Chaochuan Group, and Fangyan.

- **Guantou Formation (K_1g)**. This is the bottom part of the Late Cretaceous Yongkang Group, which can be divided into upper part and lower part with missing middle part. The low part is gravel, sandstone, and thin powder-like deep-gray mudstone, charcoal-like shale, with random-espousing power-like fine sandstone. They mainly distribute on Jishang, Qianding, and the east part of Changtai Town.

  (i) The first part of K_1g: little gravel, sandstone espousing on the bottom; powder-like deep-gray mudstone, charcoal-like shale making up middle upper part, with random power-like fine sandstone.
  (ii) The second part of K_1g: a complete deep-gray dacite, produced in extravasation status, distributing around 1 km^2, limited in the north fringe of Changtai Town. The thickness is about 50 m.
  (iii) The third part of K_1g: a complete of light-purple red-spotted rhyolite, produced in extravasation status, distributing in Kaijing, Maoxing, exposing area is about 10.6 km^2. The thickness is about 150 m.

- **Chaochuan Formation (K_1c)**. Grouan stone with gravel and middle grouan compose the bottom part. Powder-like purple red sandstone, powder-like mudstone compose the middle part, with gravel in river status, fine sandstone, siltstone, and aleuritic texture mudstone. There are gravel powder-like red purple mudstone, gravel, and conglomerate alternate with each other. The main part of Chaochuan Formation is a complete thick purple red fine fragment. Less thick coarse fragment can be seen in the bottom part. A complete very thick conglomerate-status and gravel-status grow on the bottom part of Chaochuan Formation in Bao'an District, belonging to alluvial fan status deposit.
- **Fangyan Formation (K_1f)**. Fangyan Formation is the main part that composes the Danxia landform in Jianglangshan, which is composed mainly by thick layer or bulk purple red, light-gray conglomerate, with malmstone, gravel and lenticle placed inside, and random extrusive rock. This is a complete river-appearance and lake-appearance deposit stratum which was river valley, lake plain, and delta deposit caused by mountain foot alluvial fan, with imposing dyke and dam from Late Yanshan or Himalaya. Fangyan Formation is mainly distributed on the edge of basin and is controlled by basin break, which is exposing in Jianglangshan–Zhangcun Village, about 16 km^2. The stone quality is light gray bulk conglomerate, gravel with little lenticle silty sand fine sandstone. There is 500 m thick gravel around Laohutou Mountain, Dongnanyuan on the southwest of the basin. They become powder conglomerate, silty mudstone with thin layer coarse conglomerate with gravel in east and northwest of the basin. It is mixed together with Chaochuan Formation in Xiafu with no more thickness of 581.2 m.
- **Zhongdai Formation (K_1z)**. It belongs to the southwest extended part of Jinqu Basin with exposing area of 22 km^2. The components of this group include: very thick conglomerate in the low part with gravel; big area of extrusive deep-gray olivine basalt grows on it; purple red siltite, argillaceous siltstone and gritstone and gravel are in the middle part interwined with each other; purple red siltite and argillaceous siltstoue in the upper part. It contacts to Wuzao Formation on the bottom with angle. It has a width more than 838 m.

Cyclicity of Basic Strata Sequence: This kind of basic strata sequence is always composed with three and more strata. Up-going clast becomes bigger or smaller with layer becoming thicker or thinner to overlap repeat. It is composed of glutenite–grit with gravel-middle fine sandstone-siltite; or glutenite-grit with gravel-fine sandstone–siltite; or sandstone with gravel-fine sandstone–siltite. The thickness of basic strata sequence is between 1.5 and 18 m. Eroding is common on the bottom of the basic layer order. Huge interwined layers are growing between them to provide features of composing and construction of cyclicity of basic strata sequence.

Non-Cyclicity of Basic Strata Sequence: This is the most popular type distributing in the basin. It is can be divided into three types by group characteristics, homogeneity basic strata sequence, rhythmicity basic strata sequence and lava-bearing basic strata sequence.

- Homogeneity basic strata sequence: this is the main part of the basin. Stone quality is simple and it is very thick. Heading area of this kind shows rarely featured layer interspace of cyclicity. It is divided into different layer by interspace, erosion surface, and ground heading face. Inner basin is composed with bulks of purple red silty, muddy silty or bulks of conglemorate and gravels. They are the production of extensive flood deposit, alluvial fan, distributing in the catchment basin, and around the fringe of the basin.
- Rhythmicity basic strata sequence: this is composed with two kinds of stone strata overlapping frequently, mainly distributing on liminal face and the top fringe of fan delta. The former is composed of thin-layer mudstone and silty or shale and marlite (millimeter to centimeter scale), showing subtle horizontal construction. The latter is composed of bulks of purple red silty and conglomerate or gravel and gritstone with gravel interlayered. Stratification cannot be seen in the single layer. The bottom is smooth. The erosion is rare.
- Lava-bearing basic strata sequence: limited distribution, seen only in Xiakou Basin Bao'an Duntou and Hushang Basin Yi Mountain. It is the production of single strata or multiple l strata of lava and deposit interlayered. Lava layers do not have the same thickness and the layer is composed with acid lava segment, rhyolite, limestone, and basalt. The location of lave strata in the basic strata sequence is not fixed. It is the same with the proportion, which shows frequent small scale volcanicity with deposition in the basin during that period.

(3) **Strata Feature after Cretaceous**

Cainozoic groups in this area are distributing in Changtai, Shimen, Jiebei, Hemu and Aopin. Layers are not fully heading, mainly Pleistocene series Lianhua Formation and post-glacial epoch Qinjiangqiao Formation. The former distributes on the hills of Chaochuan Formation and the latter is distributing side plain of Xiandai River with heading area of 68 km^2.

- Pleistocene series Lianhua Formation (Qh*l*) Lianhua Group distributes limitedly with less area of heading. Gray-yellow and bronze-yellow clay sand loam can be seen on the top of vertical section with loose soil. Gray-yellow sand loam with gravel and little bronze-yellow grid clay can be seen in the middle part. Mottle, gray-purple grave, conglomerate layer are with unstable sand loam in the bottom. It width is 2.5–5 m.
- Post-glacial epoch Qinjiang Formation Distributing on the grand plain beside the Xiandai River. Light-gray clay, sand loam is on the top of the vertical section with great sand quality and bad plasticity. Gray-yellow sand soil with gravel and gravel sand can be seen in the middle part. Gray-yellow soil and conglomerate layer are in the bottom with good psephicity. It width is 1.9–4.9 m.

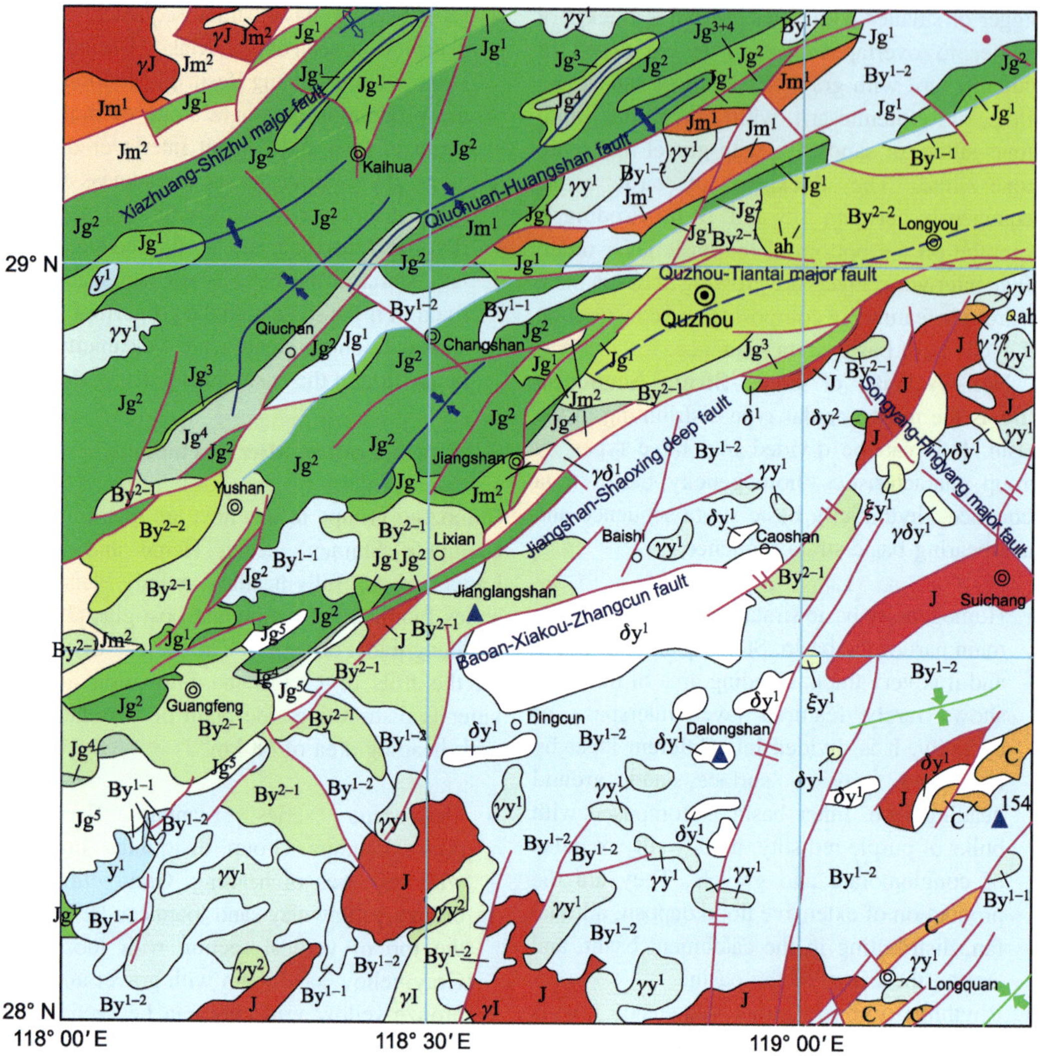

Geologic structure diagram of Mt. Jianglangshan and close region

Clumpy conglomerate with huge thickness at Peak Lang Tianyou

Diabase dike exposed at Peak Ya

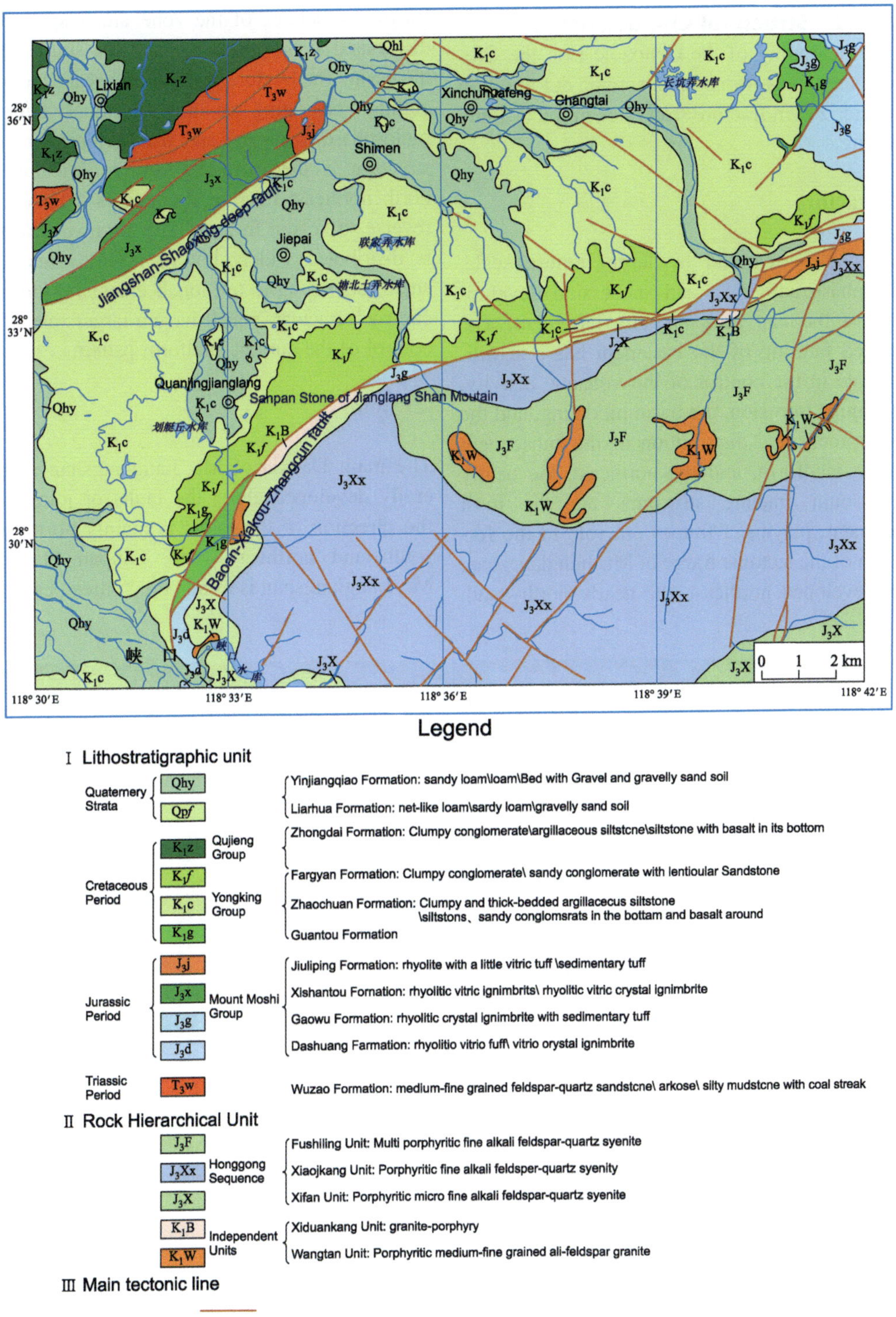

Geological map of Mt. Jianglangshan area

17.2.2.3 Structural Characteristics

At the beginning of the Cenozoic Era, the Cretaceous red beds formed the tectonic deformation characterized in fracture structures in the tectonic uplifting.

(1) **Fracture**

The zone is featured by large rupture scale, multi-phase activities, and succession development of the cap rock and the basement. The renewed faulting of the basement has a significant role in the control of the volcanic activity, the magma intrusion, the basin modeling, and the development and distribution of the subsequent Danxia landform, and the northeast, the nearly north–south fracture structures are the most prominent, playing a framework role in the formation of the tectonic frame of Mt. Jianglangshan. The developed northeast, the nearly north–south fracture structures of the zone are, respectively, distributed at the northwest and the southeast of the basin. The fracture feature large scale and deep cutting, belonging to the basement fault of multi-phase activities, and the northeast fracture structure is the extruding reverse fracture and north northeast fracture structure is the shearing reverse strike-slip fracture. The northwest fracture is distributed widely but its scale is relatively small, belonging to tensional- extension torsional cap rock fracture, which have a larger role in the control of the current landform pattern.

(2) **Joint**

The main Danxia of Mt. Jianglangshan is generally developed along the faults or joints, and the direction of the joints is mainly northwest–south and northeast–south, Sanpan Stone of Mt. Jianglangshan is in fact the "Three peaks and

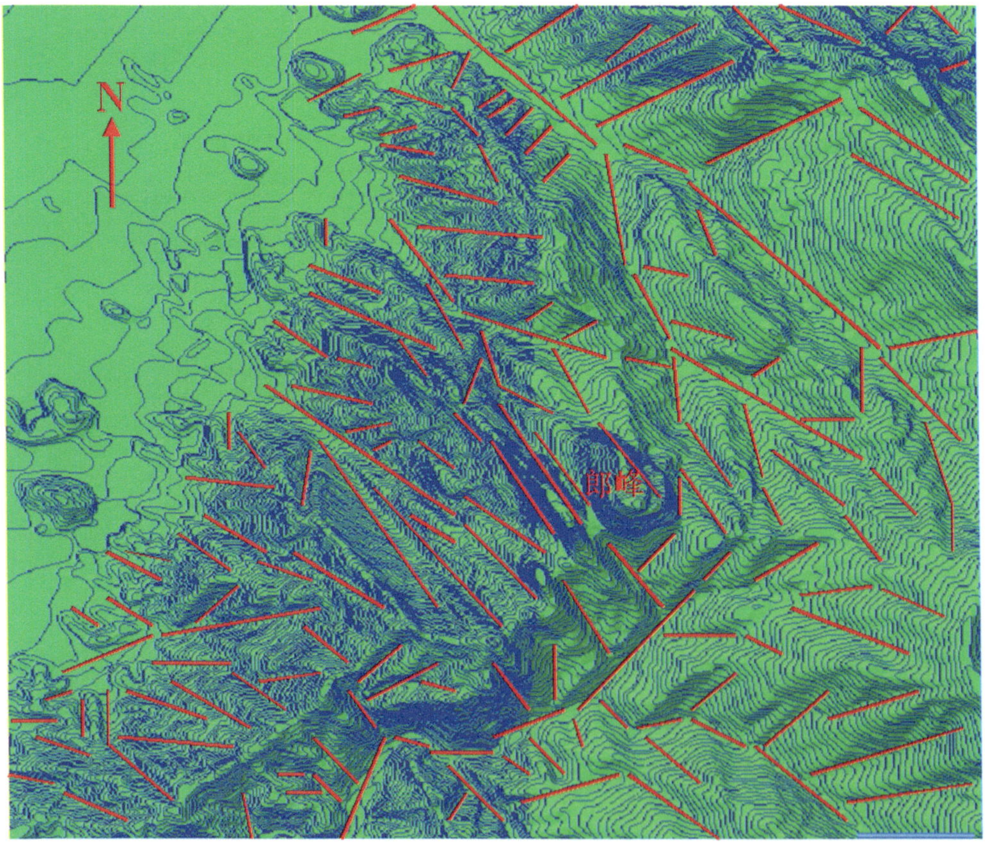

Diagram of Joint Distribution within Mt. Jianglangshan

"X" joint on megathymiinae cliff of Mt. Jianglangshan

Oblique joint of Peak Ya seen from Peak Lang

Oblique joint at one side near to Peak Ling of Linear Sky Valley of Mt. Jianglangshan (shown by red arrow)

Oblique joint at one side near to Peak Ya of Linear Sky Valley of Mt. Jianglangshan (shown by red arrow)

two valleys" developed due to the control of the northwest–south joint set. The statistics on 66 fluorite veins and quartz veins in the narrows basin shows that: majority of cracks' direction is between 300° and 320°, indicating the narrows basin is affected by the northwest–south extrusion stress, therefore, there appears a large number of northwest tensional and extension torsional fractures and joints.

In addition to groups of large vertical joints, the "X" joints and oblique joints can be seen in Mt. Jianglangshan, and many "X" joints can be seen from the megathymiinae cliff between the Peak Lang and Peak Ya, displayed in the significant diamond-shaped cutting landscape, and the nearly 20 large scale oblique joints can be seen from the cliff between Peak Ya and Peak Ling in the Linear Sky Valley, some runs through the peaks of Peak Ling and Peak Ya vertically. The thick-bedded grit in the narrows basin, due to the segmentation of the above-mentioned joints, forms many giant nubby rocks, accelerating the late weathering, erosion, and collapse.

17.2.3 Landform Features and Types

17.2.3.1 Landform Features

The most unique features of Mt. Jianglangshan are the Sanpan Stone in the isolated peak from the Linear Sky Valley and two Danxia Stone Walls.

(1) **Unique high Danxia isolated peak**

The most surprising landscape of Mt. Jianglangshan is the Sanpan Stone known as "Grandest, most unique and most beautiful mountain" comprising of Peak Lang (an altitude of 824 m, a relatively height of 369.1 m), Peak Ya (an altitude of 737.4 m, a relatively height of 287.4 m), and Peak Ling (An altitude of 765 m, a relatively height of 298 m). The isolated peak is the highest among different peaks, which at the same time seems to reach the sky and cloud, features different views at different places and grand views, known as "The Most Unique Peak Danxia."

Close Shot of Sanpan Stone of Mt. Jianglangshan (Peak Lang, Peak Ya, and Peak Ling from left to right)

Sanpan Stone overlooked from the northern side of the foothill of Mt. Jianglangshan

(2) **Unique combination of cliffs and valley**

The straight Top Valley between Peak Ya and Peak Ling has the length and the height of over 300 m, respectively, and there are Peak Great man, and steep Langfeng Tianyou, Baibu Cliff, Tiangong Cave, Heaven Bridge, Waterfall over treetop, and other beautiful scenic spots.

(3) **Space structure features**

On macro level, Sanpan Stone is developed on the hill platform of gent ups and downs, forming obvious space contract and bringing strong visual impact. At different places and with different light and weather condition, there appears mysterious and eyeable view. On medium-view level, Sanpan Stone, integrated with the natural factors such as stones peaks and valleys and others, and the human sites such as Xuxiake mountaineering booth, Kaiming Temple, Jianglang Academy, Yuankong Academy and others, forms the different scenic landscape, thus, highlighting the uniqueness of Sanpan Stone. On microlevel, Sanpan Stone itself has unique styles, providing the views of different stone peak styles, surface images, appearances and colors, and other landscape factors. Seeing Sanpan Stone from different angles can realize the magic power of the nature due to the various physical changes, such as The Top Valley, Peak Great man and cliffs, stones, gorges, caves, booths, and bridges and other small scenic spots along the way, all the above can bring visitors the eyeable views.

Sanpan Stone overlooked from the northern of Mt. Jianglangshan (Dim Sanpan Stone seems like the home of immortals)

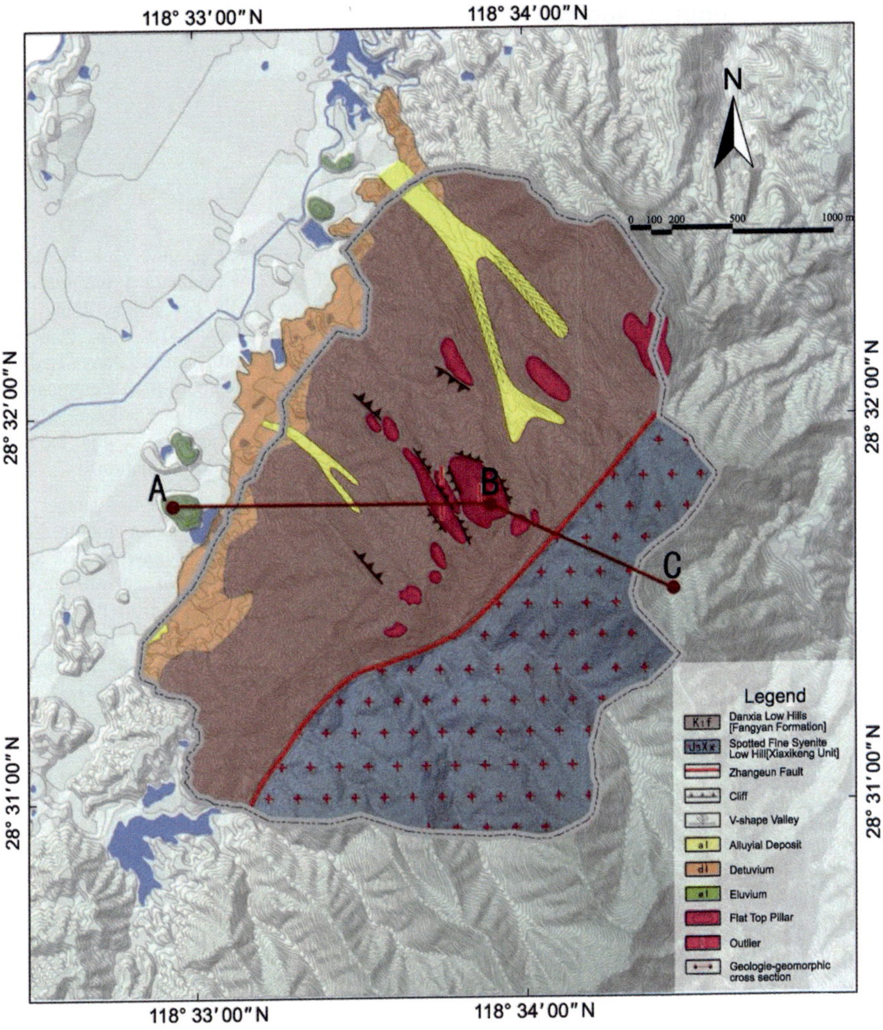

The scatter diagram of the central topographical features

17.2.3.2 Main Geomorphological Landscape

Introduction of Main Geomorphic Scenic Spots of Mt. Jianglangshan

No.	Name	Location	Feature description
1	Sanpan Stone	Mt. Jianglangshan	Sanpan Stone consists of three gigantic wall-like stone peaks that tower on top of a 500 m mountain. From northeast–east to southwest–west perspective, the first is Peak Lang (824 m), Peak Ya (737.4 m), and Peak Ling (765.0 m). Peak Lang is the largest of the three and around it are all Danya Red cliff. The height of Danya, except for the 225.0 m of Dengtianping, ranges from 300 to 369 m, towering, and rare among Danxia physiognomy at home. Experts agreed that it is "the first wonder peak of the country" after careful evaluation
2	Linear Sky	Between Peak Ya and Peak Ling	Linear Sky is 308 m long, 298 m high, and gauges 3.5 m at its narrowest point. The overall shape looks smooth. Looking up from the ravine, the limitless sky assumes the look of a crescent, giving off a mere, linear light. One of the well-known remarks goes like this: Sanpan Stone is moved from the sea of Bo, to cut the Silver River to give a linear sky." The views here vary according to seasons. In summer, when the scorching sun is above, easy and cool wind starts blowing day and night, offering a perfect place for those who want to get away; in winter, when cold wind blows, the place is all coated with snow and ice, decorated with icicles, just like a crystal palace dragons used to occupy. Linear Sky is always appealing in sunny or rainy days. When fog rises and you look up from the place, you will find clouds and fogs, just assume the shape of a silver dragon, rolling and twisting from the lower end of the valley up to your presence. This is the wonder that inspired many legends: A silver dragon ascending from the sea. When it rains on and on for days, however, rainwater accumulated on top of the cliff would find their, by leaps and skips, down to the valley, like a curtain of pearls hanging down from the sky. This is what usually called "waterfall from tree tops." When it rains cats and dogs, the curtain of pearls transforms into a cloth of water hanging down 300 m long, 200 m wide, just like a heavenly-made immense curtain
3	Eighteen Curves	At the foot of Peak Lang	The tour stairs are stages that meander upward into Mt. Jianglangshan. It measures 500 m long, 120 m high, there are altogether 461 stages. Its name derives from its many twists and turns. At the beginning of the tour stages, there towering a huge rock which is inscribed with "Mount Jianglang" written by Zhao Puchu. The calligraphy is powerful and impressive and a good match for the scene. Traditional sighting spots like sister stones, "Yidengpankong (one leap would send you into sky)" and inscriptions from Professor Huang, a Danxia physiognomy expert. On both sides of the tour stair, there planted a lot of pines and bamboos whose leaves offer cozy shelter from a scorching sun, and brooks running down along quietly. Passing through the forests of bamboos and pines, seeing Sanpan Stone from afar, looking down on the view below, the visitor would feel as if he were in heaven

(continued)

No.	Name	Location	Feature description
4	Fairy, Meeting Rock	At the foot of Peak Lang	The tour stairs are stages that meander upward into Mt. Jianglangshan. It measures 500 m long, 120 m high, there are altogether 461 stages. Its name derives from its many twists and turns. At the beginning of the tour stages there towering a huge rock which is inscribed with "Mount Jianglang" written by Zhao Puchu. The calligraphy is powerful and impressive and a good match for the scene. Traditional sighting spots like sister stones, "Yidengpankong (one leap would send you into sky)" and inscriptions from Professor Huang, a Danxia physiognomy expert. On both sides of the tour stair there planted a lot of pines and bamboos whose leaves offer cozy shelter from a scorching sun, and brooks running down along quietly. Passing through the forests of bamboos and pines, seeing Sanpan Stone from afar, looking down on the view below, the visitor would feel as if he were in heaven
5	Bell and Drum Cave	At the foot of the Peak Lang	Below Peak Lang there is a precipitous piece of rock hanging to form a cave. The wall of the cave is miraculous in that the upper wall gives the sound of bells and the lower drums. Hence, the cave got its name. The cave differs from Fairy Meeting cave in that it is a lot more spacious, with a capacity of several hundred. Zhou Wenxing, a celebrity in Ming Dynasty, used to live here. There are some inscriptions on the walls, one is "little cave heaven," the other is "cliff stiff," made by Zhan Ruoshui, a scholar in Ming Dynasty
6	100-Pace Ravine	At the foot of Dengtianping on Peak Lang	In front of Yanxia Pavilion and next to the Fairy Meeting Rock, the pace used to be a crevice in stone walls covered in bushes. Looking up, the passage is solitary, down you will see a different scene. The ravine is 20 m high, and no more than one meter wide. You have to squeeze through its narrowest point. At the end of the stone stair is a stone wall, turning to a different direction a new world suddenly opens up. A platform that protrudes out, on which Dongshan Thatched Hamlet was once built, leaving only ruins now, though. Looking afar, you will find terraces below, mountains ahead and sights around
7	Peak Great man	Villa Mt. Jianglangshan	Looking upward at Peak Lang from the gate of Jianglang Villa, the silhouette of Peak Lang assumes the shape of a great man leaning the head upon a chair, lost in thinking. The eyes that are full of confidence and the dignity of a thinker all bear resemblance to a great man. When the moon is bright and stars sparse, the silhouette looks even better, adding to the wonder of this mountain a touch of modern social life
8	Peak Jar	The front of Mt. Jianglangshan	It is said that with the help of a fairy, Jiang Lang got a pen, which endows him with a talent. But he became lost in politics and the fairy took back the pen. He lost the talent, and the fairy kept the pen here
9	Peak Bull-nose	Pictographic	A peak that bears semblance to the nose of a bull
10	Lingshi Huifeng	Between Peak Ya and Peak Ling	On the west edge of Ascent Platform and the cliff of Peak Ya there inscribed "ling shi hui feng." When in storm, the leaves gathered at the top of the peak would fall off and circling around in the little and great alley beside Peak Ya. Flying, they are like thousands of birds frolicking among the three peaks, hence the name "ling shi hui feng," which is a description of the scene

(continued)

No.	Name	Location	Feature description
11	Danxia Red cliff	South of Peak Ling	It is a layer of arenosols and conglomerate. Due to long-lasting geographic forces, the slant of geographic layers and the breaking of drapes, the erosion of streams, and collapses, the steep red cliff of conglomerate rock is formed. Queer peaks and weird stones, and their sizes, are rare nationwide
12	Peak Lang Tianyou	Peak Lang	824 m above the sea level and the vertical distance of the largest piece of Peak Lang is 370 m away from the bottom of the peak. There used to be no passages, and human activities are even rarer. The passage that leads to the top was cut out in 1990. Counting from the Ascent Platform, there are 3500 stages which adapt very well to the lie of the mountain. The narrowest point of the stage is less than 0.3 m; the steepest point seems to be impossible to pass. Climbing along the stage, tourists can see towering cliffs and fathomless abyss; with the wind blowing by the ear and in the coat. Each step forward brings a new sight, indeed wonderful. There are sky bridges that link up, and natural caves for one to sojourn, indeed a wonderful place in the middle of the air. The beauty of the views along the way is stunning, but only the strong and the brave can make it to the top to enjoy the grandeur of Peak Lang
13	The Top of Peak Lang	On top of Peak Lang	The top of Peak Lang has accumulated both cragginess and wonder. There is a "lawn" that measures several thousand square meters; there are orchids and herbs and other thick vegetation that gives the place a look of forest park. Inside the "lawn" there are many primitive plants, rare and huge trees, Sabina chinensis, juniper, etc. Family Fagaceae, Magnoliaceae, Cornaceae, Elaeocarpaceae, Lauraceae, and other valuable tree species, and herbs like Shigu, dendrobium, gyrophora, and many more others. On top of the peak, there built a pavilion called "Wentian," which means questioning the heaven. Standing on the pavilion, wind blows strong on the face and sights rushing to the eyes: mountains afar meander to nowhere, peaks towering and noiseless, terraces divided evenly, lakes reflecting the sun, indeed wonderful, and a reminder of famous poetic lines like "on top of the peak, the rest of the mountains seems weak," or "how wonderful our country is!"
14	Lion and Elephant Guards	The back of Mt. Jianglangshan	Two mountains, one in the front, the other behind, like a lion and an elephant, are guarding the three peaks
15	The Heavenly Palace	On top of Peak Lang	2 m in length and width, 6 m in depth. A natural cave formed due to collapses
16	Carp Stone	Collapsed pictographic rock	A large plate of stone that resembles a skipping carp. What is amazing about this stone is its verisimilitude to a real carp skipping the dragon gate. Truly, it is about to be successful. Once over the gate, the scenic sights of Jianglang are right before the eyes
17	Jiugu Cliff	In the middle of the valley	A protruding cliff, below which water is always running and infuses the cliff with a dignity of an abyss. Legend has it that there was a woman named Jiugu used to wait for the three brothers of Jianglang for all her life. Failing that, she became a cliff. Upon the cliff, there is temple well adapted to the lie of the place. It is as though the temple is floating in the air, gaining a dignity that is rare elsewhere

(continued)

No.	Name	Location	Feature description
18	Dragon Ravine	The end of Xunv Lake	It is a valley that is extremely narrow, just like stone cut open by god with a huge axe. It is said that there used to be a giant dragon hidden here, and one day, it suddenly broke out of the stone and fly to a place of nowhere. The ravine, therefore, became narrow and steep and weird and craggy, and hence, the name
19	Stone gate	Opposite to the Xianju Temple in Shimen Town	The Stone gate, also known as Waterfall Cave, is located at the mountain side that is opposite to the Xianju Temple. The Stone gate is 10 m in width and 20 m in width and height. The top of the stone is arc and the Linear sky is formed at the crack. There are millenary rattans overhanging on the stone gate and fountain flowing down by the cave ceiling. On the inner wall of the stone room, there is a round gate with two stone doors closed tightly. The transverse dent of a lock is left on the rock tunnels. Seen at several miles away, the waterfall cave is so alike a stone gate. On both sides of the stone gate are Danxia red cliffs that set out the imposing and magnificent impression on visitors
20	Moon rock	In scenic spot of the Xianju Temple	Over the top of silver platform there is a rock known as Shangyao rock. The cave is 20 m in depth, 5 m in height, and 15 m in width. There is a stone mortar in the cave. Story has it that there lived a monk with unsurpassed military arts who can bounce from the ground with a broad-rimmed hat (made of indocalamus splints and leaves) in hands. Two miles downward from the Shangyao rock there is the Xiyao rock, which is the Moon Rock with side back to the White Rock. On top there is a stone that bears resemblance to the head of a dragon, from which water drips out all the year around. On top, there is a space that looks like a crescent, hence its name "Moon rock." The temple was brought back to the Moon Rock and the figure of Buddha was repaired in Ming Dynasty by the officer, named as Mao Kai. And a front palace hall was built in 1936. The whole temple is substantially in existence today and becomes a busy and crowded place of worship for Buddhists

Three overlapping peaks seen from the northwest

The west of the Peak Ling

Linear Sky, 308 m in length, 298 m in height, and 3 m in width

Valley

Passage of Peak Lang Tianyou

Danong Valley

Sanpan Stone seen from Xiake Booth

Running water incision on the upper part of Peak Lang

Peak Ling and Peak Ya, two stone walls over 300 m in height seem to be stone columns from northwest

Stone wall, Peak Ya (in the middle), and two linears on both sides seen from Dengtianping

Stone wall, Peak Ling seen from northwest

Crest of Peak Lang

Peak Ya (middle) and crest of Peak Ling

Peak Great man

Crest cliff on the north of Peak Ling

Linear Sky, Zipao Valley developed in the direction of 20°NE joint

Linear clough below Wentian Booth

Drum and Bell Cave at half-way up the Peak Lang of Mt. Jianglangshan

Fairy Meeting Cave at the foot of the Peak Lang and residual falling stones

Falling stone at Linear Sky, Mt. Jianglangshan

Recess of Jiugu Rock, Longqing Ditch formed by collapse after weathering

Collapsed Fengxi Stone between Peak Lang and Peak Ya

17.2.3.3 The Effect of Geological Structure on Development of Jianglangshan Danxia

(1) **The effect of tectonic line on landform pattern**

This area belongs to Cenozoic era, which is featured with massive structure. Lots of ruptures were formed when the area was arising. All the structures have profound effect on the development of the Danxia landform in Jianglangshan. Among the structures, the large northeast, north–northeast, and south–north raptures mainly affect the edge lands and the main valleys to form the Danxia landform; while the effect of large numbers of small raptures on mountain blocks of Danxia landform is obvious, especially the effect of the large perpendicular joints. The rapture to the northwest direction distribute broadly with small scale, which results in the main strike direction of northwest of the main mountain blocks, along with the development of valley landforms such as lane valley and canyon which were formed by the tension–torsional raptures and joints to the northwest direction. The boxwork line formed by the compressional tortion raptures and the tensi–trosional raptures became the basic structure of the maintain blocks and ravines.

(2) **The effect of attitude of rock formation on the topographic form**

Most of the stratums in this area were flatly inclined occurrence with angle less than 10°, which were cut by the perpendicular joints and formed the flatted mesa, stone fortress, stone peak, stonewalling, stelae and the flatly inclined Danxia landform such as the horizontal rock tank and rock cave on the Danxia.

(3) **The effect of stratum lithology on Danxia landform development**

The reason why the height of Sanpan Stone is of hundreds of meters is that, the flatted inclined rock stratum is stable and more importantly the red bed to form Danxia in Jianglangshan is homogeneous and hard. According to the determination, the content of volcanic gravels and cuttings of Fangyan formation in Jianglangshan is about 60%–70%, in which the content of SiO_2 is between 44.90 and 75.80%. The main contents of cement is silicious, ferruginous and fluvial mud, and sand. The rocks formed are very hard and of great anti-weathering ability. Besides,

Over 70% content of Fangyan Formation in Mt. Jianglangshan are volcanic agravels and cuttings

The hard conglomerate-type rocks of Fangyan Formation in Mt. Jianglangshan

Conglomerate-type rocks of Fangyan formation in Mt. Jianglangshan without clear layers

Almost no difference between layers in the cliff of more than 300 m

most of the stratums of Fangyan Formation in Jianglangshan are homogeneous massive sandstone without clear layers and they have few weak fine sand- siltstone- argillaceous interlayer, so the large bedding cave does not develop. In the other hand, the short of collected water in the residue rock block resulting in weak hydrodynamic make the large collapse impossible.

(4) **The effect of neotectonic movement**

Three summit surfaces of Mt. Jianglangshan with the altitudes of 800–900, 500–600 m, and 200 m separately represent the third-grade denudation surface of different times of Mt. Jianglangshan. We calculate that the first-grade denudation surface formed in the end of Oligocene; the second-grade denudation surface formed in the end of neogene, and the third-grade of denudation surface formed in the middle of Quaternary (middle Pleistocene–late Pleistocene).

Sanpan Stone in Mt. Jianglangshan is on the hilly surface with altitude of 500 m around, the second-grade denudation surface, which indicates that the Danxia isolated peak in Sanpan Stone was formed at least in the Pliocene. When the Danxia landform developed to the old stage, some isolated residue peaks were distributed in the peneplain gentle hill. The tectonic movement at the end of the Pliocene raised the basin with about 300 m, and the third-grade denudation surface developed in the early stable period of Quaternary. The neotectonic movement after that raised the basin with another 100 m and the modern valley plain developed.

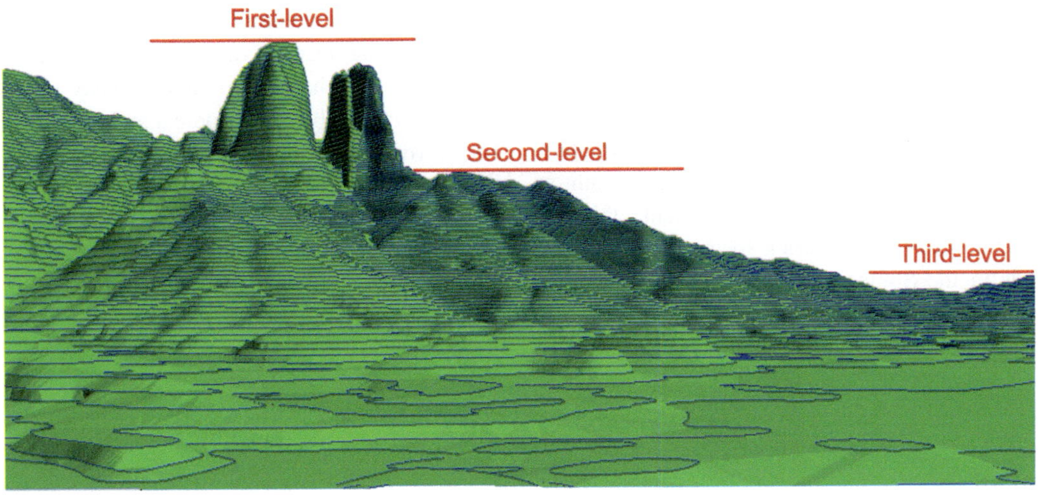

The third-grade summit surface of Mt. Jianglangshan in longitudinal profile of three-dimensional topography

Three different grades of summit surfaces of Mt. Jianglangshan seen from the entrance of Tianban Jianglang

17.2.4 Species Diversity

(1) **Biogeographic region**

According to the Udvardy (1975) biogeographic system, Mt. Jianglangshan is in the south of Biogeographic region, the ancient Oriental Deciduous Forest with rich species diversity, and it is in the Southeast China-Hainan Moist Forests, one of 200 biogeographic regions of World Wide Fund for Nature. With consideration of formation and structures, the tropical zone and temperate zone features of flora are obvious and the floristics of the region has obviously transitional characters between tropical and temperate. The vertical distribution of vegetation in Mt. Jianglangshan is obvious, the aquatic vegetation, evergreen broad-leaved forest, deciduous-evergreen broad-leaved mixed forest are distributed from the foot to the top of the mountain in sequence.

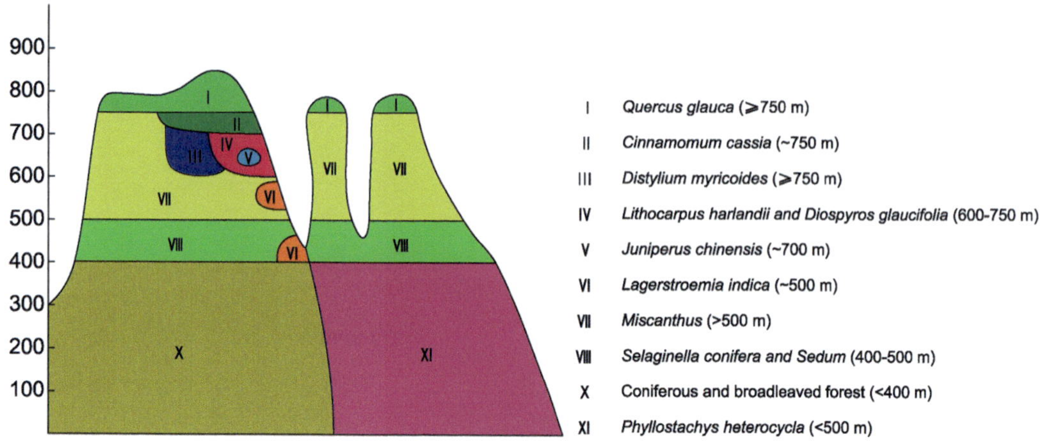

Schematic diagram of vertical distribution of vegetation in Sanpan Stone of Mt. Jianglangshan

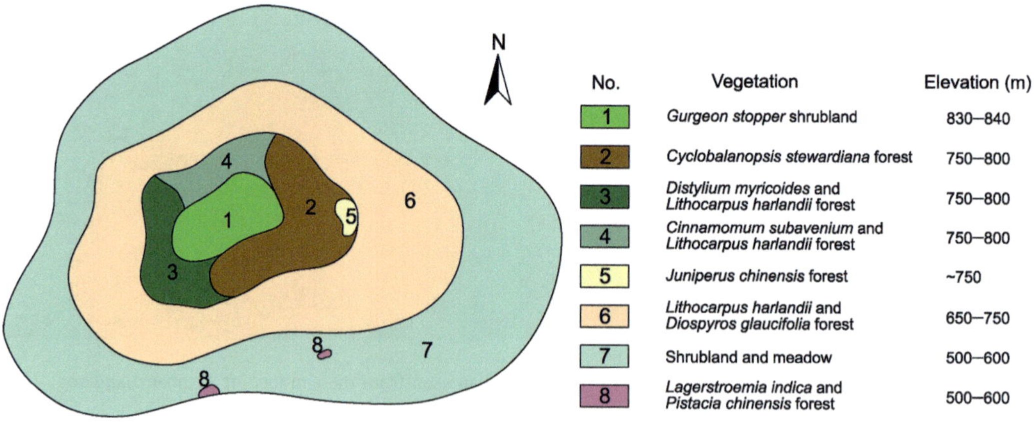

Schematic diagram of horizontal distribution of vegetation in Sanpan Stone of Mt. Jianglangshan

(2) **Species**

There are higher plants of 252 families, 969 genus, and 2116 species all over the Mt. Jianglangshan (ca below species included, similarly hereinafter), with bryophytes of 57 families, 124 genus and 236 species, vascular plants of 195 families, 845 genus and 1880 species, pteridophyte of 33 families,61 genus and 122 species, seed plants of 162 families, 784 genus, and 1758 species, among which gymnosperm of 8 families, 23 genus and 34 species, angiosperms of 154 families, 761 genus, and 1724 species, wild woody plants of 87 families, 242 genus and 681 species, and cultivated plants of 73 genus and 144 species. Among the bryophytes above, 23 families, 33 genus, and 54 pieces are hepatica; 33 families, 89 genus and 180 species are moss and 1 family, 2 genus, and 2 species are Anthocerotopsida. In addition, *Cynodontium* Bruch. et Schimp. ex Schimp, and *C. gracilescens* (Web. et Mohr.) Schimp are the new genus and species listed in the bryophyte distribution record of Zhejiang Province. Among all the species above, 34 species are listed in the China Red List; 51 species are listed in the ICUN Red List; 28 species are listed in CITES; and 47 species are listed in State Important Conservative Plant List, two of which are under the first-grade state protection plants.

The Mt. Jianglangshan has got 12 vegetations, 35 plant formations, and 80 associations.

The wild animals belong to the central China region of the oriental realm, which is adjacent to the northern China realm of the palaearctic realm. There are terrestrial vertebrate of 4 classes, 29 orders, 63 families, and 195 species, with mammals of 30 species, birds of 119 species, reptiles of 32 species, and amphibians of 14 species. Among all the species above, 175 species are listed in the China Red List; 142 species are listed in the ICUN Red List; 31 species are listed in CITES, four of which are collected in Appendix I; and 29 species are listed in State Important Conservative Animals, four of which are under the first-grade state protection animals, and 25 of which are under the second-grade state protection animals.

(3) **Ecological process and features**

With being affected by the topographic factors, vegetations of the Mt. Jianglangshan are vertically distributed with evergreen broad-leaved forests covering the middle and lower parts of the mount; evergreen and deciduous broad-leaved mixed forest covering the middle and upper parts of the mount; and epipetreous meadow and moss vegetation covering the crag, all of which are stable zonal plants. Due to the man-made sabotage, local area is covered by deciduous broad-leaved forest, mixed broadleaf–conifer forest, bamboo groves, and cultivated coniferous forest, which are eventually turned into the deciduous broad-leaved mixed forest or evergreen broad-leaved forest as time goes by. What affects the horizontal distribution most is heat and moisture factor, manifested by forest cover covering the peak (green hat), moss vegetation covering the crag, the meadow, and secondary scrub and the crag surrounded by the evergreen broad-leaved forest (green skirt). As Danxia geomorphy covers a small area of the Mt. Jianglangshan, owing to the orogenetic movements, the Danxia vegetations are the special vegetations that are formed, remain, go through the long-term succession and become differential from the surrounding plants gradually. In addition, solitary island effect is another key factor to foster the growth of the vegetation of this kind.

Special plants on Mt. Jianglangshan: *Phlegmariurus,* Adiantaceae, *Adiantum juxtopositum,* and *Shibataea chiangshanensis*

17.3 History and Development

(1) **Basin evolution**

Development for Mt. Jianglangshan Danxia landform has experienced the formation of Xiakou Basin, red bed sediments, tectonic uplift, fracture development, erosion from the external force, development of Danxia landform, etc.

The gateway basin was formed by secular subsidence after Middle Jurassic Indosinian Movement. Large scale magmatic activities occurred within the scope of the basin in Late Jura, formed Upper Jurassic volcanic suite. Stress field in Early Cretaceous region transferred to extension mechanism and early deep fault structure pulled apart and rifted, formed Lower Cretaceous Yongkang Group Guantou Formation (K_1g), Chaochuan Formation (K_1c), and Fangyan Formation (K_1f) sediment and piled up a land-based detrital sediment combination with characteristics of massive delta, alluvial fan, stream, and lacustrine deposit. Above-mentioned two fault structures had intensive extrusion again in Late Cretaceous Period, and the gateway basin stopped sediment course but began to rise slowly and be eroded accompanying with rupture development.

At the end of Oligocene age, the gateway basin was in the state of low peneplain. Massive rising occurred during the tectonic movement in Miocene Period. Danxia landform of the gateway basin where Mt. Jianglangshan located was aged in Pliocene Epoch, and the isolated peaks such as Sanpan Stone were dispersed over peneplane hummocks. Earth crust in this region uplifted another 300 m since Pliocene Epoch, and isolated peaks such as Sanpan Stone rised and became activated high peaks, and the third grade of Planation Surface grew in the stable early Quaternary Period. And then, neo-tectonic movement in Medio-Pleistocene made this region rise another 100 m intermittently and isolated peaks such as Sanpan Stone reached the height now synchronously.

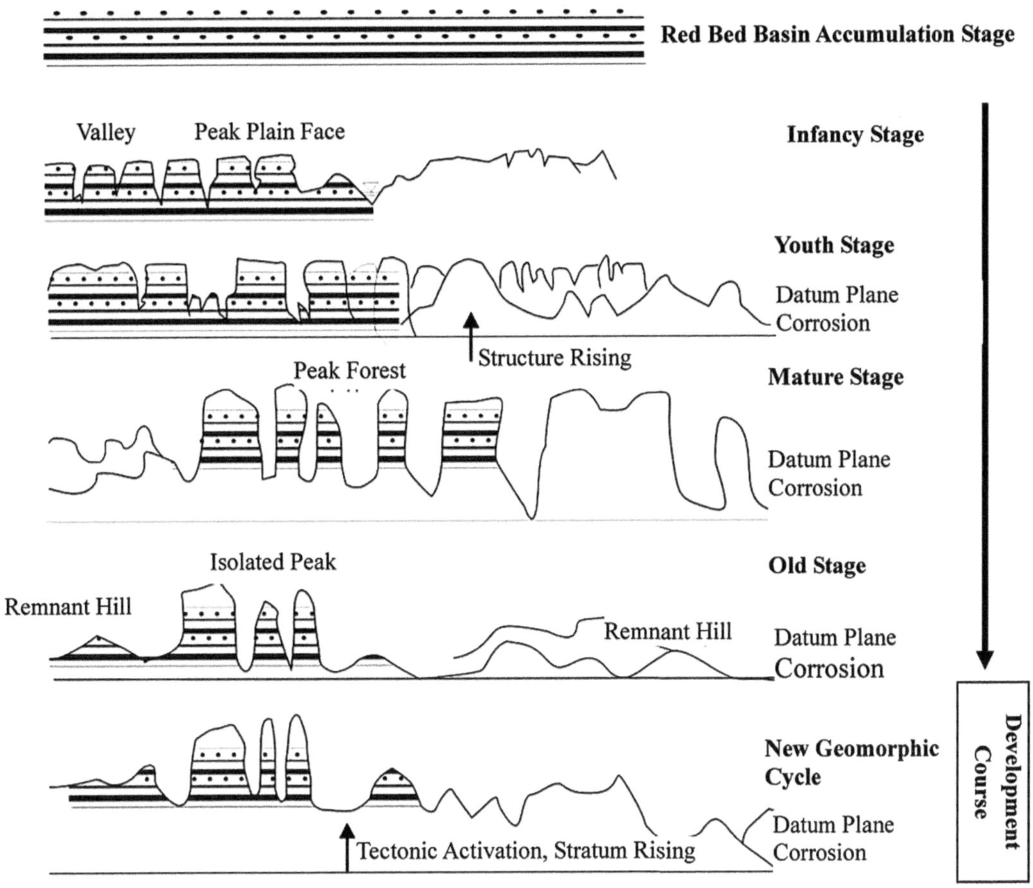

Diagram of Formation of Mt. Jianglangshan Danxia Landform

Therefore, Mt. Jianglangshan is an activated isolated high Danxia peak formed in tectonic movement of a typical aged Danxia landform, has unique geoscience value and physiognomy evolution mode sense.

(2) **Evolution of history and culture**

With the temples traceable to Tang Dynasty, Mt. Jianglangshan was firstly exploited in Jin Dynasty, emperors of all dynasties ever conferred inscriptions, tablets, and titles upon this place. Famous for delicate and majestic appearance and the wonderful scene in wind and snow, this mount attracted numerous celebrities, including the notable prime ministers Yao Chong and Zhang Jiuling who ever wrote poetries to praise the mount; the great poet Bai Juyi wrote a poem to express his feeling about the mount that "I wish to have wings to fly in the mist and clouds forever." Zhu Dongshan, a learned and lofty scholar from Jiangshan lived in the mount for a long time and established a school there to teach students.

The culture of Mt. Jianglangshan arrived at the peak and attracted more and more celebrities since Song Dynasty. Xin Qiji, a ci Writer of Southern Song left a poem to praise the gorgeousness of this mount, i.e. "extremely green and precipitous, as high as thousands of meters, towering to the skies without support, the three peaks are the pillars between the heaven and the earth." And the renowned poet Lu You expressed his endlessly patriotic passion as an old hero when he facing Mt. Jianglangshan with "three peaks strike into the clouds." Wang Anshi and

Mao Pang ever studied in Xianju Temple in this mountain and other celebrities such as Wang Yucheng, Lv Gongzhu, Wang Dan, Zhao Bian, Wen Yanbo, and Shen Jiuru, etc., ever sang poems to glorify the scene of Mt. Jianglangshan.

Mt. Jianglangshan landscape was not cultivated during the initial post-liberation period until 1985 when it was listed among the first series of scenic and historic interest areas of Provincial level and began to accept purposive protection and planned development.

Mt. Jianglangshan was authorized as a National Key Scenic and Historic Interest Spot in 2002. The People's Government of Jiangshan City established "Interim Management Procedures on Scenic and Historic Interest Areas of Jiangshan City" in April, 2006 and implemented it. In addition, relevant department started to draft protection ordinance for Mt. Jianglangshan landscape of Zhejiang Province in March, 2007 to meet the demand of heritage conservation and application conditions of World Natural Heritage, which made protection procedures for Mt. Jianglangshan more legalized and standardized.

(3) **Cultural heritages**

Mt. Jianglangshan is rich in cultural heritages, especially famous for Kaiming Temple traceable to the Era of Tianxi in Northern Song Dynasty and Jianglang Academy with the history of more than one thousand years that established in the Era of Xining, Northern Song Dynasty.

With the original name of Jianglang Temple, Kaiming Temple locates at the terminal of the eighteen bends in Mt. Jianglangshan, is the start point to climb Sanpan Stone peak and is a famous temple in the western part of Zhejiang Province. In 994, Yiyan, an eminent monk of Five Dynasties piled rocks to cover the mouth of a halbhohle in Mt. Jianglangshan and sat in meditation in the halbhohle. Later, the King of Yue conferred him the title "Master of Open Mind (Kaiming)," and the temple's name was changed to "Kaiming Temple." Accompanying with worldly things changing, this temple was rebuilt for many times and destroyed for many times, and present temple rebuilt in 1990 attracts pilgrims in an endless stream.

With the area of 780 m^2, Jianglang Academy locates on the northern peak of Mt. Jianglangshan, including a calligraphic tablets corridor, a showroom of Zhu Dongshan history and accommodations for tourists. This academy was established by the scholar Zhu Wei in the Era of Xining, Northern Song Dynasty, which lasted for hundreds of years and trained numerous talents during Northern Song and Southern Song Dynasties. More than ten people from this academy were listed among successful candidates in the highest imperial examinations in feudal China early or late. The academy now was rebuilt in 1996.

17.4 Natural Characteristics and the Value for the Serial Heritages

(1) **Natural characteristics**

Mt. Jianglangshan locates at a gateway basin whose foundation was volcanic rocks formed by magmatic activity in the region in Late Jurassic epoch. The stratum contains quite significant information about volcanic events in the early term of Early Cretaceous Period, fault events in the middle term of Early Cretaceous Period, and raising events in the early term of Late Cretaceous Period, is a key base for researching Danxia landform and historical evolution of the earth.

Mt. Jianglangshan mainly includes eight types of Danxia landforms including wall-like landform, columnar landform, linear structure, massive landform, gulch and valley, tafoni and landslip, etc., wherein, Sanpan Stone and Xianggu Valley are most notable.

(2) **The status in and value to serial heritages**

Mt. Jianglangshan has the same or similar architectonic background and evolution history as well as common geomorphic features and physical geography characteristics with other heritage areas in the serial heritages, comply with the conditions for nomination of World Heritage series (ref. article 137, III.C), and has non-replaceable status and effects in the series. Its outstanding values include:

- Mt. Jianglangshan is the typical representative of Danxia landform of aged isolated peak type and is a non-replaceable component element in serial nomination.
- Geologic history reflected by lithological character, structure, basin development, extinction, and formation course of Mt. Jianglangshan Danxia landform represents an important stage of earth history.
- Landform and characteristics of Mt. Jianglangshan are typical examples in Danxia landform types, so the mount represents a special development stage of Danxia landform, conformation as an exemption.
- Three huge rock peaks with the height about 300 m stand closely on the mount top 500 m above sea leave, including Peak Lang, Peak Ya, and Peak Ling, forming Sanpan Stone with the title that "most grand in the world and most beautiful in Southeast." With the length of 308 m and the height of 298 m, and the width of the narrowest place 3.5 m, Xiaonongxia, the gorge between Peak Ya and Peak Ling is praised as "the First of Linears Sky in China" by 56 geological specialists.

17.5 Overview of Protection Management of the Heritage Site

The Jianglangshan World Natural Heritage Site is national scenic area (national park) of China, and is under statutory protection of the state laws and regulations including Constitution of the People's Republic of China, Law of the People's Republic of China on Forest, Environmental Protection Law of the People's Republic of China, Water Law of the People's Republic of China, Law of the People's Republic of China on the Protection of Wildlife, Regulations of the People's Republic of China on Wild Plants Protection, Enforcement Regulation of Forest Law of the People's Republic of China, Forest Park Management Measures, and Regulations of Scenic Spots of the People's Republic of China. The Jianglangshan Heritage Site is currently in good condition.

The Jianglangshan World Natural Heritage Site has established an effective multi-level management system to undertake environment protection works, with collaboration of relevant departments, support of the public, and unified coordination. The relevant resource management departments and enterprises have also established environment protection organizations to control pollution, protect the ecological environment, and managed to keep the air, water, soil, and noise in ideal condition.

The boundaries of the Jianglangshan World Natural Heritage Site have been clearly defined with establishment of key monitoring parameters. Sufficient personnel, institutions and funding are provided to ensure prompt reaction to emergencies and problems.

The outstanding universal values of the Jianglangshan world natural heritage, such as the Danxia landform, the esthetical value, the ecological system, the endangered species, and their habitat are preserved in good condition; the integrity of the evolution of species, the ecological system, and the natural environment are fairly maintained.

The Jianglangshan World Natural Heritage Site is in certain extent affected by natural factors and human activities. Natural factors include landslide and mudslide, forest fire, forest plant diseases, and insect pests. Human activities include agricultural activities, lumbering, and engineering projects within the Heritage sites. Accordingly, the relevant management departments implemented effective measures to solve these problems, including returning land for farming to bamboo and production forest on hilly areas with Danxia landform, especially hillside lands, adjust the agricultural structure, develop diversified economy, etc., with satisfactory results. Therefore, the outstanding universal values of the Heritage site are not radically affected either by natural factors or human activities.

Appendix

Decision: 34 COM 8B.1

The World Heritage Committee,

1. Having examined Documents WHC-10/34.COM/8B and WHC-10/34.COM/INF.8B2;
2. Inscribes **China Danxia, China,** on the World Heritage List under criteria (vii) and (viii);
3. Takes note of the following provisional Statement of Outstanding Universal Value.

Brief synthesis

China Danxia is a serial property comprising six areas found in the subtropical zone of Southern China. The term "China Danxia" describes the physical landscape developed from continental (terrestrial) reddish conglomerate and sandstone, also known as "red beds," in a warm, humid monsoon climate. These landscapes are developed on continental red terrigenous sedimentary beds influenced by endogenous forces (including uplift) and exogenous forces (including weathering and erosion). The process of development is characterized by a particular rock sequence, tectonic background, warm and humid climatic conditions, and resulting erosion processes and landforms.

China Danxia is unrivaled in its rich warm and humid climate red beds sandstone landform geomorphologic features characterized by spectacular red cliffs and a range of erosional landforms, including dramatic natural pillars, towers, ravines, valleys, and waterfalls. China Danxia is also noted as a natural esthetic landscape comprising red rocks, green vegetation, blue water, and white clouds.

The rugged landscapes in the nominated property have helped to conserve subtropical broad-leaved evergreen forests, and these forests are found within all six serial sites. A range of important microhabitats are also found. The nature of the Danxia landforms leads to intensive fragmentation and isolation of ravine and mountain top habitats. The natural habitats host many species of flora and fauna including endemic, endangered, and threatened species of conservation significance.

Criterion (vii): China Danxia is an impressive and unique landscape of great natural beauty. The reddish conglomerate and sandstone that form this landscape of exceptional natural beauty have been shaped into spectacular peaks, pillars, cliffs, and imposing gorges. Together with exuberant forest, winding rivers, and majestic waterfalls, China Danxia presents a resplendent natural picture. The sharp contrast of red rock against green forests and blue rivers is a striking feature of China Danxia and renders great scenic appeal. China Danxia sites have long been appreciated by both the general public as well as the academic world and further celebrated by artists. It is one of the most important scenic identities of China and has even attained significance as religious shrines. Its significance is further elaborated by the countless paintings, poems, and articles eulogizing these unusual beautiful sites since ancient time.

Criterion (viii): Compared with other similar areas, China Danxia is the outstanding example

H. Peng, *China Danxia*, Springer Geography,
https://doi.org/10.1007/978-981-13-5959-0

of warm and humid climate red beds landform in the world. As a result of favorable geological, hydrological, and climatic conditions since at least late Mesozoic period, China Danxia areas preserve and display much richer geomorphological, ecological, biological, and scenic features for warm and humid climate red beds landform than any place in the world. The component parts represent the best examples of "least eroded" to "most eroded" Danxia landforms, displaying a clear landform sequence from "young" through "mature" to "old age," and with each component site displaying characteristic geomorphologic features of a given stage. China Danxia contains a wide variety of well-developed red beds landforms such as peaks, towers, mesas, cuestas, cliffs, valleys, caves, and arches. Being shaped by both endogenous forces (including uplift) and exogenous forces (including weathering and erosion), China Danxia provides a range of different aspects of the phenomenon of physical landscape developed from continental (terrestrial) reddish conglomerate and sandstone in a warm, humid monsoon climate, illustrating both the range of landforms in relation to the forces and processes that formed them.

Integrity

The nominated property of China Danxia satisfies the requirements of integrity, protection, and management set out in the Operational Guidelines for the Implementation of the World Heritage Convention. The nominated property encompasses all the elements in sufficient size necessarily to reflect the natural beauty and earth science values of Danxia landform from young stage through mature stage and to old stage. The boundaries of the component parts and their associated buffer zones were adequately defined on maps as well as on-site. The boundaries of the China Danxia itself are adequate in relation to the nominated earth science and esthetic values, and the buffer zone boundaries are also clearly defined. The level of management commitment appears adequate to the main challenges and threats that could face the property.

Protection and management requirements

The component parts of the nominated property are all state-owned and have national protected status including national park, national nature reserve, national forest park, and national geopark. They are protected under relative laws and regulations at both the national level and regional level, which ensure the adequate long-term legislative, regulatory, institutional, and traditional protection of the outstanding universal values. On the basis of the laws and regulations, the protection activities are well practiced in the nominated sites.

Efficient management systems at different levels have been built with enough qualified staff in China Danxia areas. Planning for the serial property is advanced. An integrated management plan has been prepared for the property as a whole, as well as individual plans for the six areas in the series. These plans identify a clear rationale for management and mechanisms for the protection of the property. Research and adaptive management techniques, including baseline condition assessment and monitoring of change for both natural values and species have been established. Local communities are aware of the World Heritage nomination, and all stakeholders are also very supportive of the World Heritage proposal, which ensures the long-term management.

1. Commends the State Party for its efforts toward protection and management of the property across different provinces of China;
2. Requests the State Party to ensure the effective long-term management and protection in the future, with a view to make all the property components meet integrity requirements for natural World Heritage properties and supported by both adequate and effective buffer zones and the protection of wider catchment areas;
3. Invites the State Party to support the organization of international meetings and to continue scientific research regarding the Danxia landform;

4. Also requests the State Party to continue its focus on the protection and effective management of the important biodiversity values;
5. Further requests the State Party to translate and make available in translation key scientific studies on the topic of the China Danxia phenomena and to actively assist the further development of international scientific knowledge of the China Danxia phenomena and red beds sandstone geomorphology more generally.

References

Ai JG, Yu L, Zhang LY et al (2007) Plant community properties in a Hemsleya zhejiangensis habitat in Wuyanling Nature Reserve, Zhejiang Province. J Zhejiang For Coll 24(6):706–710

Bai K, Wu CJ, Tao YK (2007) Tourist cognitive behavior of national geo-park based on the signification of geological science—a case of Cuihua National Geo-park. Arid Land Geogr 30(3):438–443

Bao J, Zhang C (2006) The TALC in China's tourism planning: case study of Danxia Mountain, Guangdong Province, PRC. Tourism Area Life Cycle 1:107–115

Bao B, Huang XW, Ma LQ et al (2008) Characteristics of vascular plants flora and vegetation types in Chaocha Nature Reserve. Sci Soil Water Conserv 6(6):92–97

Cai JB, Zhang K, Liu SN (2004) Size analysis of red sandstone, quartz surface morphology and its environmental significance of Danxia formation in Guangdong Pingshi. In: Physiognomy·environment·development. Environment Science and Technology Press

Cao QX, Guo FS, Cao YT (2007) An introduction on study of geopark. Water Conservancy Sci Technol Econ 5:317–319, 341

Chen GD (1935) Red rock series in Guangdong. Neo-confucianism Q J Guangzhou Sun Yat-sen Univ 6(4):755–784

Chen GD (1938a) Division of red beds in Southeast China. Chronicles Chin Geol Soc 18:301–324

Chen GD (1938b) On the subdivisions of the red beds of South-Eastern in China. Bull Geol Soc China 18:315–316

Chen QS (1989a) The Fangyan formation in Zhejiang. Geol Bull China 1:52–58

Chen ZJ (1989b) Study on Gansu Danxia. J Northwest Normal Univ (Nat Sci) 4:68–71, 82

Chen JG (2003) Study on the forming conditions in first class waterfall of Danxia in China—Shizhangdong of Chishui. Guizhou Sci 21(4):63–67

Chen AZ (2005) Issues on Danxia landform. In: Symposium of the 19th national annual meeting of tourism geography and the development strategy of Shaoguan City

Chen S, Zhu C, Peng H et al (2010) Anti-pressure experimental study on rocks in level cave of Jinshiyan of Danxiashan in Guangdong Province. J Anhui Normal Univ (Nat Sci) 33(2):170–174

Chen SJ (2010) Comparative study on the regional characteristics of Danxia landform in southeast and northwest China—a case study of Danxia landform in Fujian Province and Gansu Province. J Anhui Agric Sci 38(29):16446–16448

Chen GD, Liu HS (1939) Geology of the Gongshui basin in Jiangxi. Collect Jiangxi Geol 2:1–64

Chen GN, Liang BH, Zhu SL (1991) The development of Danxia landform in Jinjiling District, North Guangdong, and its relationship to geological structure. Quat Sci 3:229–236

Chen CH, Cao J, Wang Y (2007) Study on classification and evaluation of Danxia landform tourism resources. J Anhui Agric Sci 35(30):9678–9680, 9779

Chen BM, Li J, Peng SL et al (2008) Preliminary study on diversity of vegetation community and ecosystem of Danxia landform in South China. Ecol Environ (Chinese with English abstract) 17:1058–1062

Deng MCH, Wang GM (1996) Analysis and evaluation of Danxia Landform Landscape in Lushan Mountain. Econ Geography 16(suppl):4–20

Deng JQ, Yan M, Peng WL et al (2007) System integration and development with information system of land resources in Hainan Province. Microcomput Inf 23(21):11–13

Dictionary of Geology (I) (1983) General geology-structural Geology volume (1). Geology Publishing House 12(1)

Dong CW, Zhu GQ, Yin W (2003) Geochemical characteristics and origin of silicolites in the early Cretaceous basin in Xinchang, Zhejiang Province. J Zhejiang Univ (Sci Ed) 30(2):230–235

Dowling RK (2006) Newsome D. Geotourism. Routledge

Du FJ, Xiong KN, Xian QP (2008) The aesthetic importance and comparison of Chishui-Xishui Danxia landscape. J Guizhou Normal Univ (Nat Sci) 26 (1):26–29

Editorial Board of Dictionary of Geography. Dictionary of geography. Shanghai Lexicographic Publishing House (1993)

Fan KF, Yang DC (2006) Geomorphologic system in Taihang Mountain are. J Changchun Inst Technol (Nat Sci Ed) 7(1):51–62

H. Peng, *China Danxia*, Springer Geography,
https://doi.org/10.1007/978-981-13-5959-0

Feng JL, Zhu XS (1928) Nanxiong geological and mineral resources in Shixing, Renhua county, Qujiang city, Guangdong. Ann Two Geol Surv Rep 1:29–51

Feng JL (1930) Some geological problems in Guangdong and Guangxi. Rec Geol Soc China 9:2

Feng JL (1939) Views on "the classification of red rock in southeast China". Comment Geol 3–4:173–192

Feng NP, Yang SL (2010) Innovations on county economy to be sustainable. In: 2010 International conference on information management, innovation management and industrial engineering (ICIII), vol 2. IEEE, pp 277–280

Feng SH, Sha R, Ou YD (2007) On the spatial characteristic on Danxia landforms tourist scenic spots and optimization in Jiangxi Province based on fractal geometry theory. J Jiangxi Normal Univ (Nat Sci Ed) 31(3):321–326

Feng SH, Sha R, Kang YW et al (2008) Study on landscape value of Danxia landform and visitor aesthetic perception of mount dragon and tiger. Geogr Geo-Inf Sci 24(2):104–112

Fu QS, Liu ZW (2002) The type and origination of Danxia Landform in Feitianshan, Chenzhou. In: Memoir of the symposium of Danxia landform and eco-tourism in Feitianshan

Gao SK, Zhu GQ, Dong CW et al (2004a) The morphological development of Danxia hillslope—a case study of Xinchang Danxia landform. Trop Geogr 24(2):131–135

Gao TJ, Liang SJ, Chen ZL et al (2004b) Research on the geological structure and Danxia landform of Taining Basin, Fujian. Map Publishing House in Fujian Province, Fuzhou

Ge YJ, Zhang RS (2004) Research on the relation of overhanging coffin with Danxia landform. J Nanjing Normal Univ (Nat Sci) 27(3):92–96

Ge ZS, Yang DY, Ren CX (2004) Danxia morphological features and the formation and evolution of Gannan areas. Physiognomy·environment·development. Environment Science and Technology Press

Ge YJ, Zhang RS, Yang GS (2007) The relation between Danxia landforms and Buddhism grottoes along the Silk Road. Geogr Res 26(6):1087–1096

Guo K, Di MH (2008) Theory and utilization of Zhangshiyan landform. Geogr Geo-Inf Sci 3:79–82

Guo FS, Liu LQ, Hua M et al (2006a) On the regional planning for Danxia landform travel resources in Jiangxi Province and the idea of "Double Red Travel Area". J East China Inst Technol (Soc Sci) 25(4):363–366

Guo JQ, Yang JY, Yang G et al (2006b) Characteristics, geomorphogeny in Jianmen landform and correlation with the Daxia Landform. Acta Geol Sichuan 26 (4):228–232

Guo GL, Guo FS, Liu XD et al (2006c) Study by empa on microcosmic chemical weathering to sandstone in Danxia Landform. Carsologica Sin 25(2):172–176

Guo JX, Zou D, Sun SC et al (2006d) Analysis on wild characters of weedy rice filial generation in Heilongjiang Province. Heilongjiang Agric Sci 6:11–13

Guo FS, Liu LQ, Yang Z (2007a) Development of Danxia landform in Jiangxi province and its regional planning for tourism. Resour Surv Environ 28(3):214–222

Guo YC, Xie Q, Wen JQ (2007b) Red beds distribution and engineering geological problem in China. Hydrogeol Eng Geol 19(6):67–71

Guo JQ, You ZP, Zhao YN (2008) A study of geoheritage regionalization in Sichuan. Acta Geol Sichuan 28 (4):327–330

Guo M, Yin M, Li Y et al (2009) Comprehensive evaluation of landscape resources in Qianshan Scenic areas. J Northwest For Univ 24(3):173–176

Han ZY, Li XS, Ren XM et al (2006) Rate estimation of scarp retreat on bank slope at Wanzhou of Chongqing. J Mt Sci 3:327–332

Huang J (1982) A basic way of Danxia landform slope development. Trop Geomorphol 3(2):107–134

Hang J (2009) Preliminary study on Danxia landform in China. In: First international symposium on Danxia Landform. World Danxia. The 1st collection. Danxiashan, Guangdong, China 28:132–146

Hao YC, Su DY, Li YG et al (1982) Stratigraphical division of non-marine cretaceous and the juro-cretaceous boundary in China. J Geol 3:187–199

Hao L, Chen JT, Zhong GP (2006) A comparison between the tourism scenery of Danxia landform and that of karst landform. Trop Geogr 26(1):12–17

He JD, Huang RJ (1979) Late Cretaceous to Paleogene strata in Nanxiong Basin, Guangdong Province. J Stratigr 1:30–39

He QC, Ren F, Liu K (2009a) The influencing factors development and evolution of Danxia landform in Longhushan. In: World Danxia—first international symposium on Danxia Landform, pp 242–248

He P, Song X, Luo Y et al (2009b) Reproductive biology of Dendrobium officinale (Orchidaceae) in Danxia landform. Zhongguo Zhong yao za zhi = Zhongguo zhongyao zazhi. China J Chin Mater Med 34(2):124–127

Hou YP, Peng SL, Li FuR et al (2008) Characteristics and scientific values of ecological succession in Danxia Landform of China. Acta Ecol Sin 28(7):3384–3389

Huang KG (1987) The palatial similar Danxia landforms in Anning Bao of Lanzhou. J Northwest Normal Univ 3:49–52

Huang J (2004) Quantitative survey of several important issues concerning the formation of the Danxia landforms. Trop Geogr 24(2):127–130

Huang S (2008) Research on geopark tourism exploration layout in the ethnic areas: a case of western Guangxi. Study Ethnics Guangxi 94:187–194

Huang YZ, Yang SY (2003) The geologic formation of Danxia geomorphologic landscape in the Three River area. Bull Mineral Petrol Geochem 22(3):270–272

Huang YZ, Yang SY (2004) Study on the features and formation of Danxia geomorphologic landscape in Liming Yunnan. J Kunming Univ Sci Technol 29 (5):23–26

Hu NY, Zhong JS, Cai RP et al (2002) Danxia landscape formation, evolution and evaluation in Langshan, Hunan Province. In: Memoir of the symposium of Danxia landform and eco-tourism in Feitianshan

Hu NY, Dai TG, Cai RP (2005) Discussion on the distribution and the governing factors of Danxia landform in Hunan Province. In: Symposium of the inaugural meeting of the tourism geography and national geopark branch of Geological Society of China and the 20th annual meeting of tourism geography and national geopark

Hu P, Quan DJ, Wang Z (2008a) The appraisement of landscape resource in ruins tourism district: based on Ningxia Shuidonggou landscape. J Northwest Univ (Nat Sci) 2:318–322

Hu XM, Xu HG, Chen MJ et al (2008b) The rhyolite landforms and development law in Yandang Mt. Acta Geogr Sin 63(3):270–279

Huang J, Huang RH, Su ZL (1994) Study on Danxia cave landform. In: Symposium of tourism geography, vol 2 (supplement)

Huang Q, Shu XB, Wang J (2002) A Study on Xu Xiake's investigation about Danxia landform in Xinjiang red-bed basin, Jiangxi Province. Jiangxi Sci 20(2):104–109

Jiang YB, Guo FS, Liu LQ et al (2006) TL Dating research of geomorphic surface on lower terraces of rives in Longhushang Danxia landform areas. J East China Inst Technol 29(3):225–228

Jiang YB, Guo FS, Sun CM et al (2008) The features and origin of the Danxia Landform in Guifeng Scenic Park in Yiyang County, Jiangxi. J Mt Sci 26(1):120–125

Jiang YB, Guo FS, Hu ZH et al (2009a) A study on the features and types of geosciences tourism resources in Longhushan World Geopark. J East China Inst Technol (Soc Sci) 28(1):23–28

Jiang YB, Guo FS, Xiong GB et al (2009b) A study on the features and development of the Danxia landform in Guifeng scenic park in Yiyang County, Jiangxi. J East China Inst Technol (Soc Sci) 28(3):221–224

Jiang AW, Su TP, Yu JJ (2010a) Survey of amphibian and reptiles in Central Guangxi karst area. Carsologica Sin 29(3):313–318

Jiang YB, Guo FS, Hu ZH et al (2010b) A study on the features of Danxia landform and its landscape types in Xinjiang Basin. J Mt Sci 28(4):505–512

Jiang YB, Guo FS, Liu LQ et al (2010c) A study on the structure features and its relationship with the Danxia Landforms in Xinjiang Basin. J East China Inst Technol (Nat Sci) 33(4):325–331

Jianjun J, Xun Z, Youfang C (2006) Geological heritage in China. Geotourism 140

Jin XF, Qian L, Lu YH et al (2008) Seed micromorphology of Sedum (s.l.) from Zhejiang and its taxonomic implications. J Zhejiang Univ (Agric Life Sci) 4:409–417

Komoo I, Patzak M (2008) Global geoparks network: an integrated approach for heritage conservation and sustainable use. Geoheritage of East and Southeast Asia, pp 3–13

Kusky TM, Ye M, Wang J et al (2010) Geological evolution of Longhushan World Geopark in relation to global tectonics. J Earth Sci 21(1):1–18

Li JX (Huang J) (1961) The landform types of Guagndong. J Sun Yat-sen Univ 4:72

Li XD (2004) Eco-environment and ecological construction proposals in Danxia mountain. In: Nationwide 19th annual meeting of geotourism, pp 42–44

Li GM (2006a) Forecast model for discharge of karst spring based on the coupling of phase space reconstruction and neural network. J Geotech Inv Surv 4:30–32

Li HF (2006b) Research on the tourism development of the Danxia Landform in Lang mountain. Yunnan Geogr Environ Res 18(5):104–107

Li F (2007) A study on the growth degree of post-development scenic spot based on double space —a case study of Mount Yun Tai scenic spot. Tourism Tribune 10:55–60

Li Xiang W, Fang JX (2006) Study on the sustainable utilization of biological and tourist resources in Mt. Huangshan. Territory Nat Resourc Study 1:69–70

Li Z, Bao JG, Qin CF (1998a) The impact of tourist development on the vegetation cover of Mt. Danxia. Acta Geogr Sin 53(6):554–561

Li Z, Bao JG, Qin CF (1998b) The impact of tourist development on the vegetation cover of Mount Danxia, Guangdong Province. Acta Geogr Sin 53 (6):554–561

Li YX, Wang YP, Ding BL (2001) New advances of the study of Cretaceous strata in southeastern China based on comprehensive stratigraphy. Acta Geol Sin 75 (4):451–458

Li DW, Wang PL, Yu JB (2004a) Micro-study of Zhejiang Xinchang Danxia weathering characteristics of rock. Prog Nat Sci 14(1):75–80

Li XZ, Dong GR, Chen FH (2004b) Ontology strata desert sedimentary and its significance in Gansu Zhangye Matisi Danxia. In: Physiognomy·environment·development. Environment Science and Technology Press

Li JX, Lin QZ, Zhao LJ (2006) The flora of seed plants in Pingjiang Mufu Mountain. J Central South For Univ 5:93–97

Li FL, Bao WK, Wu N (2007a) Relationship between leaf traits in *Jasminum humile* and environmental factors in the dry valley of the Upper Minjiang River. J Trop Subtrop Bot 15(4):315–323

Li XF, Chen XS, Wang XM (2007b) Investigation on the resource of containing cantharidin insects in Guizhou. Hubei Agric Sci 46(2):300–302

Li JX, Luo KW, Huang HQ et al (2008a) Additions novelity to the seed plants flora of Tianmenshan, Zhangjiajie, Hunan Province. J Trop Subtrop Bot 6:573–576

Li YH, Yu L, Liu HX (2008b) The geological features of Danxia landform in the east part of Qinghai Province. Hydrogeol Eng Geol 4:126–128
Li XH, Chen SD, Cao K et al (2009) Paleosols of the mid-cretaceous: a report from Zhejiang and Fujian, SE China. Earth Sci Front 16(5):63–70
Li CL, Wang XD, Shen NH et al (2010a) On characters of plant community of rock vegetation of Danxia landform in Xinchang of Zhejiang. J Hunan Agric Univ 36 (4):470–473
Li ZX, Yan H, Zhu C et al (2010b) Genesis on two valleys of the Mt. Jianglang in Zhejiang Province and its implications on geomorphology. J Xinyang Normal Univ (Nat Sci Ed) 23(4):546–549
Liang BH (1992) Discussion on sedimentology control of Danxia landform in Guangdong. Papers of 3rd sedimentation and lithofacies geography conference of China
Liang BH, Zhu SL, Chen GN (1992) Petrologic analysis of Danxia landform in Jinji Ridge, North Guangdong. Trop Geogr 12(2):133–140
Liang SJ (2009a) Characteristics of Danxia Landform of the cretaceous Redbed Basin in Yongan City, Fujian Province. Geol Fujian 28(1):1–9
Liang SJ (2009b) On the characteristics and genesis of the channel erosion microtopography in Baiyun Mountain of Fu'an City. Geol Fujian 28(3):213–227
Liangyi R, Jinzhao Z (2007) Hydrological effects of forest litter and soil in the Simianshan Mountains in Chongqing, China. Front For China 2(2):157–162
Liang J, Xie S, Zhang W et al (2010) Investigation and analysis on growth condition of seven clumping bamboo species in Chishui City. Guizhou Agric Sci 6:161–167
Liao RJ (2003) Basin-margin types and basin-fill types of Cretaceous-Neogene terrestrial red basins in Jiangxi. Geol Bull China 22(9):680–685
Lin JQ (2004) The slope development of Huoyan Mountain-another Danxia terrain evolution. In: Physiognomy·environment·development. Environment Science and Technology Press
Lin D (2006) On the formation of the senior vocational textbook of Chinese tourist geography orientated towards occupational capacity. J Sichuan Vocat Techn Coll 16(2):80–82
Lin HB, Tang SQ (2006) Floristic analysis of seed plants in Longhu-shan Nature Reserve, Guangxi Province. Subtrop Plant Sci 35(1):57–59
Lin H, Huang F, Lu G (2009) Development of virtual geographic environments and the new initiative in experimental geography. Acta Geogr Sin 64(1):7–20
Liu SR (1992) Red bed karst and hydrogeology features in Guangdong. Guangdong Geol 3:9–17
Liu SR (1994) Red bed karst and its mechanism in Guangdong. Carsologica Sin 13(4):395–403
Liu SR (2004) The concept of Danxia and profile of parts of foreign Danxia. In: Physiognomy·environment·development. Environment Science and Technology Press
Liu SR (2005) Introduction to foreign Danxia landform. Principle and practice of tourism geography (vol 11). China Forestry Publishing House, Beijing
Liu SR, Liu RH (2003) On the conception of Danxia landform. J Mt Res 21(6):669–674
Liu HF, Chen LJ (2004) Danxia landform and Yardang landform. Manag Strat Qinghai Land Resour 3:47–48
Liu SR, Huang RH (1991) Red bed karst and Danxia landform of Guangdong Province. Carsologica Sin 10 (3):183–189
Liu SR, Peng H (2006) An introduction of Danxia landforms in Australia. Econ Geogr (Suppl) 26:222–232
Liu Y, Wang ZZ (1986) The petrographic characters and environmental analysis of the Cretaceous-Tertiary Boundary strata at the Nanxiong Basin Guangdong. J Stratigr 10(3):190–203
Liu WQ, Li ZH, Liu LF (1999) A preliminary study on the flora of the tourist landform of Danxiashann, Guangdong, China. Guangxi Foliage 19(1):15–21
Liu RW, Xiao X, Wu DT et al (2006) The research on the phenomena of "first impressions are strongest" and "accomplished Fact" in the process of the tourism destination image apperceiving—a case study of Sand-lake of Yingchuan City and Hongjiannao Destination of Yulin City. Hum Geogr 1:34–40
Liu ZW, Zhang ZL, Wang BC et al (2006) Paleoecology and sedimentary environment of ostracods in the Fuyu and Yangdachengzi oil horizons in the Sanzhao area, Songliao basin. Geol China 33(5):1059–1066
Liu LQ, Guo FS, Zeng XH (2007a) Study on the Danxia geomorphologic landscape with remote sensing technology. J East China Inst Technol 30(3):247–251
Liu TG, Ye L, Wang XL et al (2007b) Chemical action is an important factor for rock weathering in arid areas Simulating experiments of evaporation and leaching. Geol China 34(5):815–821
Liu K, Gao QZ, Jiao SL et al (2008) The double level structure of the weathering terrace base and its significance: example of the 6~(th)river terrace of Pingshi Basin in the North of Guangdong Province. J Mt Sci 26(2):137–144
Liu ZH, Zhang B, Wang WJ et al (2009) Advantages and countermeasures on the development of eco-tourism resource in Xingwen World Geo-park. Sichuan For Explor Des 3:51–53, 61
Liu H, Luo Y, Liu H (2010a) Studies of mycorrhizal fungi of Chinese orchids and their role in orchid conservation in China—a review. Bot Rev 76(2):241–262
Liu YS, He ZW, Long XJ et al (2010b) Characteristics and geological significance of geological relics in Qijiang Geopark, Chongqing city. Chin J Geol Hazard Control 21(2):118–124
Lu YT, Lu HS (2005) Study on the ancient Danxia culture in Danxia landform areas. In: Symposium of the 19th national annual meeting of tourism geography and the development strategy of Shaoguan City, pp 63–84
Luo BS (1991a) Rock mass mechanical mechanism of the formation of Danxia landform in Jinji ridge. Papers of Tourism Academy

Luo BS (1994) Rock mass temperature stress effect in the process of the formation of Danxia landform. Ann Guangdong Ind Coll 11(4):1–7

Luo HF (1991b) Geological structure characteristics and formation and evolution mechanism of Meso-cenozoic basin. Jiangxi Geology Science and Technology

Luo BS (1993) The development of Danxia Landform in Jinjiling District and its relationship to unloading effect. J Guangdong Univ Technol 10(1):83–90

Luo CD (1999) Danxia landform in south Sichuan basin. Econ Geogr 19:37–43

Luo CD (2005) Regional difference of Sichuan Danxia landform. J Leshan Teachers Coll 20(12):134–138

Luo PS, Zhang MF (2010) The temporal characteristics of tourist flows to Danxia Landform scenic spots in Fujian Province and its influence factors. Sci Geogr Sin 30(3):377–382

Luo H, Chen JT, Zhong GP (2006) A comparison between the tourism scenery of Danxia landform and that of karst landform. Trop Geogr 26(1):12

Luo KW, Yu XL, Deng YY (2009) A floral study of Danxia Landform in Shiniuzhai, Pingjiang County Hunan Province. J Central South Univ For Technol 29 (2):93–97

Luo D, Xiao YP, Zhan L et al (2010a) The tectogenesis control of Danxia landform in southeast China. J East China Inst Technol (Nat Sci) 33(22):147–153

Luo Y, Guo FS, Jiang YB et al (2010b) Design and implementation tourism resources information system of Shangrao City based on VB+ MO. Water Conservancy Sci Technol Econ 16(1):108–110

Lv XB (1989) Analysis of structural landform in Fangyan, Zhejiang. Ann Zhejiang Normal Univ 12 (2):129–133

Lv FD, He HX (2006) Determination of seed vigor in Carya nutt species. Nonwood For Res 24(3):11–14

Ma CM, Zhu C, Peng H et al (2006) Danxia landform genesis of the Qiyun Mountain, Anhui Province. J Geogr Sci 16(1):45–56

Margottini C, Spizzichino D (2009) The management of cultural and environmental heritage sites: a pivot for conservation and enhancement. In: First international symposium on Danxia Landform. World Danxia, The 2nd collection. Danxiashan, Guadong, China, pp 26–28

Meng HH, Wang LC, Su WC et al (2009) Development of a karst sinkhole-based semi-distributed hydrological model and its application. Sci Geogr Sin 29 (4):550–554

Migon P, Andrew G Sandstone geomorphology of south-west Jordan, Middle East, and its relation to Danxia Landform of China. Quaestiones Geographicae 33(3):123–130

Ou ZR (2004) Usage of GIS and RS in geomorphic evolution research of the three parallel rivers area in northwest of Yunnan. Normal University of Yunnan, Yunnan

Ouyang J, Zhu C, Peng H et al (2009) Types and spatial combinations of Danxia landform of Fangyan in Zhejiang Province. J Geogr Sci 19(5):631–640

Partin C, Robinson S, Meade B (2006) Geological heritage in Chinese parks: balancing protection and development. Focus Geogr 49(3):10–16, 37

Peng H (2000a) A survey of the Danxia landform research in China. Sci Geogr Sin 20(3):203–211

Peng H (2000b) Danxia geomorphology of China and its progress in research work. Sun Yat-sen University Publishing House, Guangzhou

Peng H (2001) Danxia geomorphology of China: a review. Chin Sci Bull 46(S1):38–44

Peng H (2004a) Research of China Danxia landform: a review. China geographical society annual meeting in 2004, 100

Peng H (2004b) World geopark: red stone park of China. Geology Press, Danxiashan

Peng H (2005) Danxiashan: red stone park of China. Chin Nat Geogr (Supplement), 110–115

Peng HM, Liu LQ, Guo FS (2001) On geological genesis of the Danxia landform in Zhejiang Province. Volcanol Min Resour 22(2):143–149

Peng H, Wu ZC (2003) A preliminary study on the characteristics and the distribution of red beds. Acta Sci Nat Univ Sunyatseni 42(5):109–113

Peng H, Wu ZC, Zhang K et al (2004) The construction meaning of Danxiashan World Geopark and the characteristics of Danxia development. Physiognomy Environment Deveolpment. Environment Science and Technology Press

Peng H, Zhang K, Liu SR (2004a) Academic values of Danxiashan Geopark and the significance of the international comparison of Danxia geomorphology. In: Proceedings of the first international conference on geoparks. Geological Publishing House, pp 123–130

Peng H, Zhang K, Liu SR (2004b) Danxiashan world Geopark's scientific value and significance of international comparisons of the topography. The first session of the world parks congress geological collection. Geological Publishing House, Beijing, pp 404–409

Peng H, Liu SR, Zhang K (2005a) Meanings and scientific values of the application for the Global Geopark of Danxiashan. In: Symposium of the 19th national annual meeting of tourism geography and the development strategy of Shaoguan City

Peng H, Zhang J, Zhou TT (2005b) Study on the development of popular science tour in Danxia landform scenic spot: a case of Danxiashan Geopark. In: Principle and practice of tourism geography (vol 11). China Forestry Publishing House, Beijing

Peng SL, Li FR, Zhou T et al (2008a) Special ecological effects of ravines in Danxia landform, China. Acta Ecol Sin 28(7):2947–2953

Peng SL, Li FR, Zhou T et al (2008b) The ecological ravine effects of Danxia landform. Acta Ecol Sin 28 (7):2947–2953

Peng YX, Wu CJ, Zhang L (2009) Literature analysis and progress on tourism earthscience in China during the period 1980–2008. Prog Geogr 28(5):723–733

Qi DL (2005) Multi-scale comparative study on Danxia geomorphology in China. Nanjing Normal University, Nanjing

Qi DL, Yu R, Zhang RS et al (2005a) On the spatial pattern of Danxia landform in China. Acta Geogr Sin 60(1):41–52

Qi DL, Zhang RS, Ge YJ et al (2005b) Comparative studies of Danxia landforms in China. J Geogr Sci 15 (3):337–345

Qiu WL, Xia H (2010) A study of raising finances in Jiangxi geo-tourism resources development. J East China Inst Technol (Soc Sci) 29(2):133–135

Ren MJ, Cao FX (2009) Application of high density resistivity method to prospecting groundwater in red bed region. Sci Technol West China 8(5):51–53

Ren CX, Yang DY, Ge ZS (2004) Qinglongyan cave's characteristics, causes, and protection in Gannan Danxia. In: Physiognomy·environment·development. Environment Science and Technology Press

Shang YJ, Yang ZF, Li LH et al (2006) Scientific and engineering highlights of an 1171-year-old weir in Eastern China. Agric Water Manag 81(3):371–380

Shan LL, Qin JX, Wei YL et al (2008) Study on the agricultural geological background of Jianmenguan scenic spot in Sichuan Province and its agricultural tourism development. J Anhui Agric Sci 13:5600–5601

Shen ZH, Hu ZW, Zhao J et al (2007) Altitudinal patterns of plant diversity on Mt. Guniujiang, Anhui, China—with a discussion on the ecological impacts of hilltop condition. J Mt Sci 2:160–168

Shen NH, Wan ZZ, Tang GG et al (2009) Community structure and species diversity of a Quercus variabilis forest on Mount Zijinshan. J Zhejiang For Coll 26 (5):696–700

Shen R J, Lin SS, Yu Y et al (2010) *Chiritopsis danxiaensis* sp. nov. (Gesneriaceae) from Mount Danxiashan, south China. Nordic J Bot 28(6):728–732

Sirenbatu E (1996) Inquire into Danxia landform in Chuanji Inner Mongolia. J Inner Mongolia Normal Univ (Nat Sci Ed) 3:74–76

Song H (2009) Spatial pattern of geological relics in western Guangxi, China. J Guangxi Normal Univ (Nat Sci Ed) 27(1):92–97

Su YQ (2008) Dynamic study on under branch height of artificial *larix gmelinii*. For Inventory Plann 1:21–24

Su X, Wan XN (2010) Cause analysis of the rock-columns in the Mufu Town of Enshi City, Hubei Province. South-to-North Water Transf Water Sci Technol 8(2):107–109

Sun YL, Tu YL (2004) The present situation and conservation about the endangered plants and animals in the Chishui *Alsophila spinulosa* natural reserve. J Guizhou Normal Univ (Nat Sci) 22(4):22–26

Sun YP, Zhang GH (2009) Issues during geoparks construction in China. Resour Ind 11(1):68–69

Sun Y, Deng TP, Liu YJ et al (2007) Inquiry into problems of groundwater chemical field in the red layer of Sichuan Province with Zigong City as an example. J Southwest Univ Nationalities (Nat Sci Ed) 32(2):408–413

Tang Y, Li YX (2008) Analysis for the features of community structure of the forest in the Dongyanggou Natural Reserve, Sichuan. J China West Normal Univ (Nat Sci) 29(1):10–14

Tao Y, Wu C (2006) A study on the construction of national geo-park of Shannxi Cuihua Mountain based on the analysis of tourist recognition. J Arid Land Resour Environ 20(6):130–135

Tian YZ (2008) Wudang landscape spectacles in ancient Wudang Mountains travelogues. J Yunyang Teachers Coll 28(4):1–6

Tong YS, Zhang YP, Wang BY et al (1976) Early-paleogene stratum of Nanxiong basin and Zhenjiang basin. Ancient Vertebr Hum 14(1):16–25, 75–76

Udvardy MDF (1975) A classification of the biogeographical provinces of the world, vol 8. International Union for Conservation of Nature and Natural Resources, Morges, Switzerland

Vlcko J, Brcek M, Varilová Z et al (2009) Deformation monitoring at cultural and natural heritage sites (in Czech and German). In: First international symposium on Danxia Landform. World Danxia, The 2nd collection Danxiashan, Guangdong, China, pp 317–324

Wang CL (1999) Formation of Luoxiao Mountains and development of its Danxia land feature. J Xiangtan Normal Univ 19(3):110–115

Wang LY (2007) Multi-designated geoparks face challenges in China's heritage conservation. J Geogr Sci 17(2):187–196

Wang C (2008) Summary of the tourism of Hexi corridor within the latest ten years. J Hexi Univ 24(4):54–57

Wang YY, Wang TY (2004) The characteristics of geologic physiognomy in Taitong-Kongtong, Natural Reserve Pingliang, China. J Gansu For Sci Technol 29 (2):27–31, 36

Wang XQ, Wei YF (2007) The preliminary analysis on formation mechanism of the earth forest of Xichang-Huanglian. Res Soil Water Conserv 1:218–220

Wang ZY, Zhu SL (2009) The Strategies of protection and characters of georogical vestige resources pertritied woods in Jiangan of Yibin, Sichuan. Urban Geol 4(1):1–4

Wang M, Tu YL, He MJ (2005) The present statues and characteristics analysis of the plants and vegetation diversity in the Chishui *Alsophila spinulosa* Nature Reserve. J Guizhou Normal Univ (Nat Sci) 32(1):19–22

Wang W, Huang S, Liang MZ (2007) Formation and development of the potholes formed in a gorge river in Guangdong. Acta Geogr Sin 62(7):691–697

Wang JW, Zhang GF, Chen HY (2008) Population pattern and community characteristics of endemic and rare plant *Magnolia zenii* in Baohuashan National forest park. Guihaia 28(4):489–494

Wei FM, Zhang XY, Liu P et al (2007a) Effects of different altitudes on pigment content, water content

and specific leaf area of *Heptacodium miconioides*. Subtrop Plant Sci 36(1):1–4
Wei YL, Qin JX, Chen LL et al (2007b) Danxia landform analysis and tourism resources evaluation in Jianmenguan tourism area. J Guilin Univ Technol 27(4):490–495
Wei P, Wang YS, Wu JF et al (2009) Experimental research on the time effect of the red beds rock-water interaction in the northeast of Sichuan. South-to-North Water Transf Water Sci Technol 7(1):83–85, 106
Wen XC (1993) Research on Mali red rock series and Danxia landform. Trop Geogr 13(2):184–192
Wen T (2007) Analysis of the life cycle of Mt. Danxiashan World Geopark. Econ Geogr 3:496–501
Wen L, Cheng Z, Hua P (2009) The contribution lithological characteristics to Danxia landform genesis in Jianglang Mountain, Zhejing Province. Bull Mineral Petrol Geochem 28(4):349–355
Wu JH (1994) Sedimentary facies of the cretaceous in the northeast of Jiangxi Province and its tectonic significance. J East China Coll Geol 17(4):313–319
Wu ZC (2004) A preliminary study of Guangdong red beds and its development contrast. Sun Yat-sen University, Guangdong
Wu SS, Zeng ZX (1946) The red bads in north Kwangtung. Ann Lingnan, 12–20
Wu SS, Zeng ZX (1948a) Geology and landform in red rock series of North Guangdong. Ann Geogr 6:13–45
Wu SS, Zeng ZX (1948b) Red basin in Pingshi, Guangdong. Geol Rev 13(Z2):274–275
Wu ZC, Peng H (2005) A study on Danxia landform classification in Guangdong. Trop Geogr 25(4):301–306
Wu JT, Liu JX, Liao ST (2001) Geological characteristic, evolution of the Danxia basin. Reg Geol China 20 (3):274–279
Wu J, Peng SL, Lin ZG et al (2008) Ecological effect on the hilltop of Danxia Landform. Acta Ecol Sin 28 (7):3390–3400
Wu WW, Yuan SB, Liu SC et al (2010) Analysis and comparison of flora characteristics of vascular plants in" Danxia' geomorphological area of Langshan Mountain in Hu'nan. J Plant Resour Environ 19 (4):84–91
Xiao WF (2009) The analysis and development of liaoning map supply system based on arc engine. Mod Surv Mapp 32(3):27–29
Xiao ZX, Shang GX, Zou WF (1997) The formation of Langshan and its development characteristics of landforms. J Xiangtan Normal Univ 18(3):65–71
Xiao ZX, Zou WF, Zhou LC (1998) The calcium characteristics of Langshan red beds and its effect on Danxia Landform. Econ Geography 1998(Supply):18
Xie L (2009) Study on the development of village travelling that depending on Danxia landform scenic areas in Zixin. Ecol Econ 2:209–212
Xu DX (1996) River capture in Danxia landform of West Fujian. Fujian Geogr 11(2):11–14
Xu SH (2006a) An integrated and harmonious way on exploitation of tourism resources among boundaries of administrative areas—on case of Chenzhou and Shaoguan boundary, Hunan and Guandong Provinces. J Shaoguan Univ 27(5):115–119
Xu CP (2006b) Guangxi dietetic culture and *Xu Xiake Travel Notes*. J Anhui Agric Sci 34(14):3562–3564
Xu X, Ge TM, Xiao H (1990) Discussion on stratum era of Danxia red beds. J Chengdu Coll Geol 17(2):79–86
Xun Z (2009) Discussion on Danxia landform. In: First international symposium on Danxia Landform. World Danxia. The 1st Collection, 26–28
Yan YB, Zhang D (2006) Research into sustainable development of red tourism. J Hunan Inst Eng (Soc Sci Ed) 16(1):34–36
Yang ZT (1997) Xu xiake's contribution to the scientific exploration of Chinese Danxia's landforms. J Hengyang Normal Univ 6:39–46
Yang ZT (2002) The journaling of travelling in Chu" and famous landscapes resource of Danxia landform in south Hunan Province—to commemorate Xu Xiake's death for 360. Fujian Geogr 17(1):38–41
Yang SY (2008a) A research for the strategy of tourist development of geological resources in Yunnan. Yunnan Geol 27(4):391–407
Yang SY (2008b) The special tourist geological resources in Yunnan. Yunnan Geol 3:278–292
Yang ZJ, Zhou MZ (1962) Vertebrate fossil in red beds of North Guangdong. Ancient Vertebr Hum 6(2):130–137
Yang ZT, Liu TZ (2009) Study on the development of leisure agriculture that depending on Langshan Mountain Scenic. J Hengyang Normal Univ 30(6):119–121
Yang ZT, Fang YF, Liu TZ (2010) Development of rural tourism in traditional settlement landscape in Chenzhou. J Hengyang Normal Univ 31(2):45–49
Yuan WU (2007) Research of drifting tourism in Anhui Province. J Anhui Agric Univ (Soc Sci Ed) 16(1):22–25, 67
Yuan JP, Yu LS, Deng GQ et al (2006) Geomorphologic division and classification of Hainan Island. J Hainan Univ (Nat Sci) 4:364–370
Yu KJ, Ji QP (1992) Study on principles and technical system of landscape planning and protection planning of Danxia scenic. J Guangdong Landscape Archit 1:6–13
Yu YW, Chen J, Jin QS et al (2003) Discovery of Faveoloolithidae dinosaur egg fossils in Yongkang, Zhejiang Province. Geol Bull China 22(3):218–219
Yu XM, Liang SJ, Chen MG (2005) On the Danxia landform and its characteristics of Ninghua County, Fujian Province. Geol Fujian 24(3):171–178
Yu XL, Liu KM, Gu ZR (2007) New recorded plants from Hunan Province of China (II). J Central South Univ For Technol 3:66–69
Zeng ZX (1943) Discussion on the thick layer red sandstone region in north Renhua County. Ann Sun Yat-sen Univ 12:19–24
Zeng ZX (1960) Rock topography. Geology Press, Beijing, pp 45–57

Zeng SQ (1991) The chemical weathering and soil forming process of aubergine sandstone and conglomerate. Ann Sun Yat-sen Univ 10(2)
Zeng ZX, Huang SM (1978) Red beds landform in Southeast China. J South China Normal Univ (Natura Sci Ed) 1:56–73
Zeng ZX, Huang SM (1980) Red beds landform and granite landform. Physical Geography and Landform of China. Science Press, pp 139–150
Zhang JF (1983) The era of red beds of Danxia-Pingshi basin in North Guangdong. Guangdong Geol Sci Technol 2
Zhang XQ (1987a) Strata classification and fossil Ostracods Sequence in the cretaceous-early Tertiary. In: Symposium on mesozoic and cenozoic geology
Zhang XQ (1987b) The combination sequence of ostracoda and division of stratum in Cretaceous—tertiary of Nanxiong basin
Zhang YR (1988) Landform types, study in genetic mechanism and the tourism value of red beds karst in the Qingchengshan. Acta Geol Sichuan 2(3):21–24
Zhang XQ (1990) The discovery and value of ostracoda in Danxia of North and Central Guangdong. Guangdong Geol 5(2):54–64
Zhang GZ (2006a) Environmental impact and countermeasure of tourism development in fragile ecology area—take tourism strategical planning of four states in Xinjiang as an example. J Desert Res 26(1):126–130
Zhang XL (2006b) An analysis of thorough exploitation for Langshan national scenic and historic area. Territory Nat Resour Study 2:408–413
Zhang Y (2006c) The research of Xu Xiake's board and lodging during his travel. J South Yangtze Univ (Humanit Soc Sci Ed) 5(6):45–48
Zhang JX (2010) Practice and exploration of practice base for Taihang mountains geological and geomorphologic field practice in Linzhou. Exp Technol Manag 27(4):150–151
Zhang YP, Tong YS (1963) Subdivision of "Redbeds" of Nanhsiung Basin, Kwangtung. Certebrata Palasiatica 7(3):249–262
Zhang ZX, Fu X (1986) Chara in red basin and Danxia landform of Danxia-Pinghi in North Guangdong. J Stratigr 10(1):75–77
Zhang K, Grapes R (2006) Relationship between large rivers and granite-cored anticlines in the lower Pearl River system, Southeast China: an example of a long-lived drainage pattern. Catena 66(3):190–197
Zhang HS, Cui ZQ (2007) Landscape character and forming of Danxia landform and color hill in Zhangye of Gansu Province. J Desert Res 27(6):942–945
Zhang JX, Tian GZ (2010) A study on the landscape features and tourism system of Hongqiqu-Linl (u) shan Geopark. Urban Geol 5(4):24–28
Zhang CJ, Zhang TG (1986) The characteristic of Aeolian sandstone within red bed of old Tertiary in Danxia Mountain Beauty Spot of Renhua County, Guangdong. J Mineral Petrol 6(3):51–54
Zhang SP, Zhou G (2007) Study on development of Dabie Mountain National Geopark in Anhui. J West Anhui Univ 5:122–124
Zhang XQ, Lü JC, Bian GG et al (2005) A preliminary study on the red beds in the Northern Heyuan Basin, Guangdong Province, China. Acta Geol Sin 79 (5):598–604
Zhang GQ, Tian MZ, Guo FS et al (2007) Study on spatial pattern and geological background of Danxia Landform in Jiangxi Province. Resour Ind 9(4):27–30
Zhang JP, Chang XL, Cao YY et al (2007) GIS-based spatial analysis of tourism resources in Yantai-Weihai region. Sci Surv Mapp 6:182–184, 210–211
Zhang YQ, Guo XY, Tai ZW (2007) "One Village one Trademark" revolution in Japan and its inspiration to the new countryside construction in China. J Northeast Agric Univ (Soc Sci Ed) 5(6):11–14
Zhang S, Guo FS, Jiang YB (2008) Landscape types and genesis analysis of Danxia Landform in Guifeng, Jiangxi province. Resour Surv Environ 29(1):70–75
Zhang YK, Jiang Y, Zhu W (2009) Resources of rosa in China and its prospect of landscaping application. Seed 28(8):68–70
Zhang C, Guo FS, Jiang YB et al (2010a) Features and tourism resources of Danxia landform in Jiangxi Province. Water Conservancy Sci Technol Econ 16 (3):243–246
Zhang FY, Guo FS, Zhang ZB (2010b) The problems and suggestions on tourism development of Mountain Danxia in Guangdong. J Shaoguan Univ 31(11):111–114
Zhang GS, Zhu C, Yu JB et al (2010) Experimental study on lithological characteristic with Danxia landform, Jianglang Mountain, Zhejiang Province. J Mt Sci 28 (3):301–312
Zhao YN (2009) On the relationship of geological landscape resources with geological evolution in Sichuan. Acta Geol Sichuan 29:294–298
Zhao PS, Liu DZ, Ma RZ et al (2007) Cretaceous jiaguansandstone in west Sichuan and tourism culture. Sci Technol Manag Land Resour 24(3):123–127
Zheng JJ, Qiu ZX (1979) Discussion on the features and relevant question of Cretaceous-Tertiary continental strata
Zheng DX, Zeng CS, Chen JB (2007) Remaining palaeotopographic surfaces of the upper reaches of Muyang River in Fujian Province—characteristics, causes and utilization value. J Subtrop Resour Environ 2(2):1–7
Zhou HJ (1990) Formation division and age assignment of Danxia basin, Guangdong. Guangdong Geol 5 (4):51–58
Zhou RZ, Yang CQ (1951) Several opinions on the red rock formation division in Southeast of China. Geol Rev 16(1):131
Zhou SZ, Liu JP, Xi ZF et al (2008) The residual top planation surfaces in the Nanling Mountains and their adjacent regions. J Glaciol Geocryol 30(6):938–945
Zhou YY, Wang XD, Fu XY et al (2010) Community features of rock vegetation of Danxia topography in

Mount Jianglangshan, Zhejiang Province. J Zhejiang For Coll 27(5):699–705
Zhu F (2010) A study on spatial structure of tourism region of south Anhui based on the aggregation fractal. Yunnan Geogr Environ Res 22(1):94–99
Zhu SL, Liang BH (1992) Analysis of sedimentary environment in Jinji Ridge Danxia Landform Group in North of Guangdong Province. In: Papers of 3rd National sedimentary and lithofacies ancient geography conference
Zhu C, Yu JB, Zhao NX et al (2000) Danxia landform genesis and scenic feature on Guanzai Mountain, Fujian Province. Acta Geogr Sin 55(6):679–688
Zhu C, Peng H, Li SC et al (2005) Danxia landform genesis on Qiyun Mountain, Anhui Province. Acta Geogr Sin 60(3):445–455
Zhu C, Peng H, Li Z et al (2009a) Age and genesis of the Danxia landform on Jianglang mountain, Zhejiang Province. J Geog Sci 19(5):615–630
Zhu C, Peng H, Ouyang J et al (2009b) Age genesis and characteristics of Danxia Landform of Fangyan in Zhejiang Province, China. Sci Geogr Sin 29(2):229–237
Zhu C, Peng H, Ouyang J et al (2010) Rock resistance and the development of horizontal grooves on Danxia slopes. Geomorphology 123(1):84–96
Zhu XF, Lin BZ, Zhou H (2006) Structure of *Cyclobalanopsis disciformis* community and analysis of features of microclimate in Jiulianshan. Jiangxi For Sci Technol 36(5):4–9
Zhu Y, Wang XD, Hu YQ (2010) Structure characteristics of rock plant community on Danxia landform in Fangyan of Zhejiang. Northern Hortic 10:107–109

GPSR Compliance

The European Union's (EU) General Product Safety Regulation (GPSR) is a set of rules that requires consumer products to be safe and our obligations to ensure this.

If you have any concerns about our products, you can contact us on ProductSafety@springernature.com

In case Publisher is established outside the EU, the EU authorized representative is:

Springer Nature Customer Service Center GmbH
Europaplatz 3
69115 Heidelberg, Germany

Batch number: 10732807

Printed by Printforce, the Netherlands